Praise for
The Emperor of Wine

"No wine book will receive more attention than *The Emperor of Wine*. . . . Scrupulous and balanced . . . McCoy does justice to the paradoxical tale of this voice of the people who became the most powerful man in wine."
—*New York Times*

"Highly anticipated. . . . The thoroughness of McCoy's reporting is impressive . . . prodigious."
—*Los Angeles Times*

"This illuminating biography of America's preeminent oenophile not only traces Parker's ascension from bored, pot-smoking 1970s government lawyer to global arbiter of vino taste, but also scrutinizes his phenomenal impact—for better or worse—on the world of wine. . . . Fascinating analysis."
—*BusinessWeek*

"At its core, *The Emperor of Wine* is a study of power. . . . The author is steely herself . . . she is both tough and fair. . . . This biography helps readers gauge an extraordinary life—and nose—at the center of all the fuss."
—*Cleveland Plain Dealer*

"Ms. McCoy chronicles Mr. Parker's rise and in the process charts the recent history of wine politics. It's a cautionary tale . . . an informative, fair, and even entertaining book."
—*Wall Street Journal*

"*The Emperor of Wine* is terrific—meticulously researched, well written, and balanced. McCoy has captured Parker in full: He comes across as a man of uncommon enthusiasm, integrity, egoism, and prickliness. . . . McCoy ends her book on exactly the right note: 'There will never be another emperor of wine.'"
—*Slate.com*

"[An] absorbing portrait. . . . As McCoy convincingly argues . . . [Parker] is probably the most powerful critic ever in any field."
—*USA Today*

"The book beautifully chronicles . . . his rise to unparalleled prominence . . . well-researched history, highlighted with gossipy intrigue. . . . For anyone with an interest in the way wine is made, sold, marketed, promoted, and critiqued, *The Emperor of Wine* is a must-read."
—*Seattle Times*

"Elin McCoy . . . [is] uniquely qualified to chart the rise of the most powerful wine critic in the world. . . . Unlike most of the wine industry, McCoy isn't afraid of Parker, and while she finds him personable, she's critical of his influence."   —*San Francisco Chronicle*

"This is an extraordinary exploration of the Parker phenomenon. It's impeccably researched."   —Decanter.com

"Balanced, timely, and brilliantly written . . . *The Emperor of Wine* is a fascinating portrait . . . a superb book."   —*The World of Fine Wine*

"McCoy offers us the best portrait yet of the man, his sins and his virtues. . . . This is a very solid book, well written, copiously documented, and presenting a well-rounded picture of the man so many love to hate and hate to love."   —*Barron's*

"An essential book for anyone interested in wine, but it would also be enjoyed by general readers, especially those interested in a uniquely American accomplishment. . . . McCoy, who has written about wine for thirty years, writes exceptionally evenhandedly and lucidly about this man, who some credit with improving wine around the world and others damn because they believe that winemakers tailor their wines to his tastes to ensure high ratings."   —*Boston Globe*

"Eminently readable . . . [an] impartial assessment of Mr. Parker's taste and influence."   —*The Economist*

"McCoy, a longtime freelance wine writer, articulates this rags-to-riches story well, with plenty of telling anecdotes."   —*Wine Spectator*

"Perhaps the most influential critic writing on any subject, Robert M. Parker, Jr., is a polarizing figure. . . . In this well-researched biography of Parker and his wine milieu by Elin McCoy, herself a respected wine writer, Parker's fans and detractors will find plenty to stew about."   —*Washington Post*

THE
EMPEROR
OF WINE

THE
EMPEROR
OF WINE

The Rise of

Robert M. Parker, Jr.,

and the Reign of

American Taste

ELIN McCOY

AN ECCO BOOK

HARPER PERENNIAL

NEW YORK • LONDON

TORONTO • SYDNEY

HARPER ● PERENNIAL

A hardcover edition of this book was published in 2005 by Ecco, an imprint of HarperCollins Publishers.

P.S.™ is a trademark of HarperCollins Publishers.

THE EMPEROR OF WINE. Copyright © 2005 by Elin McCoy. All rights reserved. Printed in the United States of America. No part of this book may be used or reproduced in any manner whatsoever without written permission except in the case of brief quotations embodied in critical articles and reviews. For information address HarperCollins Publishers, 10 East 53rd Street, New York, NY 10022.

HarperCollins books may be purchased for educational, business, or sales promotional use. For information, please write: Special Markets Department, HarperCollins Publishers, 10 East 53rd Street, New York, NY 10022.

First Harper Perennial edition published 2006.

Designed by Jessica Shatan Heslin

Library of Congress Cataloging-in-Publication Data has been applied for.

ISBN 0-06-009368-4

ISBN-10: 0-06-009369-2 (pbk.)
ISBN-13: 978-0-06-009369-3 (pbk.)

06 07 08 09 10 ❖/RRD 10 9 8 7 6 5 4 3 2 1

To John and Gavin
With love

CONTENTS

THE
EMPEROR
OF WINE

PROLOGUE

S *SPLAT!!!!!* THE MOUTHFUL OF WINE STRIKES the side of the small stainless steel sink, adding a few red splashes to the basin before disappearing down the drain. The taster frowns at the aftertaste that lingers briefly on his gums, his tongue, and the inside of his cheeks. Too tart and herbaceous—he's sure it's been acidified—and stripped of flavor. The wine is the fourth in a lineup of twenty-six California Cabernet Sauvignons. He knows he'd be generous to give this one a score in the low 80s. He'll taste it again, but right now it looks like a disappointing 81.

The next wine has plenty of saturated purple color—it's almost opaque—but he knows many of these California Cabs do. Eyes narrowing, he blocks out all distractions—the hum of arriving faxes, the sound of the rain on the window, the barking of his basset hound outside, and a ringing phone and the murmur of voices in the next room—and sticks his nose into the glass, then inhales quickly with a short, silent snort. A panoply of scents, he thinks. Is that a cassis-like note? Blackberries, or is it black cherries? To confirm what his nose tells him, he tips the crystal glass, sucks in a generous mouthful of wine, and lets it spread across his tongue. After lowering the glass, he swishes the liquid around his mouth for a few seconds as though gargling mouthwash, pauses for a moment while staring into space, his eyes unfocused, then spits it into the sink with expert aim and force. It forms a perfect arc that pings sharply against the metal.

He is still tasting the wine, his mind completely concentrated as he savors the persistent echo of flavor and lingering scents and sorts out his impressions: explosive richness, yes; oodles of powerful ripe black cherry fruit . . . definitely some cassis there; a thick, voluptuous texture, and a wonderfully spicy note in the 30-second finish. It's a hedonist's dream.

The number comes to him. This is a 95—he is certain of that.

When these scores are published a month later, the producer of the 95-point wine will celebrate. If his is a new tiny winery, people will call to congratulate him. He's made the 90+ club. His wine will be sold out by the end of the week—if, indeed, it takes that long. The recipient of the 81-point wine will gnash his teeth; it may be hard to convince some retailers to buy this particular bottling. Too many scores like this and his fortunes will start to decline.

By noon, the taster will have passed judgment on several dozen wines, judgments that will create overnight successes, deflate long-held reputations, and move markets around the world. They will be anticipated eagerly by retailers, who will quickly order the wines with scores of 90 and above, and with trepidation by winery and château owners, who know the scores can cause a rise or fall in their prices. These judgments will be regarded as Holy Writ by thousands of serious wine consumers.

Why? Because the taster, Robert M. Parker, Jr., whose palate has been called the oenological equivalent of Einstein's brain, is the most powerful wine critic in the world. But even more than that, right now he is the most powerful critic in any field, period. If a New York film critic pans or praises a film he may influence its reception in that city, but his view won't have the same effect on moviegoers in Paris or Tokyo, nor will film directors around the world create movies to appeal directly to his taste. But over the past twenty years Parker's passions and ideas have influenced how wine is made, bought, and sold in virtually every wine-growing and wine-drinking country on earth, and there are winemakers who consciously aim to make a wine that will seduce him.

Despite the world's attention and unending acclaim, Parker is a controversial figure in the wine world. He has saved wineries from bankruptcy and turned winemakers into millionaires and unknown wines into sought-after collectibles, but his low scores for other wines have meant lower prices and diminished reputations. He has been sued and received death threats. Parker is blamed by some for helping to turn the world wine industry into a single global market, causing prices to skyrocket, and for reshaping the taste of wine to his own personal preference for dark, high-alcohol wines with lots of power and intensity, and in the process killing tradition and reducing great wines to mere numbers. Others revere him, claiming he is largely responsible for the vastly improved quality of wines made across the globe and is the wine consumer's best friend.

But for all, Parker's scores are the wine industry's report card.

How did an American raised on soft drinks, who never tasted fine wine until he went to Europe during college, become such a colossus? What were the times and circumstances that made his extraordinary rise possible?

1

AN
AMERICAN PALATE
IN FORMATION

*M*ONKTON, MARYLAND, POPULATION 4,615, seems an unlikely hometown for the world's most important wine critic. When Robert McDowell Parker, Jr., was born in Baltimore on July 23, 1947, the landscape north of the city around Monkton was much as it is now—a typical rural American mix of working dairy farms, white clapboard farmhouses and modest brick ranch houses, and patches of second-growth woods and rolling fields. As I drove to his house to spend a day with him, though, a few large, white-fenced horse farms alerted me to the presence of the affluent elite of hunt-country blue bloods who have long presided over point-to-point horse races each spring and fall and attended the annual Blessing of the Hounds at the local Episcopal church.

But Parker didn't grow up on one of those grand estates, where the occasional bottle of wine may have graced the table even back in the 1940s and 1950s. His parents married at eighteen and never went to college. For the first few years of his life, home was the family dairy farm in Monkton, only a 10-minute drive from where he lives today. Some of the first smells to hit his now famous nose were fresh milk, cows in the barn, and hay warm from the sun.

Three hundred years ago, when barrels of Bordeaux wines like Château Haut-Brion and Château Margaux were being regularly auctioned off in London, Monkton and all of northern Baltimore County was still the domain of the Susquehannock and Piscataway Indian tribes. In 1634 it was granted to the king's representative, Charles Calvert, the third Lord Baltimore, who founded Maryland and leased thousands of acres to settlers before giving a choice parcel that included Monkton to his fourth wife. Besides dairy farming and creameries, grist- and sawmills were the earliest industries here, now long gone; the old stone mills survived as antiques shops and homes. Back in the 1950s, when Parker was growing up, Monkton and neighboring Parkton (where he now lives) were mere clusters of houses and churches; Monkton had only a tiny food market, no library or dry cleaner, and the nearest drugstore was 15 miles away.

Today the town is bigger, encompassing all-American nondescript shopping plazas with supermarkets as their centerpieces; a main street with the standard town businesses—banks, pizzerias, auto repair shops, and garden centers; and a bland brick high school set in a field with a sign that reads Hereford High School, Home of the Bulls. This wasn't a fine wine and food place in the 1950s, and it still isn't, despite the growing number of "executive-style" development homes and the presence of two tiny wineries a few miles away. The distance to Baltimore is only 27 miles but must have seemed farther before Interstate 83 was built, linking the city to points north.

Forget the wide range of gourmet staples urban Americans take for granted. The store nearest Parker's house in Parkton turned out to be just a country place with an old-fashioned Drink Coca-Cola sign and out front, a barrel of cabbage flowers edged with American flags. The local give-away paper has ads for popular wines he ignores, like Beringer White Zinfandel, and lists of tree farms where you can cut your own for Christmas. The score of the latest Hereford Bulls basketball game is the big news.

Parker's childhood was a typical one in small-town America of the 1950s, which meant that food was plentiful but unimportant. His mom's kitchen repertoire was unsophisticated farm cooking of the meat loaf and fried chicken variety, but apart from her wonderful banana cream and lemon meringue pies, Parker remembers having no particular opinion about what appeared on the family table. Wine was never served; the beverages of choice were milk, soda, or coffee, and the occasional cocktail. As a teenager Parker wasn't the least interested when his father, Robert Sr., who drank Bourbon and loved inhaling its rich scent, held out a glass and suggested, "Oh, you should smell this." And when Robert Sr., an enthusiastic hunter who kept bluetick hounds, told Robert Jr. on a hunting trip that you could tell the breed of a dog by its smell, the boy thought he was joking. Only much later in life did he realize that his father had an unusually acute sense of smell, and that he had inherited this same ability.

It was a happy childhood steeped in the middle-class normalcy of riding bikes, roughhousing with the family cocker spaniels, playing soccer, hanging out with friends, and enjoying the security of a stay-at-home mom, Ruth "Siddy" Parker, who never left her only child with a baby-sitter and didn't work outside the home until he was in college. When he was four his father gave up the dairy farm, the family moved a few miles to a newly built ranch house, and Robert "Buddy" Parker, Sr., went to work selling construction site equipment for a Baltimore-based oil company, eventually becoming a vice president. He traveled extensively for his job, but Bob Jr. didn't go much farther than Baltimore, Washington, D.C., and once by train to New York City. This cluster of towns in northern Baltimore County, now known as the Hereford zone, was his world.

A soccer star in high school—he twice made the all-county team as goalie—Parker was popular, handsome, and easygoing, with the boundless energy that would one day carry him through marathon wine tastings and 12- to 14-hour days visiting vineyards. His intensity, competitive spirit, and fierce determination emerged on the athletic field, where, his mother once observed, "he'd kill himself to win." During one faculty–student soccer game, he was carried off the field and rushed to the hospital with caved-in ribs and bruised lungs after a teacher butted him in the chest. While most of the students at this "redneck" high school were 4-H

Club future farmers, Parker was in the college-bound program for smart kids, but was widely popular anyway.

His doting but strict parents encouraged him to talk about anything—no subject was taboo—and his mother emphasized two apple-pie American principles: be true to yourself and be honest with others. In turn, like many only children whose parents lavish affection on them, Parker adored his parents, despite differences in what he terms "philosophies of how to live life." He and his father were, as he says, "completely different people." Robert Sr. loved to ride motorcycles; Robert Jr. didn't. His father loved hunting and fishing, and decorated his office with bear pelts. But when they went on a few hunting trips together, Parker found he didn't enjoy "killing just for the fun of killing."

If all this sounds completely unremarkable, it was. The only clues that Parker, who was known as Dowell (a nickname derived from his middle name that is still used by his wife and closest friends), would become the premier taste arbiter of the world's wines were his deep self-confidence and his conviction that he could go his own way with impunity.

———

PARKER IS SINGLE-MINDED IN MANY THINGS, AND LOVE IS ONE OF them. He never tires of telling the story of how he met Patricia Etzel when they were both twelve years old, fell in love with her in high school at age fifteen, married her in 1969, and is still married to her more than three decades later. She and her family, he told me with evident emotion, changed his life.

In 1959 Pat's parents had moved to a ranch house in Parkton—the same house Parker and his wife live in now—from suburban Long Island, where they had kept a boat. Although they didn't drink wine, "they entertained a lot and their cooking was more sophisticated," as Parker remembers noticing when he came to their house to pick up their pretty, lively daughter for dates. It was in their house, at Pat's eighteenth birthday party, that he had what he now calls "my first taste of what I thought was wine"—André Cold Duck, a bubbly sweet mix of red and sparkling wine bottled by Gallo that was all the rage in the mid-1960s. (Though it later fell out of favor, this bottom-of-the-barrel wine is still available today, nearly forty years later, for a mere $3.99.) Parker's first experience with wine was not a success. He got so drunk he could barely walk, and Pat's father had to drive him home. In the middle of the night Parker woke up feeling queasy, staggered over to what he thought was the toilet, wrapped himself around it, threw up until he felt weak and exhausted, and immediately fell asleep. In the morning he discovered he'd clutched an open dresser drawer and vomited all over his clean clothes. His mother was furious. It wasn't an experience he wanted to repeat.

After graduation in 1965, both headed off to college, Parker on a soccer scholarship to Randolph-Macon in Ashland, Virginia, Pat to Hood, a then all-women's college in Frederick, Maryland. Petite, dark-haired Pat was the star student and the adventurous one. She'd dreamed of going to Europe on a big passenger liner since she was eleven years old and saw ships docked at the piers in Manhattan. And she was focused on mastering French, a skill that would prove a critical help to Parker in the years to come. In contrast Parker, now a rangy 6 feet 1 inch with the sideburns and longish hair then gaining popularity among students, had no idea what he wanted to do with his life. So far he hadn't shown passion about anything other than soccer and Pat. A year later he transferred to the University of Maryland in College Park, joined a fraternity, and like many of his friends became opposed to the war in Vietnam. He held off sending in his student deferment form after knee surgery for a soccer injury temporarily put him on crutches. When the inevitable induction notice arrived, Parker hobbled down to Fort Holabird for the physical. His injury earned him a permanent 1-Y deferment. At college he smoked his "share of joints" and enjoyed getting high, but drinking alcohol still held little appeal.

That would change in December 1967, when love for his girlfriend motivated twenty-year-old Bob Parker to travel to Europe for the first time. That fall Pat had sailed on the S.S. *Statendam* for France, where she would attend the University of Strasbourg during her junior year of college. Ships in the Holland-America line hauled many American students to Europe back then and I'd made the crossing myself on one of them, the *Seven Seas*. It carried only students and served spaghetti and meatballs every night, but Pat's was a luxury liner, she remembers, "with real tourists and good food." Spending a year abroad wasn't yet the common experience it is for college students today, and Europe seemed very far away to Parker. Telephoning was difficult and expensive, and their few conversations were short and awkward; and while they exchanged letters, Pat didn't write often. Though Parker's parents had given him permission to meet her in France over Christmas vacation, it had been a long four months, and he brooded over their lack of contact. He thought his pretty girlfriend looked like a young Natalie Wood, with her long, dark hair, slim, graceful figure, and a smile that lit up her face, and he worried that she would be swept off her feet by some suave French student she'd meet in a café. Besides, Parker knew that her parents weren't thrilled by their relationship and that they hoped Pat might forget about him and find someone else. Determined not to let that happen, he dropped out of school to spend six weeks traveling with her.

For Parker the trip was momentous, a turning point in his life, and he still delights in recounting the experience and pointing out his youthful naiveté. By now he's told the story so many times that it has become a well-polished narrative, like a favorite family tale handed down through the years. "I wanted to look as good as possible," he told me nostalgically as we sat in his cluttered, bottle-strewn office. "I put on a brand-new three-piece suit and a white shirt—which was ridiculous to wear on a plane—and my father insisted I shine my shoes." Unfortunately, a pocket of air turbulence on the short flight to Idlewild airport (now JFK) spoiled that idea as he spilled coffee all over the suit. It seemed a bad omen.

On the Pan-Am transatlantic flight Parker was in awe of his seatmate, a sophisticated, well-traveled Harvard student who spoke fluent French, whose father worked in Paris, and who found it hard to believe his flying companion was someone who had never been on a plane. "I felt like this little yahoo," Parker said, "Elmer Fudd sitting next to James Bond." His other concern was Pat, who was supposed to meet him at Orly airport when the plane landed at 10:30 P.M. About an hour before the scheduled landing, all the worries of the past months crowded into his thoughts—that she wouldn't be there, that she wouldn't be as much in love with him as he was with her. He tried to still his nervousness by guzzling a couple of drinks, then promptly fell asleep.

Parker vividly remembers waking up, looking at his watch, which said 10:45, leaping from his seat, and shouting at the stewardess, "Damnit, I missed my stop!"

"Just calm down," she replied. "Airplanes are not like trains. We'd never let a passenger sleep through a scheduled stop," this last delivered with a smirk.

When Parker sat down, the Harvard student looked at him in amazement. Then came the announcement from the captain that due to heavy fog at Orly, the plane would be diverted to Rome. Parker was devastated. "I can't believe my luck," he moaned to his seatmate. "I don't speak a word of Italian, I don't know anything about Italy. How in the hell am I going to get from Italy to France?"

His worldly seatmate explained that they would stay in Rome only overnight, that Pan Am would fly them to Paris the next day and foot the bill for all of it. Parker, relieved, decided the unexpected Italian detour "sounded great."

The first thing that struck Parker about Rome were the smells—even on the bus into the city they seemed so foreign, especially the acrid smell of horse urine from a gypsy encampment next to the Colosseum near his hotel, the Napoleon. They arrived at 1 A.M., and since they had to catch the bus back to the airport five hours later, there was no time to lose. Parker joined several other young men from the ill-fated flight to look for an open bar. As they wandered through the dark streets, he took in the smells, the dawn rising over a strange city, the ancient buildings and aura of history with enthusiasm, and after a few drinks learned to pronounce "lire" correctly.

His European education had begun.

The next day Parker flew to Paris, which was still so foggy that he couldn't see the tarmac when they landed. He didn't know enough to be as anxious about the conditions as the businessman next to him. His worries about Pat, he quickly discovered, were groundless. She'd slept at the airport—by the time they'd canceled his flight there was no transportation back to Paris—and was waiting there, as beautiful as he remembered. During high school Pat had viewed her boyfriend as quite a catch. Now that he'd come all the way to Europe to pursue her, she was more in love with him than ever. Pan Am had refused to tell her whether Parker's name was on the passenger list, but she had been convinced he was, sure he wouldn't let her down.

Just about everyone remembers his first view of Paris, and Parker's was romantic and dramatic. He and Pat boarded the Métro to the Place du

Trocadero, where she made him walk up the steps backwards with his eyes closed, then playfully told him to turn around. And there was picture-postcard Paris with the Eiffel Tower across the River Seine, right in front of him. "Oh my God, wow!" was Parker's response to this and much that followed.

That first day, Pat took him to an Alsatian brasserie and ordered "all these wonderful weird things I'd discovered." If she was testing his willingness to be adventurous, he passed, embracing everything with gusto. There was a bucket of mussels and a plate of snails, neither of which he'd had before. Since Coca-Cola was so expensive, a dollar for a tiny bottle, Pat insisted he try *un verre du vin*, the first dry wine Parker had ever tasted. For someone raised on meat loaf and soda, these tastes were all new and wonderful, a revelation, and it didn't hurt that he was so much in love. Parker couldn't get over the different aromas and flavors in the food and wine, and he wanted to taste everything—frog's legs, pâté, Camembert—and much more wine.

The couple spent their first night at the Hôtel du Midi, a $3-a-night fleabag in the Quartier Latin. Their room was ice cold, with a floor so slanted it was disorienting, and they wouldn't take off their socks for fear of what they might step on. But thirty-four years later, Parker fondly remembered the wonderful smell of chestnuts being roasted on the street right below their window.

The next day they moved to the nearby Hôtel Danube on rue Jacob, a step up at $5 a night (it even had an elevator) and much cleaner, and for several days they wandered around Paris, in love with the city and each other. They spent their evenings eating and talking for hours in "dirt cheap" neighborhood bistros where they carefully calculated the cost of the *plat du jour* to make sure they could also afford to order wine. Parker had brought $600 with him, which seemed like a lot of money, but they schemed to make it go as far as possible. The simple carafe wines they drank each night with their prix fixe dinners had no names or labels and were basic regional *vins de table*, doubtless from the Loire or Beaujolais. They wouldn't pass muster with Parker today, but from the first they fascinated him as no other drink ever had. He noticed they all tasted different and enhanced the food, promoted conversation, and gave him a comfortable buzz. Wine's effect, he observed, was very mellow and gradual; it didn't make him drunk or blur his vision as liquor did or feel bloated and heavy the way beer had. He didn't have a clue how to describe or analyze the wines' differences, but he knew they tasted good.

For Parker, as for the many students who visited and lived in France during the 1960s and early 1970s, falling in love with French food and wine was a key part of falling in love with France and French culture, which seemed so much more sophisticated than anything back home. Thanks to the introduction of large jet airplanes—Boeing 707s that could carry 160 passengers—into commercial service in the mid-1960s, more and more Americans were traveling there. In today's world of cutting-edge American cuisine, it's hard to imagine the enormous impact on all these middle-class students of seeing wine on every table in restaurants, of eating a simple *omelette aux champignons* or a basic *poulet rôti* three-course dinner at a tiny Left Bank bistro, of buying a long baguette still warm from the oven and choosing among fifty completely different cheeses at a *fromagerie* for a picnic lunch in the Luxembourg Gardens or the Tuileries.

What a contrast with the food most Americans had grown up on! In the 1950s and early 1960s bread was white and squishy; cheese was Velveeta and thick packages of "Swiss" with big holes in the slices; vegetables meant frozen succotash, green beans cooked with Campbell's cream of mush-room soup, and canned peas; salads were thick wedges of iceberg lettuce covered with pink Thousand Island dressing. On the dinner plate were steak or hamburgers, or sometimes pork chops, with baked or mashed po-tatoes, and occasionally spaghetti with meat sauce. Often the center of the meal was the ubiquitous hot dog, an American obsession. None of the dishes had the subtle flavors of herbs like tarragon or thyme, or a hint of fresh garlic. Shrimp cocktail seemed luxurious and exotic. Dessert was red or green Jell-O cut into squares, or perhaps Tollhouse chocolate-chip cookies and Sealtest ice cream. The best part of a meal was often—as it was in Parker's home—a delicious homemade pie.

France seemed like another, better world, one where people were more attentive to sensual pleasures, kissed in cafés at breakfast, and knew how to live the good life. Parker found the idea of lingering two or three hours over a meal, eating a whole series of courses rather than chowing down one plate of meat and potatoes, and sipping wine along with it "as civilized as spending a day in the Louvre."

The couple spent Christmas in Germany with friends of Pat's parents who served a sweet wine that unfortunately reminded Parker of his cold duck experience, but on New Year's Eve they were happily back in France celebrating at a bistro, Pot d'Etain (the only restaurant name from that trip that has stayed in their memory). On their six-week itinerary, of course, was northeastern France, and Strasbourg, where Pat had been living. There, Parker's experience of food and wine expanded even more, when a local doctor whom Pat had met introduced the couple to some notable restaurants and treated them to several expensive dinners that included far finer wines than those they'd been able to afford. This generous gourmet clearly enjoyed sharing his knowledge about wine and food with someone so keen to learn.

Though Parker was beginning to pick out some of the different flavors in these wines, which were tantalizing enough that he knew he wanted to learn more, simply drinking them was part of his delicious immersion in the food, wine, and culture of France. Only a "horrific" meal of boiled brains on an excursion to Germany marred his palate impressions. Other than that he ate French food, and drank French wine, and he loved it.

His world was opening up in other ways, too. Though he was majoring in European history in college, Parker had arrived with only an abstract understanding of how World Wars I and II had devastated Europe and how much people had suffered under bombardment and occupation. But now the evidence was in front of him—in the 1960s the landscape was still scarred with bombed and burned-out ruins that had never been rebuilt, not to mention the plethora of monuments and graveyards, and seats on buses for *les mutilés de guerre*. Parker realized he'd led a very sheltered life and seen the world narrowly. His natural sympathies were aroused, and he was filled with admiration for the French, who had, despite their recent history, been able to preserve their joie de vivre and their enjoyment of fine food and wine.

The couple had traveled so frugally that Parker had a couple of hundred dollars left when they arrived back in Paris at the end of their holidays. "Let's blow it all," he insisted, so they went to the Lido, an expensive nightclub on the Champs-Elysées popular with tourists, which offered a Ziegfeld-type spectacle of dancers, show girls, and three orchestras. And they decided on a grand final dinner at the then world-famous, three-star restaurant Maxim's at 3, rue Royale, noted for its luxurious fin-de-siècle décor and elaborate haute cuisine. Though it was definitely on the tourist circuit at the beginning of 1968, Maxim's was still one of the city's top restaurants and a culinary shrine. To the American gastronome Waverley Root, who dined there the same year, however, Maxim's was more a bastion of chic celebrities and "the richer variety of snobs," who went there to see and be seen rather than eat. Although the food "can be top-level," he wrote, it was also uneven and often overpriced, as Pat and Parker were about to discover. But their experience was worse than that.

In their cheap hotel the couple did the best they could to bring their clothes up to the standards of a grand restaurant. Pat remembers using two books to press the collar of Bob's white shirt; Parker remembers shining his shoes. In a photograph taken that evening at the restaurant, Pat looks young and beautiful in a simple black spaghetti-strap dress and Parker appears serious enough in his three-piece suit to critique anything served him. But this was not enough to impress the restaurant staff, who were disdainful and arrogant in a way the French of that era had polished to perfection.

The evening started badly and went downhill from there. The couple were immediately banished to "Siberia," a dining room full of unchic tourists, probably by the famous head waiter Roger Viard, whom Root described as being able "to separate the U's from the non-U's with a single piercing glance." An undaunted Parker complained to a haughty waiter that the electric lamp on their table was flickering. The waiter's efforts to fix it gave him such a strong shock he stumbled and fell to the floor, to the couple's barely concealed amusement. The food and wine were disappointing. The dessert had a pastry crust so hard that when Parker tried to cut it, the whole thing shot off his plate, struck a passing waiter, and stuck to his pants. The maître d' even asked them to leave the tiny dance floor because Parker's shoes were the wrong color. The final blow was the outrageous bill. Still, the experience didn't dampen his new enthusiasm for French food and wine.

It was "one of the saddest days" in Parker's life when he put Pat on the train back to Strasbourg at the Gare de l'Est and waved goodbye, and then flew back to the United States. His life was forever changed. He was determined to continue two romances: with Pat and with wine.

BACK AT THE UNIVERSITY OF MARYLAND, PARKER LOST NO TIME IN pursuing his newest passion. The French word *formation* means training or molding, and usually refers to the formal education or apprenticeship an individual undergoes to develop his knowledge of a subject or to shape his taste in, for example, literature or art. But the shaping of Robert Parker's palate was undertaken without professional training or tutoring from a mentor as had been the tradition in Europe, where wine experts came out of the wine business. Instead it began quite informally, with like-minded amateurs and a self-organized curriculum.

Eager to learn everything he could, Parker first bought the wine books available at the time: several volumes by the English writer André Simon, Alexis Lichine's *Wines of France* and his recently published *Alexis Lichine's Encyclopedia of Wines and Spirits*, and Frank Schoonmaker's *Encyclopedia of Wine*.

Schoonmaker and Lichine were then the two most important wine authorities in America. In fact, they'd been transforming the country's wine business for decades, crisscrossing the United States to introduce the fine wines they imported to consumers and at the same time educating the American palate when no one else was doing so. Neither came from families with any connection to the wine industry. Schoonmaker, like Parker, fell in love with wine at age twenty, having dropped out of Princeton University in the 1920s to travel to Europe and take up what he thought would be a much more adventurous and interesting life. While writing guidebooks (one, *Through Europe on $2 a Day*, was the model for the later Frommer guides), he met Raymond Baudoin, the much feared and influential editor of *La Revue du Vin de France*, and regularly accompanied him on his tasting trips to various wine regions. Baudoin, known for his uncompromising palate and fierce temper, "could cast judgment on an entire lot of wine with a quick stroke of his pen," as wine writer Frank Johnson put it. Schoonmaker learned at his side, recording his wine impressions methodically in a little black notebook just as Parker would do decades later. He was confident that Prohibition would eventually be repealed, and planned to import wines to America using Baudoin's palate to help him select the wines.

Lichine, on the other hand, was born in Russia just before the Revolution, grew up in Paris, and was exposed early to wine's pleasures, but only became convinced his future lay in wine after visiting France's major wine regions to write a series of profiles for the Paris *Herald Tribune* (later the *International Herald Tribune*). The editor hoped this would attract wine advertisers eager to reach the American public when Prohibition ended.

Schoonmaker and Lichine joined forces for a time after Prohibition's repeal, at Schoonmaker's wine shop in New York and as importers. Anticipating World War II and the difficulty of bringing in wine from Europe, they championed California wine and on a trip to the state persuaded a few wineries to label their wines by grape variety (they coined the word "varietal"). As a result, their sales of California wines soared. After the war, during which both were in intelligence and Lichine provided wines for Churchill and Eisenhower, the two again imported fine European wines. Consumers who didn't know much about wine often relied on those that carried a "Frank Schoonmaker Selection" or "Alexis Lichine Selection" label. Both men tirelessly promoted wine; the urbane, glib Lichine was an especially effective propagandist who once suggested to the governor of Wisconsin that he lay down a wine cellar, then ended up selecting the wines for him and selling thousands of bottles in the state. Schoonmaker reached budding gourmets through his articles in *Gourmet* and *The New Yorker*, and his book *The Wines of Germany* reportedly impressed German chancellor Konrad Adenauer, who sent Eisenhower a case of one of the recommended wines. Lichine became such a celebrated authority on French wines that he was dubbed the "Pope of Wine." He bought a Bordeaux château and was eventually given the Legion of Honor. Parker read their books voraciously, sometimes two and three times, as if he were trying to cram for a college exam.

He also formed a wine tasting group, recruiting willing students from his fraternity and his classes. "It was a time when everyone was game to try anything—I could probably have found people to taste marigolds," he recalled. Soon he'd assembled five or six regulars, and sometimes others whom no one had ever seen before would turn up at tastings attracted by rumors of alcohol. "About half of this crazy group came totally smashed on dope or hash or something," Parker said, but already he drew the line at smoking during the tasting itself. He didn't have any idea how to structure a wine tasting, but his sensitive nose told him that strong odors interfered with tastes and smells. No one thought of spitting out what they sampled as professionals do. Part of the fun of tasting wine a couple of times a month was "getting a good buzz on."

The group kicked in contributions of several dollars each to a wine budget that seems laughably low today. But simple Côtes du Rhônes and Beaujolais were very cheap, good Burgundies could be had for $5, and top Burgundies rarely cost more than $10 or $15. These were the wines everyone wanted to taste; Bordeaux was considered too grand, and though the group occasionally included wines from California, no one was really interested in them. Most of those available were jug-level wines like Almadén Mountain Red Burgundy selling for about $1 a bottle. Inglenook North Coast Counties Burgundy, at $1.50, was slightly better and vintage dated. Parker bought all the wine, shopping regularly at big shops in Washington, D.C., such as Pearson's and Central Liquor (which had the lowest prices) because "there were no real wine shops in Maryland."

Washington D.C., with its cosmopolitan population of foreign ambassadors and well-heeled, well-traveled government employees, was a town of serious wine lovers, and local wine shops responded with a good selection at reasonable prices. Calvert Liquors in Georgetown, owned by Marvin Stirman, had brought in nearly one hundred cases of every *grand cru* Bordeaux from the great 1961 vintage and sold them all. Another top store, Addy Bassin's MacArthur Liquors (now MacArthur Beverages), had opened ten years before, in March 1958, and was devoting more and more of its floor space to wine. Both would play a key role in Parker's future.

The point of the tasting group was to have fun, not to study wine, but as it evolved, with members dropping out and others joining, Parker was clearly the one who took it seriously, and he quickly became an early example of the wine geek. It's easy to imagine him in his jeans, brown hair curling around his collar, carefully lining up the bottles, tasting them intently one by one and trying to sort out his impressions. It became increasingly apparent to him that he had a more sensitive sense of smell and taste than the others and that when it came to wine, he had powers of focus and concentration that he'd never had for any other subject. He was beginning to "see" the wine when he tasted it, as though the taste and smell and texture were architectural. He tried to capture what he thought of each wine by writing down a note in clear, direct language and short sentences, but Parker barely knew what to look for. The books he was reading weren't much help. They were packed with fascinating historical details on regions and châteaux, statistics on vineyards and wine production, and long lists of the best vintages and traditional hierarchies of producers, but they didn't say much about the ways in which one wine tasted different from another. That was what really interested him. Instead writers waxed rhapsodic about a wine's taste, often comparing it to women. To explain the difference between two Bordeaux first growths, Lichine quoted Jean-Paul Gardère, the manager of Château Latour: "Lafite is a svelte and elegant Madame Récamier, while Latour is more like a Rubens heroine." André Simon was equally unhelpful. He described a young Château Margaux as a "girl of fifteen, who is already a great artist, coming on tip-toes and curtseying herself out with childish grace and laughing blue eyes." Sometimes writers mentioned a wine's "refinement and great breeding" or "goût de terroir," or called a Beaujolais "charming." Parker found it frustrating. How did any of it help you know whether you should buy a wine or not?

By the time Pat returned from France for her senior year at Hood College, Parker was living off-campus in an apartment. This made it harder for him to pull a tasting group together very often, but he persisted. One of the people who joined the tastings was Victor Hugo Morgenroth III, who was studying to be a toxicologist. He was "a brilliant, brilliant taster," Parker recalled, who was as serious about wine as Parker was, and the two quickly became friends. Although Morgenroth loved German wines and Parker didn't, they were in definite agreement that many famous wines were not up to their reputations. One of their first major disappointments was a 1957 or 1958 Château Lafite-Rothschild—Parker doesn't remember which—for which they paid a whopping $12.50. That hardly seems expensive compared with the cost of Lafite today (the famed first-growth Bordeaux starts at about $150 a bottle), but a double-digit price tag seemed a huge sum to the group, and they were expecting greatness. In anticipation, they flipped through several books to read about the château's illustrious three-century history. They learned it is situated on the highest knoll in the commune of Pauillac, that the vineyard's good, gravelly soil is one of the reasons the wine has such extraordinary finesse and complexity, and that it was reputedly a favorite wine of Madame du Barry in the eighteenth century. Then they tasted it. The wine was awful.

Parker turned to everyone and asked, "Do you all think what I'm thinking?" Morgenroth exclaimed, "Yeah! This tastes like cat piss." All of them agreed it was an acidic, horrible wine. "If I had tasted the wine alone," recalled Parker, "I'd probably have thought it was just me, but the reaction was unanimous." Everyone was outraged.

Neither 1957 nor 1958 was a particularly good year in Bordeaux, but the possibility that this might have been a bad bottle of what at best would have been a lean claret, or one that hadn't been stored properly, didn't occur to Parker at the time. What stayed with him was the idea that a wine as famous as Lafite could be vastly overrated and could coast on its reputation.

═══

IN THE SUMMER OF 1969, AFTER PAT GRADUATED FROM HOOD, the couple were married in an old brick church in Monkton. Parker had just turned twenty-two; Pat, ten days younger, was still twenty-one. Parker knew that her parents were not happy at their "pride and joy" marrying "this farm boy." The reception for 150 family and friends at their house was something of a mixed celebration. No Champagne—or wine of any kind—was served, and there was no photographer to record the happy occasion. The newlyweds quickly left for a five-day stay at the beach before flying to Europe for a four-week honeymoon, a week each in Lisbon, Madrid, Paris, and London, which was a wedding gift from Parker's parents.

Parker was looking forward to advancing his wine knowledge, but an abscessed impacted wisdom tooth interrupted their stay at the beach. After it was pulled, the painkillers turned him into a "zombie" and kept him from drinking during the first few days in Lisbon. For the rest of their stay in that city the couple downed quantities of fresh, fruity, slightly sweet and lightly sparkling chilled rosés, the only wines they could get. The most recognizable brand they drank—Mateus—was wildly popular in 1960s America, especially on college campuses. No restaurants or wines stand out in their memories of the rest of the trip, but it's safe to say that none was very grand as they went armed with a copy of the 1960s bible of cheap travel, Arthur Frommer's *Europe on $5 a Day*, and closely followed the book's recommendations.

The day after they arrived home, Pat started teaching French at a local public high school at a salary of $6,900 a year. It was their only income. Parker wouldn't graduate from the University of Maryland for another year, and although he'd become a vastly improved student, he still had no idea what he wanted to do with his life. The couple lived just outside College Park in a large basement apartment that had enough space for a Ping-Pong table. They had a wine cellar of one hundred bottles, which were stored in the living room on a makeshift rack built with bricks and planks of wood. Parker insisted they keep the apartment at a chilling 55 degrees, which he'd read was the perfect temperature for storing wine. Cold-sensitive Pat was forced to walk around the apartment bundled in sweaters. To save enough money to afford summers in Europe, the young couple subsisted on vegetable soup, hot dogs, and macaroni and cheese. Their budget was so tight that Pat's lunch at school was a Coke and vending-machine peanut butter and cheese crackers; coffee, fortunately, was free. At the end of their first year of marriage, she weighed 99 pounds. It was a starving-student existence except for one fact: they bought good wine, and drank it every night.

The best wine in Parker's starter collection was a 1966 Vidal-Fleury Côte Rôtie, a red from the northern Rhône Valley. It was his favorite wine, one that he had discovered at a tasting, where "it just blew people away." No one knew anything about Côte Rôtie (some in the group pronounced it "koh-tee roh-tee"); Rhônes weren't well known or fashionable in the United States then the way Bordeaux and Burgundy were. But Côte Rôtie was such a "seductive, voluptuous wine," Parker would write later, that it "takes little experience to appreciate." Parker himself would be partly responsible for the rebirth of interest in these wines.

Another favorite was a Beaujolais, Château des Jacques Moulin à Vent, which Parker considered "phenomenal." Moulin à Vent, the top *cru* of the Beaujolais region, is noted for its depth and richness, and Château des Jacques made several special cuvées. These and the Côte Rôtie were their house wines. Parker never intended to age any of them; his philosophy was "if it's good, drink it." The couple entertained frequently—serving more ambitious menus than their standard macaroni and cheese—and generously and readily poured their wine for guests. In fact, wine had become such an important part of Parker's life that he "began to choose friends on the basis of whether they liked wine or not." If they didn't, he didn't want much to do with them, and, he says, "it was sometimes a problem." Pat, however, was starting to complain about how much money they were spending on wine, as well as the temperature of the apartment. Though she was young and in love and willing to put up with a lot, there were arguments.

Getting a job, the obvious answer, seemed a grim prospect to Parker, so he applied to law school. He wasn't particularly interested in the law, but the idea of becoming a lawyer to change the world was in the air at the time and at least it would put off the decision of what to do with his life for another three years. (It didn't occur to him that he might seek employment in the wine industry.) After graduating in June 1970, with a B.A. honors degree in liberal arts with a major in history and a minor in art history, he planned to attend University of Maryland Law School in the fall.

That summer he and Pat left for Europe and their first visit to a wine region. Renault was offering a plan whereby tourists could buy a car at the factory, then sell it back at the end of their stay, and Parker and Pat signed up.

En route to Spain's beaches, they drove to Bordeaux and stayed for several days. Parker hadn't made any appointments, assuming they could just "knock on the door" of any château. They made their way north on the D-2, the main road through the Médoc, eagerly watching for château names they recognized, impressed with the sweep of vines fronted by low grey stone walls and the grand, formal châteaux. "Hey, I know that name, let's go visit them!" Parker would say, then pull in and park in front of the château's *bureau*. Their first stop was Château Lascombes, north of the village of Margaux, which was then owned by a group of wine-loving Americans that included Alexis Lichine. Château Lascombes and Lichine's Château Prieuré-Lichine were the first châteaux in the Médoc to post signs welcoming visitors, and Lascombes was high on Parker's list of must-visits because he'd tasted a bottle of the 1966 vintage and thought it was "a beautiful wine with great complexity."

"Just tell them I love their wines and ask if we could visit and taste," he impatiently instructed Pat. Few tourists dropped by the vineyards in France's wine regions in those days, and the *maîtres de chai* (the men in charge of the winemaking and wine cellars) were clearly taken by the novelty of this pretty young American speaking impeccable French. She chatted and charmed, and soon Parker, who wanted only to taste, would be holding a glass for the *maître de chai* to dribble in a few ounces of the astringent but fragrant young wine drawn from a barrel with a glass pipette. Pat heard the details of the vintage, which she translated for Parker. It was a formula that worked wonders. They visited about a dozen top estates and the staff at all, including first growths like Châteaux Lafite and Latour, were "wonderfully sweet and nice" to them, Parker recalled. No one turned down their request to taste. Pat's French and her warm, engaging personality, Parker knew, had been essential to their friendly reception.

Parker wanted to know what the places he'd read about looked like, so they drove through all the important wine areas of Bordeaux. He and Pat found the walled medieval village of St.-Emilion, surrounded by vineyards, especially beautiful. They even managed to make a pilgrimage to tiny Château Pétrus in neighboring Pomerol, which had a cult following in the United States. Parker was not yet so obsessed with wine, however, that he wanted to spend his entire summer visiting wine regions, so they headed for the beach.

On their return, law school loomed. In September, on the first day of classes, Parker realized to his dismay that "these other students were really serious." He, on the other hand, was barely going through the motions. Later, the fact that he appeared only in classes in which the professor took attendance earned him the nickname "the phantom." A night owl, he was addicted to staying up late to watch Dick Cavett on television and usually slept until noon. Parker remembers that most of the classes bored him, except for a course on conflict of interest taught by Watergate counsel Sam Dash. The consumer movement was growing at the time, and Ralph Nader's task force of young people, "Nader's Raiders," were making the news with a series of important exposés, which also made a deep impression on him.

Unlike the other law students in his class, Parker had no intention of spending his summers clerking in law offices. Though both professors and student friends thought he "was out of his mind," as Parker put it, he took the view that they were the ones who were crazy. They were all going to be working like dogs as young lawyers when they graduated, he reasoned, and none of them would ever again have two months off every summer to see the world. He intended to take advantage of this freedom and visit all the wine regions he could.

The next two summers were a continuation of the young couple's first summer in Europe. They traveled light and lived cheaply, but shopped carefully for restaurants, scrutinizing the posted menus, and they always, always drank wine. "We got high every night and had a blast," Parker recalled. Eager to learn and experience everything, they crammed as much as possible into every day—"at least three museums and a cathedral," Pat remembers. Nineteen seventy-one was the year they first wandered through the vineyards of Burgundy and the Rhône Valley. From his reading Parker had become fascinated by Provence and Avignon, and a visit there introduced him to the full-bodied wines of Châteauneuf-du-Pape. He became an immediate fan, and the region became Pat's favorite in France.

But equally important to Parker was a goal common to many in his generation: getting to Tangier and Marrakech "to score some good hash." The trip proved more adventurous than he'd anticipated. First came the painful cholera shots in a filthy clinic in Spain, which were required to reenter the country because of an epidemic in Morocco. Then the couple had trouble getting into Morocco, even though they had chosen a supposedly more liberal entry point, because Parker refused to cut his then very long hair. Pat, as usual, smoothed the way with immigration officials. And finally there was the Moroccan political situation. A group of dissident army officers stormed King Hassan II's palace during his grand birthday reception on July 10, 1971, while 10 miles away, hundreds of rebels assaulted the radio station, army headquarters, and Interior Ministry. The attempted coup was bloody (ninety-two people killed) but ultimately unsuccessful, and within three days the ten leaders were executed by a firing squad. Pat and Parker, unaware of the situation, were on their way from Tangier to Marrakech—Pat was driving while Parker was trying to light up some hash he'd purchased after haggling with a street dealer—when they suddenly came upon a series of barricades manned by machine-gun-toting police. By the time they reached Fez, Parker, worried about the military presence and nervous about his stash under the car seat, was ready to leave.

France continued to draw the couple, but they also visited other wine regions—Rioja, Piemont, and Tuscany. Wine wasn't Parker's only passion at the time. He lavished a similar intensity and focus on photography. On these European trips he would drive for miles and spend hours to set up a particular shot exactly the way he wanted. Once he even got into a fight when another photographer moved his tripod and took his chosen spot in front of the Acropolis while he and Pat were eating lunch. (A determined Parker won.) Surprisingly, he never took photographs of châteaux or vineyards. Later, he would speculate that was because wine was about the taste in the bottle, not about what the vineyard or château or label looked like.

Instead of traveling to Europe in the summer of 1973, Parker crammed for the bar exam. He hadn't been a very diligent student and was consumed with fear that he would fail. To his surprise he pulled through, passing it on the first try. To celebrate, he and Pat toasted their future with 1964 Dom Perignon and sole amandine, and a brief vacation in Montreal and Quebec, so they could spend some time in a French-speaking environment. As a graduation present his parents gave him a case of 1970 Château Lafite-Rothschild. He was overjoyed to receive it, he recalls, "knowing the château's reputation and the excellence of the 1970 vintage."

PARKER ABHORRED THE CUTTHROAT COMPETITION AMONG HIS fellow law students (friends had refused to share their notes for classes he'd missed), and the dog-eat-dog atmosphere and long hours required to become a partner at a corporate law firm held no appeal for him. Banking law, he concluded, might be more relaxed. In fall 1973 he began his law career at Farm Credit Banks in Baltimore. Pat was relieved. Parker's biggest worry, that he wouldn't be able to get up at 7 A.M., proved groundless. But the workload of bankruptcy and commercial law, while consuming at first, soon bored him.

In truth, he found wine much more fascinating than any aspect of the law. Parker wanted to discover who made the best wines and why their wines turned out better than those of their neighbors. By now he was sure he was a good taster, which he credited more to his ability to focus and concentrate than any innate ability, and he started systematically training his brain to isolate smells. As he walked down Baltimore streets he would practice identifying the ones that were similar to those he'd detected in wines: "That's melting asphalt, that's a hot summer day, that's exhaust fumes, that's decaying garbage." He read everything he could on wine, and on Wednesday, on the way to his office, he always picked up *The Washington Post* to read the wine column written by the paper's food editor, William Rice. He continued his tasting group, met frequently to try wines with his friend Victor Morgenroth, and he and Pat traveled to Europe whenever they could.

Wine had become Parker's all-consuming interest, but he had no idea how to make it his life.

2

THE
WINE BOOM
AND THE
OTHER BOBS

WHILE THE FUTURE FAMOUS WINE CRITIC WAS reading books about wine and tasting as much as he could afford, the landscape of wine and food in America was changing rapidly. Robert Parker was just one of hundreds in the baby boom generation who experienced a gastronomic epiphany in France and came home determined to learn more. Many were so smitten that they abandoned their original ambitions and completely changed the direction of their lives, opening restaurants, working at wineries, becoming wine importers, or just writing glowingly about their experiences for any paper that would print their work, becoming proselytizers for a new consciousness.

The world into which Parker was taking his first steps was wide open and his timing was all-important. The beginning boom in wine that coincided with his new-found interest meant there were many more wines at hand to try and new ways to learn about them. As general enthusiasm accelerated in the late 1970s, he would be ready to impart what he'd learned to an audience who still knew much less than he did and were hungry for guidance. But of course it was more complicated than that.

American fascination with food and wine was fragmented in the 1960s and early 1970s. Those with a passion for food didn't always have an equal interest in wine, and vice versa, and there was a definite split between wine lovers and writers on the West Coast, who were beginning to champion California wine, and the East Coast, where only European, especially French, wines reigned.

The boom in the appreciation of fine food came first, starting in the early 1960s with Julia Child. Despite the fact that she had a very different background than Parker—she came from a wealthy Pasadena, California, family and had attended fashionable eastern schools—she'd had a similar inspirational experience in France, but earlier and at a more sophisticated economic and social level. Child's husband, Paul, was transferred to Paris in 1948 by the U.S. Information Service, and their first meal in France, a "plate of oysters, followed by sole meunière and accompanied by a bottle of pale Chablis," opened up her "soul and spirit." The wonderful smells of butter and wine inside La Couronne restaurant in Rouen stayed with Child for more than fifty years. She, like Parker, took the route of gaining knowledge and set herself the task of learning to cook, attending Le Cordon Bleu Cooking School. Her desire to introduce sophisticated French cooking to Americans led to the publication of *Mastering the Art of French Cooking*, which she wrote with her two French partners in a cooking school, and then to her popular television show, *The French Chef*, which debuted in 1963. Already a craze for things French was in the air thanks to First Lady Jacqueline Kennedy, who brought a French chef and wines to the White House and dazzled Charles de Gaulle and the French as well as Americans with her sense of style.

Aspiring gourmets were glued to the TV set while Child demonstrated how to roast a chicken the French way—the message was understood as slather on the butter, stick in some garlic, anyone can do it—and her show was so popular that when she cooked broccoli, the vegetable was said to have sold out within 200 miles of the broadcast station. It began a transformation of the way the middle class cooked; fine French cuisine, previously enjoyed only by a tiny privileged group, was now something anyone could learn to make. Child was well on her way to becoming a celebrity and an American icon when Robert Parker came on the scene.

Another well-known food revolutionary, Alice Waters, was also an American struck by the pleasures of wine and food during a junior year in France. In the mid-1960s, she too fell in love with the inexpensive but delicious bistro fare eaten by average French working people and romanticized the way of life and culture that placed so much emphasis on food and wine. Waters found what she ate so authentic, so different from the bland and boring American food she'd known, which seemed to reflect the American conformity she was trying to escape. Her revelation that food and wine could "nourish the heart and soul" came at a tiny restaurant in an old stone country inn in Brittany, where she ate trout fresh from the stream outside and a simple raspberry tart made with fruit from the garden. Like so many other Americans she wanted to take this experience, this way of life, home and was determined to make her revelation part of her future life. And so, jettisoning her original plan to be a Montessori teacher, Waters opened a restaurant, Chez Panisse, in 1971. Her goal was to make it an American re-creation of that tiny inn, where everyone had clapped for the chef and announced the dinner had been "*fantastique*."

Both stories seem so completely American: when you find something revelatory elsewhere, there is no reason not to change your life to follow your new star, your new dream. Just because you didn't grow up in a restaurant, just because you didn't grow up with wine, just because you learned about something only last week doesn't mean you can't enter that new world and make it your own. Of course, it's Bob Parker's story, too.

So began the decade that would soon see culinary trends trumpeted in the mainstream press.

———

A TINY GROUP OF WEALTHY AMERICANS HAD ALWAYS DRUNK WINE and eaten very, very well, starting as far back as Thomas Jefferson. An English traveler to the United States in the 1880s observed that "fine cellars of wines are to be found in the houses of millionaires, but otherwise they are rare in the States and not to be met with, as in England, in the more ordinary households." In the late nineteenth century, French wines, and especially Champagne, were powerful status symbols to the newly moneyed class that made their wealth after the Civil War. The fabulously rich J. P. Morgan had three thousand bottles of Bordeaux in his 11,000-bottle cellar when he died in 1911. Many others, like John Randolph Bland, uncle of author Upton Sinclair, continued to drink imported wines throughout Prohibition. Anticipating what would happen he wisely stocked up, spending $6,000 on wines for his cellar to see him through.

After Prohibition's repeal, when André Simon came to New York to try to extend the franchise of his recently founded Wine & Food Society, he found both upper- and upper-middle-class Americans were eager to join, and international branches of the Society were quickly established in New York, Chicago, San Francisco, Los Angeles, and New Orleans. These were people who were already comfortable wearing black tie to the opera and at home with the rituals of elaborate dinners, and they saw the Society as social as well as educational. The serious wine lovers eventually joined organizations like the Commanderie de Bordeaux and the Confrérie des Chevaliers du Tastevin, too, and most subscribed to *Gourmet* magazine, which started in 1941. Since the country was soon to enter World War II, this wasn't the most auspicious time to launch a food magazine, but an affluent elite soon made up a loyal readership for articles that for the next twenty-five years would be mostly a romantic glorification of the past.

But in the mid-1950s, a market research survey of attitudes toward wine found that 90 percent of the public still didn't think table wine was a drink for Americans—only for rich people or foreigners. Wine was a luxury, and in the 1950s the middle class was making a beeline for the growing number of inexpensive fast-food restaurants dotting the American landscape—McDonald's, Kentucky Fried Chicken—and greeting the appearance of Gallo's first pop wines, Thunderbird and Ripple, and later, cold duck, with enthusiasm. Although wine appeared on the tables of recent immigrants from Mediterranean countries, until the 1960s fine wine was mostly consumed by the rich. There were inklings of future wine sophistication in several San Francisco tasting clubs, but elsewhere sweet wines were often served with dinner, even with roast beef.

Nineteen sixty-eight, when Robert Parker returned from Europe and formed his tasting group, was the first year that sales of dry table wine surpassed those of sweet fortified wines. The boom in American wine drinking was about to begin.

───────

EVERYONE SEEMED TO BE GETTING INTO WINE. CONSUMPTION WAS growing by a fast-paced 10 percent every single year in the late 1960s, and between 1960 and 1973 it more than doubled. People who'd never even thought about drinking wine before were buying and tasting. The big sellers in liquor stores were American jug wines like Almadén Mountain Red Claret, which cost only $1, and Gallo Hearty Burgundy, which was even cheaper, and those with a sweet edge, like fizzy Riunite Lambrusco, launched in 1967 specifically to appeal to the American palate (by 1984 it was selling 130 million bottles a year). But as Americans became more and more interested in premium dry table wines, they looked to Europe for the finer bottlings, and that meant France. On the East Coast wine snobbery demanded French imports, and on the West Coast even those who knew that some good wines were being made in Napa and Sonoma still believed that French wines were the pinnacle.

Ten years before the American market had hardly counted in French eyes, so small was the amount of wine being imported, at least in comparison with other countries. The British were much, much more important, as were the Belgians, who bought consistently year after year. The first clue that things would be different in the future was after the 1959 harvest, when reports of the "year of the century" in Bordeaux made news in America. Art Buchwald's syndicated column in early October and simultaneous articles in *Time* and *Newsweek* a month later all acknowledged that 1959 was indeed the "vintage of the century." Predictably, voracious demand in major cities pushed prices "up to the skies," and as one English wine merchant wrote, "the Bordelais suddenly realized they could not only open wide their mouths, but get away with it, too." For seven and a half centuries the English had bought the lion's share of Bordeaux's best wines, but in 1959, for the first time, the Americans took over "the historic role as principal buyers of the finest wines, and consequently, as pacesetters for prices." In retrospect the price jump doesn't look quite so exorbitant—Château Lafite went from $75 a case to $85 to $90—but it was the beginning of an upward spiral of prices that never really reverted to amounts once considered reasonable, and, grumbled the British, it was all the fault of the Americans.

Prices for Bordeaux soared even higher with the newspaper articles and general hype for the next "vintage of the century," 1961, which deserved the accolade even more. Because of weather conditions a much smaller amount of wine was produced in 1961, and as a Washington, D.C., retailer observed, "Wine is a law of supply and demand. Back in the early 1950s, the French couldn't give all their wine away. Only when they saw that they couldn't supply enough did prices really start to escalate—that was the story in 1961."

The Americans had the money, and now they were interested in wine. It was a heady combination. Because they concentrated their interest on Château Lafite and other first growths, the wines that had previously been the favorites of the English, their effect on the Bordeaux market was dramatic. The high-priced Burgundies from the Domaine de la Romanée-Conti had found a ready home in American cellars, too. Many in the British wine trade, used to purchasing those wines at much lower prices, wondered what would happen if something went wrong with the growing American market. The international market was now being built on these high prices, and if the Americans dropped out it would surely crash. Who else but the Americans would pay so much money for wine?

The French, on the other hand, seemed oblivious to what all this meant, and their attitude toward the Americans remained a mixture of condescension and greed. They circulated the usual unflattering jokes and stories demonstrating how ignorant Americans were about wine, firmly held the assumption that their pocketbooks and bank accounts were limitless, and gleefully anticipated raking in the cash. And the prices went up. But at the same time a huge number of Americans, like Robert Parker, were energetically and enthusiastically educating themselves about wine in ways other countries, including France, never had. Within a decade, tasting, comparing, and discussing French wines would, as Nicholas Faith wrote, "spread more widely through the American social system than it had over the previous two hundred years in England, let alone in France." The chauvinistic French suffered from a bad case of assumed wine superiority, and were in for a very rude awakening before the decade was out. They didn't reckon with the possibility that American wine enthusiasts would look elsewhere, or that the desire for the old and traditional would eventually give way to the cult of the new.

In 1969 sales of Bordeaux jumped a whopping 50 percent, and by the following year people were remarking that wine drinking in America was swelling torrentially. And once again a Bordeaux vintage, the 1970, was being hyped as the "vintage of the century." The wine was good, and there was plenty of it. The fine-wine buyer for Austin Nichols, the ingratiating but shrewd and calculating Abdullah Simon, had made massive profits buying up the 1966 and 1967 vintages, and he jumped on the 1970 with aggressive new marketing tactics. He brought in Gerald Asher, a young Englishman who would later become *Gourmet*'s wine columnist, to help him. Simon had rolled through Bordeaux with an open checkbook, scooping up vast quantities of wine. His idea was to get Americans excited enough to buy the wines as futures, while they were still in barrel, or *en primeur*, which meant that consumers would pay a retailer to reserve wines that would be delivered two years later, after they were bottled.

To drum up enthusiasm among both retailers and collectors, Simon organized an enormous tasting in New York City only six months after the grapes had been harvested and invited the sophisticated wine world. Barrel samples of the unfinished wines were flown in, something that had never been done before—even in England, where importers prudently waited to see the size and quality of the next vintage before deciding how much to buy—and for good reason. Barrel samples are difficult to taste because their harsh, bitter tannins aren't yet smoothed out and softened by time. But the samples of the 1970s were unusual, Asher remembers, ripe and fruity, already attractive. Nelson Rockefeller showed up as well as wine merchants like Sam Aaron from New York's Sherry-Lehmann, who was anxious to get into this new futures game. For retailers, it was enticing: by law they didn't have to pay the importer until the wines actually arrived, so they would essentially have a two-year financial float. If they sold ten thousand cases at $300 a case, they would have $3,000,000 to play with for two years without paying any interest. By the time they had to pay Austin Nichols, there would, it was hoped, have been another *en primeur* sale and another significant bit of money rolling in. It came close to a pyramid scheme and seemed a sure thing. With all the worldwide publicity for the vintage, Americans were clamoring to buy.

Simon is generally credited with igniting the Bordeaux futures market in America, and he was also held responsible for its debacle two years later. But once the concept had been successful in America, it was only a matter of time until it would be resurrected by another great vintage.

———

MEANWHILE, THE CALIFORNIA WINE COUNTRY WAS ALSO POISED for change, thanks to a band of ambitious pioneers determined to make wine to compete with the best of France. In most consumers' minds, especially on the East Coast, California wine was synonymous with inexpensive jugs, not prestige bottlings. The transformation began with another Bob, winemaker Robert Mondavi, who had finally made his first trip to Europe in 1962, the year after Child's *Mastering the Art of French Cooking* was published. He was forty-nine. And he, too, had a revelation in France, at one of the country's greatest restaurants, La Pyramide in Vienne. The food and wine, he wrote later, "transported us to a world of gentleness and balance, of grace and harmony . . . it inspired in me both a vision and a vow." Although he had been in the wine business in America for more than three decades, he had now found his calling, his purpose, his way to the future—like Alice Waters, he would bring his vision back to California, in his case to create "wines that have grace, and style, harmony and balance." Mondavi was impressed with the traditions at the great French châteaux and convinced that their secret was new oak barrels, but with typical American optimism he "felt absolutely certain we could learn quickly and make significant leaps in the quality of our wines." Ultimately, to make this happen, he would fight with his brother, get thrown out of the family winery, and launch a new one of his own, in 1966, at the age of fifty-four.

During the late 1960s and 1970s, a new tiny winery dedicated to making fine wine seemed to be opening just about every week in the Napa Valley, and the joke was that Mondavi's new winery, a hotbed of experimentation, was the best place to pick up used equipment that had been cast aside when experiments hadn't worked out. Prior to the mid-1960s only a handful of boutique wineries like Stony Hill, Hanzell, and Ridge existed, but dozens were founded in the next ten years.

In the spring of 1971 my husband, John Frederick Walker, and I made our first trip to the wine country of California, a three-week-long journey from Napa to the Santa Cruz Mountains, and became instantly smitten with a world on the edge of revolutionary change. We wandered from winery to winery, from sleepy, long-established family dynasties to start-up enterprises bursting with ambition and passion for what could be achieved with grapes on California soil. It was a small world then, but the following year saw the founding of Stag's Leap Wine Cellars, Silver Oak, Caymus, Carneros Creek, Stag's Leap Winery, Clos du Val, Château Montelena, Jordan, and Dry Creek, and that was just the beginning. Fifty thousand acres of vines a year were being planted in California, as wineries also sprang up in neighboring Sonoma and Mendocino counties and south of San Francisco Bay. To distinguish their premium wines from the generic screw-capped jugs labeled with French names like Chablis and Burgundy, the new fine wineries labeled their wines with the name of the grapes from which they were made, as Frank Schoonmaker and Alexis Lichine had urged decades before. They knew that historically all the best and most expensive wines in America, like Beaulieu Vineyards' top red, Georges de La Tour Private Reserve Cabernet Sauvignon, were varietally labeled. So it's not surprising that in the next decade the amount of varietally labeled wine soared from 5 million to 25 million gallons. Even giant Gallo finally released varietally labeled wines in 1974, anxious by then to compete in the new fine wine market.

For consumers beginning to learn about wine it was a godsend, so much easier to understand than the complicated nomenclature of châteaux and villages and classifications on French bottles. But these new wineries were releasing tiny amounts of wine, and they didn't have the advertising muscle of the ubiquitous Gallo, which in 1972, made one-third of all the wine sold in America.

They knew they had to get attention for their wines.

———

FORTUNATELY FOR BOTH THE NEW BOUTIQUE WINERIES IN CALIfornia and French imports, the new boom in wine drinking among the middle class inspired newspapers and magazines to add wine columns, writers to launch their own wine newsletters, and entrepreneurs to found wine magazines. More and more wine books hit the shelves. The cast of characters in the American wine-writing scene of the 1970s came from a surprising range of backgrounds—from police dispatcher to ex-marine to graduate student to professor to sportswriter to a couple of wine merchants who actually knew what they were talking about. Just about anyone could (and did) call him- or herself a wine writer. All it took was a serious interest in wine, a little tasting experience, perhaps a trip or two to châteaux in France or wineries in the Napa Valley, and reading a few books to put someone a few steps ahead of the majority of wine drinkers, who were more enthusiastic than knowledgeable. That, and the ego to believe that you had a good palate. A few writers were wildly eccentric, others imperious prima donnas who expected to be courted by the wine trade, but for all of them wine writing was a way to justify the amount of money they were personally spending on wine and to expand their own wine horizons.

In England, on the other hand, many of the wine writers were also wine merchants or brokers with some training and regular access to many different wines, or from a tradition of connoisseurship at home, at university, or later at the gentlemen's clubs that had formed a large part of upper- and upper-middle-class male social life. These clubs often had fantastic cellars that made it easy and pleasant to refine one's taste. Drinking fine wine in England had a venerable tradition—after all, the whole Bordeaux wine trade had originated with the British centuries before. So at Brooks's, they drank claret with grouse and partridge by candlelight shed from tall tapers in the club's famous Georgian silver candlesticks. At White's, members consumed it in the large, lofty red-walled dining room under huge portraits of past monarchs and the Duke of Wellington. At the St. James's, they dined and drank by the tall windows looking out on the verdant grass and beautiful trees of Green Park. Later there would be glasses of rich port served in a library with leather armchairs and sofas, rustling newspapers, and the smoke of good Havana cigars twirling upward toward the ceiling.

Among the British writers who weren't in the trade, Hugh Johnson, a member of the Wine & Food Society at Cambridge University, was emerging as the most important. At age twenty-four he'd become editor of the journal *Wine and Food*, taking over from the illustrious André Simon. His first book, *Wine*, had appeared in 1966, when he was only twenty-seven, and in 1971 he published *The World Atlas of Wine*, which would later be hailed as the best book on wine in the twentieth century. His graceful writing style seemed a modern, less florid version of the purple prose penned by British wine writers of an earlier generation, such as George Saintsbury and H. Warner Allen.

By contrast, U.S. wine writing developed as a "uniquely American cottage industry," as David Shaw of the *Los Angeles Times* saw it, a kind of hobbyist self-expression. In the 1970s, most wine writers were amateurs with a passion, like Parker. And by the time he joined the cast of characters, the stage was already very crowded.

The need for wine writers in America arose after Repeal for the simple reason that no one but the very well-to-do drank wine or knew anything about it, and most of them didn't know much either. For thirteen years, during the time an entire generation was coming of age, people had been downing booze in speakeasies and getting drunk on bootleg gin, Canadian whiskey, and cocktails. Schoonmaker and Lichine knew that to sell wine to this generation, they had to educate them, too. That goal lay behind the books and articles they published in the 1950s and 1960s. But both remained firmly in the wine trade; writing about wine was only a side occupation.

Then came the 1970s and a whole new group of enthusiasts, like Parker, who wanted to know what to buy. Books like Hugh Johnson's *Wine* gave fine background information for those who wanted to study the subject, but they weren't timely enough to keep up with the hundreds of wines showing up on retail shelves. What was needed were current recommendations. Many English writers suggested the way to the best bottles was by developing a relationship with a good wine merchant, but few American retailers fit the image of an old-fashioned British wine firm like Berry Brothers & Rudd of St. James's Street in London, who guided their clients' taste and gave well-informed advice on the appropriate wines to lay down for future drinking. For every Esquin's in San Francisco helping collectors train their palates, there were hundreds of retailers in America whose shops were still dominated by fifths of Smirnoff and Johnnie Walker Red, and who were scarcely aware of anything other than the most famous names. One veteran wine sales representative observed that fewer than 1 percent of retailers ever tasted the wine they sold back then, and that many knew less about wine than their customers. Only a handful had set aside a small wine section in their shops. Besides, many Americans had a caveat emptor mindset when it came to recommendations from those trying to sell them something.

Two of the decade's most influential wine writers also happened to be named Bob, but both resided on the other side of the country, in California, where the American wine scene was heating up. One was the flamboyant, often outrageous Robert Lawrence Balzer, who'd been a celebrity wine merchant, had hung out of a helicopter in Cambodia to capture Angkor Wat on film, and in 1956 had been ordained a Buddhist monk, an event that made headlines in the *Los Angeles Times*: "Millionaire Los Angeles Grocer Becomes Buddhist Monk." Balzer, who has always claimed movie star Gloria Swanson was "the love of my life," was also the first wine writer–journalist in America—as far back as 1937 he wrote a weekly column for a small newspaper in Beverly Hills, and in 1964, after authoring several books, he became the wine columnist for the *Los Angeles Times*. It was the first regular wine column in a major American newspaper. A few years later he launched what was probably the very first wine newsletter in the country, *Robert Lawrence Balzer's Private Guide to Food and Wine*. With his newsletter, newspaper column, plus a column in *Travel-Holiday* magazine, he was considered the most powerful wine writer in America in 1970, and he was helping to usher in the new era of American wine.

Unlike Parker, Balzer gained his knowledge in the trade—his father put him in charge of the wine department in the family's gourmet grocery business near Beverly Hills, and Hollywood stars like Ingrid Bergman and Marlon Brando and director Alfred Hitchcock were frequent customers—but it was a haphazard education, learned on the job. As Balzer was gregarious and charming he soon made friends not only with a Hollywood crowd, but also with all the people we now think of as America's fine wine pioneers. An early believer in the quality of California wine, he was one of its biggest boosters—the "seldom is heard a discouraging word" kind of critic, or as some said, propagandist, promoter, even industry flak. Though Balzer was occasionally critical, he wrote mostly favorable reviews. Many winemakers, some of whom first learned about wine in one of his wine appreciation classes, thought of him fondly as someone who did almost as much for the California wine industry as Robert Mondavi or the Gallos. They sought his advice, or at least listened politely (he claimed that in 1974 he told Ernest Gallo that his red wines would never be taken seriously unless they were aged in oak), and in the same way new wineries rush their bottles off to Robert Parker today, they made sure Balzer was one of the first to taste their new releases. He was lavishly wined and dined in the California wine country and took it as his due. Balzer had a lot of ego, and plenty of self-involvement, but also an endearing, enthusiastic, entertaining personality.

The concept of journalistic ethics or the problem of conflict of interest simply didn't apply to wine writing then, even at the *Los Angeles Times*, and to a certain degree it still doesn't. Though Balzer had sold his family business, he had strong, almost familial friendships with many winemakers, and he didn't hesitate to publish a wine book underwritten by winemaker Paul Masson, or, later, tout a wine futures venture backed by his best friend. But he was so universally known and had such a following on the West Coast that Ronald Reagan asked him to oversee an exclusive wine and food fete at his inauguration in 1981. Balzer was an American character, an eccentric, who had a considerable amount of knowledge and loved telling people about wine, and who at over ninety years of age still conducts wine appreciation classes and trips for a cult following.

People would later say that Balzer's wine writing belonged to the era when the main role of wine writers was to give people some basic information and enthuse and excite them about buying and drinking wine, period. After all, wine writers' reviews sold bottles, and in the end, despite all the romance and pleasure of tasting, selling wine was what the wine industry in both Europe and California was all about.

The other writing Bob was the influential Robert Finigan, a San Francisco resident who founded his own monthly newsletter, *Robert Finigan's Private Guide to Wines*, in 1972. From the beginning it had a consumer-oriented aim, and over the next ten years it would become the most respected wine newsletter in America. Like Parker, Finigan hadn't grown up in a home where wine was served, but while attending Harvard University the soft-spoken, sandy-haired Virginian met others who had— one of his classmates came from a family with a great cellar, and another's owned a château in the Bergerac region of France. They and three other classmates all shared a large house, pooling their resources to hire undergraduate women to cook for them, and it seemed obvious they should be drinking wine with those nicely cooked dinners. By the time Finigan went to work for a San Francisco–based international management consulting firm in 1967, he had been "bitten by the wine bug." While Parker was traveling in Europe on $5 a day in jeans and pooling his wine funds with other students, Finigan was wearing a suit and tie, carving out time to visit wine regions on European business trips, getting to know the wines of the Napa Valley, and buying 1962, 1964, and 1966 Bordeaux from one of San Francisco's best wine merchants, Esquin's.

The idea that he might do something professionally with wine hadn't occurred to Finigan any more than it had to Parker, but fate intervened in 1970, when, like Parker, he first visited Bordeaux. The 1969 vintage in Bordeaux was being hyped by the trade yet again as the greatest since 1961, but Ken Kew, Esquin's director, wasn't sure whether to believe it or not. Since Finigan planned a swing through Bordeaux on a coming business trip, Kew offered to give him letters of introduction to the châteaux if he would taste the 1969s in barrel and advise Kew whether to buy. Finigan, flattered, accepted the challenge, elated to be seeing up close those châteaux whose names he knew only from labels. Though he'd never tasted wines from the barrel before, at Château Palmer, his first stop, he sniffed, swirled, and spat the young wine onto the floor in the cellar in what he hoped was a very professional fashion. The wine seemed a disappointment, tannic with little character or flavor, and the story was the same at the other châteaux he visited. Despite the hype, Finigan decided, these wines were nowhere near as good as the Bordeaux he'd been drinking. He had no hesitation about advising Kew not to buy, and as it turned out, saved his wine merchant a lot of money.

This heady experience encouraged Finigan to take his intense interest in wine much more seriously. He, like Parker, wished there were a source of reliable information to tell him what bottles would taste like so he would know what to buy. Since there wasn't, he took his own education in hand. In every spare moment he was dashing up to the Napa Valley to cajole owners into opening old bottles, interspersing business in Europe with weekends in the vineyards of Germany, France, Spain, Portugal, and Italy, and methodically setting up tastings in his apartment of wines he'd purchased. Always he recorded his impressions in a notebook, and, like

Parker, he was frequently disappointed by wines that didn't match the descriptions he'd read in books. Finigan subscribed to a restaurant newsletter that gave candid reviews of the local scene and as more and more wine brands appeared on the shelves, he became convinced he could make money by publishing a similar newsletter on wine to guide confused consumers to the better bottles. In his first eight-page issue, which appeared in September 1972, he organized his assessment of the wines he'd tasted that month into four carefully thought-out categories: Outstanding, Above Average, Average, and Well Below Average. There were no numbers. He would cover both California and European wines, including wines on restaurant wine lists in the Bay Area, and he would be critical, he assured his readers, buying all the wines for full price in retail shops and assessing them through blind tastings. In the first issue he listed seven Beaujolais as Well Below Average, complaining one had an "acetone-like nose," while another was dismissed as "dreadful." Later issues would call much inexpensive red Bordeaux "barely drinkable swill" and criticize many California jug whites for reeking "repulsively of sulphur." With month two the newsletter was profitable, and by 1977 Finigan had added a national edition and developed an international following.

Where Balzer was a combination of booster, prima donna, and charming dinner companion, Finigan was quiet and restrained, with a studied manner, someone who liked the good life, played a good game of tennis, felt at home with the coteries of well-to-do, well-connected, and cultured San Francisco collectors, and who seemed to have his personality and passion for wine completely under control. His newsletter was a model of carefully modulated writing, as though he had filtered his opinions on wine through a Harvard writing program. Finigan was also a powerful player in the San Francisco restaurant scene, as he had bought *Jack Shelton's Private Guide to Restaurants* and was now writing that, too. He was living the high life, traveling to the châteaux of France, eating regularly at spots like the Berkeley Pot Luck, which had the best list of old California wines in the country, and hanging out at Chez Panisse because he was having an affair with Alice Waters.

The two Bobs were only two of the dozens of wine writers on the scene in the early 1970s. On the West Coast, Charles Olken and Earl Singer, both passionate collectors of California wine, had started their own newsletter,

Connoisseur's Guide to California Wine, in 1974, and it, too, was aimed at the consumer. A panel sorted out the best of the growing number of the state's labels through blind tastings, rating wines on a system of one to three "puffs," three being the equivalent of outstanding. Soon this newsletter, too, had a following on the West Coast.

Back east, at the *New York Times*, executive editor Abe Rosenthal had decided the paper needed a weekly wine column and tapped Frank Prial, a veteran general reporter who knew something about the subject, to write "Wine Talk," also in 1972. Prial's influence was felt quickly in New York City; the power of the *Times* ensured that a wine recommended highly in the Wednesday paper would often be sold out by the end of the week, sometimes by the end of the day, at big wine shops like Sherry-Lehmann. Stores helped sales along by putting signs in the window highlighting Prial's picks and used his name in their ads and catalogs. But Prial didn't think of himself as a wine expert or critical taster; what interested him were the stories about the people involved with wine, and he told them well. That same year wine merchant Gerald Asher began his "Wine Journal" column in *Gourmet*—reflective and literary, it was more in the style of British writers than the new consumer-oriented writing found elsewhere. And two years later, Alexis Bespaloff, who'd also worked in the trade, became the authoritative columnist for the new and trendy *New York* magazine—which just showed how fashionable wine drinking was becoming.

In fact, by 1974, New York and California were swarming with wine writers, inspiring the founding of both a Wine Writers Circle and a Wine Media Guild in New York alone. Retailers, importers, and California wineries were thrilled—so many people to educate consumers and recommend the wines they were trying to sell! They were all anxious for the attention of the writers, giving tastings and lavish lunches and dinners at the city's best restaurants where the wine press could meet important wine producers, château owners, and winemakers when they came to town. For wine writers it was an invaluable shortcut to a wine education at the highest level. New York wine writers would later look back on the decade from 1974 to 1984 as a golden age when they were feted and courted as they never would be again.

A number of wine periodicals started in the 1970s, such as *Vintage, The California Grapevine*, and the *Quarterly Review of Wines*. Some were short-lived, others are still publishing. They were embraced by a wine trade eager to reach what one called the "army of wine enthusiasts." Importers regularly sent sample bottles to the magazines, hoping for reviews, but wines weren't ranked or rated numerically. At the time, even offering descriptive tasting notes seemed revolutionary.

But by far the most successful wine publishing venture was started by Wall Street real estate and investment banker Marvin Shanken in 1973. The stocky, mustached, intense Shanken became passionate about wine, especially California Cabernet Sauvignon, while regularly putting together vineyard deals in California. Ever the businessman, he began seriously studying the industry as an area for investment and was surprised to find there was no source of accurate information or reliable forecasts, like Standard & Poor's. Then he heard that *Impact*, a struggling four-page newsletter on the alcoholic beverage industry, was for sale and snapped it up for $5,000. What he'd bought as a hobby business turned out to be a list of 250 subscribers and a file box, and then the editor quit. Shanken knew nothing about writing or publishing, but now he was responsible for both, and to the dismay of his parents he left Wall Street two years later to devote himself to the newsletter. It was a big risk, but by the end of 1975 he was predicting—accurately, as it turned out—a boom in white wine consumption and thinking of other ways to capitalize on the hunger for wine information. His biggest venture was a wine magazine, *The Wine Spectator*, which would become the only rival to Parker's influence.

─────

WINE EDUCATION WAS BOOMING AMONG THE MIDDLE CLASS, TOO, because of an innovative organization started in Washington, D.C. In 1953, Marvin Stirman had returned from military service in Korea ready to take his place in his father's business, Calvert Liquors, in the Georgetown section of the city, and eventually planned to prove himself by adding a wine department. His first big problem, he decided, was that his customers were convinced that only the French knew anything about wine. The key to sales, he reasoned, was to hire someone who had a French accent, preferably a Frenchman who actually did know something about wine. So Robert Gourdin, who'd worked for importer Brown-Forman, was installed as the shop's first "sommelier," impressively attired in the vest typically worn by sommeliers in restaurants and a silver saucer-shaped *tastevin* on a chain hanging around his neck.

But that didn't solve Stirman's other big problem: His customers didn't even understand basic wine terminology. How could you talk to them? "Try this, it's a lovely dry table wine," Stirman would suggest. "How can it be dry? It's liquid," was a common response. To educate their customers, the two began a wine club in the early 1960s, which Gourdin named Les Amis du Vin, the friends of wine. They put a large sign in the shop's window inviting customers to join the club at a free wine tasting at the nearby Mayflower Hotel, whose food and beverage manager, Jim Nassikas, had donated the space. Hundreds of people showed up, but unfortunately most were freeloaders more interested in downing as much wine as they could for nothing than in learning anything. The event was a disaster, as attendees chug-a-lugged the wine and then staggered home. Charging $5 for future tastings weeded out those looking to get drunk on free wine and drastically cut the number to a more manageable thirty.

Building an actual membership for Les Amis du Vin by offering tangible benefits seemed a better long-term strategy. For $5, members would receive a free subscription to England's popular *Wine* magazine, the chance to attend regular wine tastings and lectures, a wine of the month that they could buy at half price, and an informative newsletter that accepted no advertising. The organization was supposed to teach people about wine, Stirman said, not to be a promotional or commercial vehicle for wine companies; and besides, Les Amis du Vin needed credibility with consumers to succeed. Clearly he was right, because within a year Les Amis du Vin had one thousand members, the newsletter was up to eight pages, and the cost to join went up to $10.

It was a brilliant way to encourage the growing number of wine enthusiasts; as they learned more, they would, of course, become even better customers, buying more—and more interesting and expensive—wine from the store. No wonder top wine retailers around the country soon approached Stirman about starting branches of Les Amis du Vin at their stores. They would plan local tastings and provide the wine of the month Stirman and Gourdin—and later Alfio Moriconi, who became Stirman's partner—had selected, but the mother chapter in Washington, D.C., would keep the records and fees and send out the newsletter. The organiza-

tion grew rapidly throughout the 1960s, attracting retailers in New Orleans, Dallas, New York City, Baltimore, San Francisco, and many smaller
cities. Then Les Amis added a food magazine, and a few years later American Express wanted to buy them out. Eventually, in the late 1970s, it would
have chapters in fifty cities, including Paris, 30,000 members, and a full-
fledged magazine. It was a national wine club of members anxious to learn
about wine, and there was nothing else like it anywhere in the world. It was
the ultimate—so far—democratic wine organization. Unlike the tonier,
more exclusive Wine & Food Society, anyone could join Les Amis for a
nominal fee, events didn't cost much, and you didn't have to wear black tie.

Les Amis du Vin wasn't completely casual, however. Though most people attended tastings because they were social and fun, these were sit-
down affairs where six to eight wines were swirled and sipped one by one
in an orderly fashion while a speaker lectured from a rostrum. The cost
was usually about $20, affordable even for students like Robert Parker, and
in Washington, D.C., two to three hundred people often showed up. The
dress code was jacket and tie, and everyone sat attentively at long tables
covered by white tablecloths, with a line of glasses holding about two
ounces of wine each and a paper score sheet on which to write notes. Food
was limited to plates of French bread to clear the palate. Tastings were frequently conducted "blind," with the labels hidden, and everyone voted on
their favorites before the identities of the wines were revealed. One of the
organization's avowed aims was to get people to discover what they themselves liked to drink and to have confidence in their own judgment. Many
people were amazed that wines from the great châteaux like Margaux or
Lafite weren't always the ones they liked best, and that they sometimes
found one from a little château they'd never heard of much more enjoyable. No one spit, so things could get boisterous during the final question
period.

But as people became more knowledgeable they wanted more, and to
keep their interest Les Amis chapters obliged with smaller, more exclusive
and expensive tastings of, say, the first growths of Bordeaux. One of the
most attentive attendees at both the Washington and Baltimore chapters
was Robert Parker. To Marvin Stirman, Parker always seemed serious, as
though he were working hard to develop his skills and not just there to
enjoy himself.

Importers such as Peter M. F. Sichel and California winery owners such as Robert Mondavi were keen to speak to any chapter for the obvious reason that these were captive audiences of people who bought wine. But one of the most popular lecturers was Harry Waugh, a very unpompous Englishman with thinning white hair combed back 1930s-style, heavy black-framed glasses, and perennially rosy cheeks that seemed to express his genial personality. His infectious enthusiasm for wine and complete lack of pretension charmed absolutely everyone who met him, and he was in great demand. When Waugh presided over a review of the Bordeaux vintages of the 1960s in Washington, D.C., in 1970, well over five hundred wine lovers turned out, including plenty of young men sporting long sideburns and beards as well as a sprinkling of women. Before one of his trips to the United States, word would go out and dozens of Les Amis chapters would sign him up for a lecture and tasting while other groups of wine enthusiasts would prevail upon him to be the guest of honor at grand dinners featuring stellar wines. Within the American wine world he rapidly became an influential celebrity—after reading Waugh's impressions of the 1966 Bordeaux in one of his regularly published diaries, a wine shop in California ordered six thousand cases of the wines he had mentioned.

Waugh had been a wine merchant, and was now a director of Château Latour and a wine consultant, and he was a great fan of California wines. In the mid-1960s he had visited the Napa Valley and been impressed with the wines he tasted, and by 1971 he knew the state's pioneers of fine wine—Ridge, Heitz, Freemark Abbey, Château Montelena, Mayacamas, Spring Mountain, BV, Stony Hill, Chappellet, Robert Mondavi, Chalone. San Francisco and Berkeley were hotbeds of tasting clubs, like the Grand Cru and Premier Cru clubs, the Vintner's Club, and the San Francisco Sampling Club, whose members were among the most serious wine lovers of the day, and Harry Waugh was a frequent guest at their events. As most were avid collectors of California wine, they introduced their friend to the emerging wine scene. Long before most Americans, especially those on the East Coast, Waugh acknowledged the exceptional quality of these wines. His California friends often sneaked a French wine or two into their blind tastings, and Waugh was often surprised by how well the California wines showed.

Les Amis du Vin flourished thoughout the 1970s and well into the

1980s, headed by a new young president, Ron Fonte, who was also editor of its magazine *Friends of Wine*, purchased for him by his father-in-law. But Fonte ended up presiding over Les Amis Du Vin's eventual demise; he left for Canada and the magazine and national organization folded.

In the early 1970s, the great wine division between the East Coast and West Coast rested on their vastly differing views of California wine. On the East Coast, French wine reigned and few writers or enthusiasts had even heard of some of the California winery names prized in San Francisco. In 1971, after a year on the West Coast, my husband and I drove back to New York in our old blue convertible with a trunk full of scarce California bottlings and what proved to be the premature idea of co-authoring a book about the small fine wineries we'd visited—Mayacamas, Heitz, Ridge, and others. The project fell flat. One prominent editor's rejection letter to our agent claimed there was about as much interest in California wine as in the white wines of the Jura. Friends to whom we served those prized bottles would say disparagingly before even tasting, "Oh . . . a California wine."

That attitude was about to change.

BOTH AMERICAN WINE AND THE AMERICAN WINE CONSUMER CONtinued to be regarded with condescension by the French in the mid-1970s, when the famous, or infamous, Paris Tasting finally put them on notice that America and the Americans might be more important than they had thought, even if they didn't speak French. The tasting pitted some of California's best wines against top French bottlings, and the American side won. It was one of those pivotal events in the history of American wine, now part of wine-country folklore, whose results have been celebrated at the tenth, twentieth, and twenty-fifth anniversaries of the Tasting's date and are still discussed and debated. The winning bottles have even been put on display at the Smithsonian Institution in Washington, D.C.

The story starts with a young, whip-thin, floppy-haired expatriate Englishman with an upper-class stammer and a languid manner, Steven Spurrier. He had opened an innovative wine shop, Caves de la Madeleine, and an adjoining wine school, L'Academie du Vin, on a tiny street not far from rue Royale in Paris. Both shop and school counted many Americans among their clientele—the U.S. Embassy and IBM were nearby, and the wine school actively recruited at the many American student programs in the city. The shop became a regular stop for visiting wine writers and California wine producers, who would drop off the occasional bottle of interesting American wine. Some of them, Spurrier noted, were surprisingly good.

In 1975, looking for a way to publicize his shop, Spurrier hit on the idea of staging a tasting of the best California wines during the American Bicentennial celebrations in Paris planned for the following year. But first, he thought, a review of the California wine scene was in order. Almost all the wines Spurrier drank and sold in his shop were French, and he didn't know enough to decide which American wines to include. His American partner in the school, Patricia Gallagher, made the initial trek, visiting, among many others, Barbara and Warren Winiarski of Stag's Leap Wine Cellars, one of the Napa Valley's new boutique-sized wineries. Winiarski was a former lecturer in political philosophy who had traded academia for a life in wine in 1964, and had purchased the property for his winery only five years before. He poured Gallagher a taste of his 1973 Cabernet Sauvignon, the first vintage of this wine, and she was impressed. The following spring, Spurrier traveled to California to make the final wine selection.

Plans for the tasting began to take shape. Spurrier chose six California Chardonnays and six Cabernet Sauvignons, and to provide a standard of comparison decided to pour some of France's finest wines made from the same grapes—four top white Burgundies alongside the Chardonnays, and four *grand cru* red Bordeaux with the Cabernets. Fortunately a group of California winemakers were traveling to France to tour the vineyards around the same time; they would parcel out the bottles in the two cases of American wine and carry them into France.

The panel of judges should include the crème de la crème of the country's wine professionals, Spurrier concluded, and the tasting would have to be "blind," with the labels hidden, to insure a semblance of objectivity. He assumed the French wines would be superior, but he was hopeful that the panel, who had little knowledge of the California wine scene, would be surprised—and impressed—by the quality of the American wines made with the image of French wines firmly in mind. The tasting would honor both countries' wines and illustrate that a young vineyard area could produce top-quality wines if they received the same care and investment the best European wines did. This kind of public transatlantic taste-off was wildly popular among amateur tasting groups and the press in America, where the American wines often won, but for the French it would be a new experience.

The nine French judges Spurrier recruited for his tasting panel were indeed the elite of the French wine world: Odette Kahn of *La Revue du Vin de France* and Claude Dubois-Millot of *Le Nouveau Guide*; the owners of two prestigious properties in Burgundy and Bordeaux, Aubert de Villaine of Domaine de la Romanée-Conti and Pierre Tari of Château Giscours; the owners of two restaurants, Le Grand Véfour and Taillevent, and the sommelier of a third, La Tour d'Argent, all of which had been given three-star ratings by the Michelin Guide; and the leaders of them all would be Pierre Bréjoux, the *inspecteur-général* of the Appellation Controlée, who was responsible for upholding French wine quality, and Michel Dovaz, of the Institut Oenologique de France.

The tasting, on May 24, 1976, took place on the patio of the elegant and luxurious Inter-Continental Hotel, which had been designed in grand nineteenth-century style by the architect of the Paris Opera. The afternoon was warm. The twenty bottles of wine, numbered and wrapped, stood on a table along with sets of glasses, scorecards and pencils, spit buckets, and French bread. The judges assembled, joined by various observers—a few tourists, reporters from the French press, wine people, and *Time* magazine's Paris correspondent, George Taber, who was a student at L'Academie du Vin. The French magazine *Gault-Millau* had been offered exclusive coverage, but had backed out. Only at the last moment did Spurrier inform the judges that French wines had been included in the lineup of bottles. They made no objection, apparently unfazed by this last-minute revelation.

The whites were poured first, then the reds. As the judges swirled, sniffed, sipped, spit, and rated each wine on a scale from 1 to 20, some were quick to pronounce smugly on a wine's origins. "That is definitely California. It has no nose," exclaimed one about a wine that turned out to be a 1973 Batard-Montrachet from Burgundy. "Ah, back to France," sighed gastronome Raymond Oliver, owner of Le Grand Véfour, after a sip of a Napa Valley Chardonnay. The judges' confusion extended to the reds. One called a Napa Valley Cabernet "certainly a *premier grand cru* of Bordeaux," evidence of "the magnificence of France."

When the results were tallied and announced, several judges behaved badly, refusing to give up their notes, and one even tried to change his numbers before Spurrier whipped away the scorecards. The top-ranked white, placed first by six of the nine judges, was the 1973 Chateau Montelena Chardonnay; the top-scoring red, 1973 Stag's Leap Wine Cellars Cabernet, just nosed out two of Bordeaux's greatest châteaux—1970 Mouton-Rothschild and 1970 Haut-Brion. In fact six of the eleven most highly rated wines in the tasting were from California.

The results caused an uproar, even making the nightly CBS News television broadcast. *Time* billed the taste-off as "Judgment of Paris," announcing that "the unthinkable happened: California defeated all Gaul." The French panel was aghast; their wine industry, no surprise, immediately declared the tasting unfair and denounced Spurrier, who was equally stunned. (Twenty years later, Spurrier pointed out that he would never have staged the tasting if he'd thought the American wines would win; after all, he was running a business in France.) After *Time* hit the newsstand, the phones at Stag's Leap and Château Montelena started ringing, and cars lined up at the wineries. Winiarski restricted sales to one bottle to a carload, but soon was sold out. A retailer who had refused to buy the Stag's Leap two weeks previously called every day, begging for a few bottles.

In New York City, Mike Kapon, owner of Acker Merrall & Condit Liquors and one of the few champions of California wines on the East Coast, had been having difficulty selling California Chardonnays. Finally he'd displayed a stack of them, including Freemark Abbey and Chateau Montelena, two of the wines in the Paris Tasting, right next to an assortment of the most famous white Burgundies. He hoped that some of the French wines' glamour would rub off on the American wines, and that when consumers compared the prices—the Burgundies cost from $20 to $30 a bottle, while the California wines were $5 and $6—they would see them as bargains. So far that hadn't happened. Then came Frank Prial's *New York Times* column on the Paris Tasting. The morning it appeared, the Chardonnays were sold out in the first hour the store was open. California winemakers were quick to understand that blind tastings and the press had the power to help them succeed. Chateau Montelena's Jim Barrett figured the tasting had been worth about $4 million in publicity.

Some critics, including Frank Prial, suggested that the results of the tasting weren't quite as clear-cut as they seemed, and over the years they continued to be analyzed and reanalyzed by wine professionals and statisticians. After the judges' scores were tallied, the "winner" was the wine in each group with the most points. But the Stag's Leap, at 127.5, was a mere 1.5 points ahead of the Château Mouton and 2 points ahead of the Château Haut-Brion, and only one judge had ranked it first. Two had placed the Haut-Brion first, and a third had placed Haut-Brion and Montrose at the top. Could wine really be ranked this way? What about all the variables? For example, how good would the Stag's Leap be in ten years compared to the Mouton, which was designed to age? Did it make sense to make tasting wines the equivalent of a horse race where every second counted?

———

THE PARIS TASTING IS STILL THOUGHT OF AS THE MOMENT WHEN American wine came of age. It was a moment savored by California winemakers and indeed, plenty of American consumers. In a few short years, with the help of American ingenuity and technology, hard work, dedication, and sheer smarts, Americans had made it to the big leagues and humbled those haughty know-it-all Frenchmen. This was an attitude that would last, and last, and last.

But the Paris Tasting had many other meanings. Yes, it gave California winemakers "a new sense of mission and confidence" and illustrated that "exceptional wines could come from somewhere other than the traditionally sacrosanct French *terroirs*." It showed that even illustrious and experienced judges couldn't always tell the difference between the French and California wines. And it brought excitement to the idea of blind wine tastings among wine consumers. But above all it was an indication that from now on wine, especially in America, would be thought about and judged in a new way.

Blind tastings, where the labels of bottles are hidden so that the tasters aren't influenced by name or reputation in their assessment of the wine in the bottle, were common in the wine trade, but they were also becoming tremendously popular among American consumers in the 1970s. Instead of learning about wine by simply drinking it with meals, as the average wine drinker did in France, pitting wines one against the other in blind tastings to see which one was the best became a new American sport. The Transatlantic Taste-off game had been played all over America before the Paris Tasting, but now it had been sanctified by professionals at an international public event. It legitimized a new way of thinking—a wine tasting became a race where all participants began equal at the starting gate, and the best won. It brought wine tastings into a global arena where they were like the Olympics, and all that counted was taste. In fact, three years later, in 1979, *Gault-Millau* staged an international "Wine Olympiad" in Paris, where a 1975 South Block Pinot Noir from Eyrie Vineyards in Oregon would best some of the finest French red Burgundies. At a rematch the following year staged by powerhouse Burgundy *négociant* Robert Drouhin, the Eyrie came in just behind Drouhin's great Chambolle-Musigny, which would persuade him that he should buy land in Oregon.

But while the French jumped on the bandwagon of the transatlantic tasting, it remained a very American idea. In France wine had never been about just taste, or who won a comparative tasting; it was about history, part of a way of eating and living, part of, well, being French. Rankings of wines had been based on rankings of pieces of land as well as prices, and grouped by quality, not singled out as "winners" and "losers" with numbers. The Paris Tasting was the first step in unhinging wine from place. As Paul Lukacs wrote, the tasting suggested that quality was not a function of history, rather that it "was intrinsic, actually within the wine, no matter what was on the label." The lessons of the tasting were that from now on, what would count in judging wine would be the taste in your mouth at the time and how a wine ranked with the competition.

Of course, once you divorce quality from everything but the taste in your mouth, you have nothing to rely on but your own palate. Classifications like the French system of ranking vineyards and properties into *grand cru* and *premier cru* would no longer be the prime source of help in understanding where wine quality lay. Instead American consumers, who were still learning and often distrusted their own palates, would have to find super-tasters, whose palates could discern the very best wines on taste alone, and follow them.

The old class approach to wine—the hierarchies of châteaux, the classi-fication of land, the primacy of France, consumption of wine by the upper and upper middle class, the old-fashioned wine merchants purveying ad-vice, the slow aging of wine in the cellars of the well-to-do, the accepted wisdom—were beginning to break down. French wines previously re-served for the British wine trade found a higher price in America, and at the same time, Americans were discovering their own wines could be just as good as those hallowed bottles from across the sea. Americans, who had known nothing, were educating themselves and realizing that anyone could learn about wine, just as Julia Child had taught them that anyone could learn to cook French food. You didn't have to have grown up with wine, or have a wine cellar, or even have a lot of money to spend, and a blind tasting was the great wine equalizer.

The time was right for Robert Parker.

3

WINE

CRUSADER

On most saturday mornings in the 1970s, Robert Parker, sprung from the soulless concrete building where he spent his weekdays as a lawyer, could be found happily wheeling a shopping cart through the crowded aisles at MacArthur Liquors on MacArthur Boulevard in Washington, D.C., his body canted slightly forward as if eagerly sniffing out the right wines to buy. He felt revitalized. His real life, wine, was again in play, his day-job legal life thankfully pushed to the background until Monday morning. There he stood, in jeans (sometimes shorts), an old T-shirt, and a windbreaker, frowning as he consulted his list; he'd already circled the wines on the store's latest mailing that he wanted to investigate with his tasting group. He looked forward to talking to owner Addy Bassin and to his assistant Elliott Staren, who had started working for Bassin in 1970 and was becoming a good friend, to find out what was new, what was a good buy, what he hadn't tasted before that he should try. To them Parker was a regular guy, pretty much like every other young enthusiast seeking knowledge, not a V.I.P. like the several Supreme Court justices who were keen wine buffs or the Secret Service detail that stopped by regularly to pick up cases of Château Margaux for President Richard Nixon. Here, as at other Washington, D.C., wine shops, Parker was just another customer serious enough to ask questions and sign up for the mailing list.

After stashing his purchases in the trunk of his car, Parker would drive to the next shop with as much determination and dedication as a butterfly collector on a dangerous remote jungle trail in search of some rare variety. Sometimes he could be found at Calvert Liquors in Georgetown or at Woodley Liquors on Connecticut Avenue; sometimes he concentrated on the round of shops in Baltimore that had expanded their wine sections.

For other lawyers, Saturday meant playing tennis, sailing in Chesapeake Bay, or cutting the lawn; for Parker it meant a return to what really motivated him after a week of work as an attorney at Farm Credit Banks, where he handled bankruptcies, leases, and the legal planning and paperwork for loans to businesses. The bank was part of a nationwide system of borrower-owned lending institutions, established in 1916 to ensure dependable funding for agricultural businesses and overseen by the U.S. government. Unlike commercial banks, the Farm Credit System's banks didn't take deposits, but instead issued debt securities in both domestic and global markets to raise funds to loan. The work was demanding— though not as all-consuming as the intense eighteen-hour days required of an associate trying to make partner at a top law firm—and to Parker it was dull.

Not that Parker didn't apply himself to his job, and not that he wasn't good at it. Though a reluctant lawyer, his keen intelligence and the all-American work ethic took over, at least for the first few years, and he treated the tasks before him as problems to be solved, even if deep down he would rather have been thinking about his favorite liquid. Influenced by the recently passed truth-in-lending law, as well as consumer advocate Ralph Nader, Parker was determined to convince the bank to let him convert their commonly used legal documents into ordinary language. When people obtained a mortgage, he thought, it should be clear to them they might end up paying $100,000 over thirty years for a $30,000 mortgage. His idea of rewriting everything from mortgages to promissory notes in plain, easy-to-read English was not popular with his conservative colleagues. It took Parker several years to achieve his goal, largely because everyone at the bank, including all the other lawyers, voiced their objections. "We can't do this," they would say. "It doesn't sound formal enough." Parker stubbornly dug in his heels, and his persistence paid off. It was one of his proudest accomplishments during his eleven-year law career. (After he left the bank, however, they quickly reverted to legal documents heavy on traditional terminology.) Hiring the bank's first black lawyer and its first female lawyer was another. Thirty years ago diversity wasn't a necessary goal of all-white, all-male legal departments, and both applicants were tough sells to a conservative bank deeply suspicious of change. Convinced they were the best-qualified people for the jobs, Parker stuck with his convictions, sure he was right. Clearly he managed all this without alienating his boss and fellow workers, because he was promoted to senior attorney and eventually became the bank's assistant general counsel, managing a department of fifteen people.

Despite the challenges, the satisfying accomplishments that squared with his strong beliefs, and the possibility of money and promotion, Parker felt trapped. He wanted to be in the world of wine, smelling and tasting and thinking about what he'd tasted, something far better suited to the pleasure-loving side of his personality.

Parker and Pat now lived in Parkton, in the house that had belonged to Pat's parents. The couple bought it in 1974, about a year after Parker joined the ranks of the employed; the Etzels had wanted to retire and needed to sell the house to do so, and the Parkers were ready to buy one. The simple three-bedroom grey ranch home on four acres at the end of a narrow dead-end road, where Pat and her two younger brothers had grown up, bordered Gunpowder Falls State Park. Beyond the screened porch a steep hillside of oaks, maples, and elms fell away sharply to the rushing Gunpowder River far below, its rocky pools home to fat brown trout. Otherwise the house was nothing special, similar to the other modest brick and frame residences on the quiet road, and it needed work, but at least it had a cellar to store wine. It was only a ten-minute drive past farms and cornfields to the tiny town of Monkton with its volunteer fire department. The network of little-traveled rural roads proved ideal for jogging, then Parker's preferred method for staying in shape.

One of the first things Parker did after their move was to form a new tasting group. Of course it included his good friend Victor Morgenroth, now employed as a toxicologist by the Federal Drug Administration, but several other serious tasters also joined the weekly gatherings around the dining table at Parker's home. When he tasted alone or with Pat, however, he simply lined up bottles on a Formica counter in the kitchen badly in need of renovation and spit directly into the sink.

———

THE WINE-WORLD BACKDROP OF PARKER'S TASTINGS IN THE MID-1970s reflected an increasingly problematic imported wine market in America, especially when it came to French wines. Nineteen sixty-nine had seen another sharp increase in interest in Bordeaux when the franc was devalued; then came a frenzy of rising prices and the inevitable feverish speculation as wine hysteria for Bordeaux gripped the American market with the 1970 vintage. A 1972 Sherry-Lehmann catalog reminded (but really warned) customers that a case of 1961 Bordeaux purchased back then for $100 was now selling for $750, if you could find it, and predicted that wines from the 1970 vintage would "achieve equal heights in another decade." *Fortune* magazine chimed in. After all, weren't the *grand cru* wines of Bordeaux textbook examples of an investment possibility? Limited production, rising demand—where could the prices go but up? The message from merchants and most of the media was clear: better jump on the bandwagon now because if you don't purchase the wines as futures, you will miss out. Yes, the prices were staggeringly high, but the implication was they were going to be even higher when the bottled wines finally arrived in 1973.

The château owners in Bordeaux had watched the *négociants* make enormous initial profits on their huge crop of 1970s, which some called a "perfect vintage"—high quality and enormous quantity, too—and so they pitched their prices for the 1971s, which were good but not great, up to even higher levels, as much as three times those of the better 1970s, and saw that the merchants they'd sold to were still making a killing. Nicholas Faith quotes a story told of a *négociant* who, when asked by a wine shop customer for six cases of ordinary Bordeaux *rouge,* replied, "Certainly, but I've had to increase the price from 3 francs to 6 francs." The reply was, "Better send me twelve cases then." The wine market became the province of wild-eyed speculators.

The consumer wine craze had flooded the American market with hordes of unsophisticated new buyers eager to snap up the top wines of Bordeaux. It was a natural fit: the château names were famous, guaranteed instant status symbols. Besides, Bordeaux was so well organized and so much easier to understand and select than Burgundy, with its tiny producers, shippers, and score of communes. For Bordeaux little expertise was required: all a brand-conscious consumer had to do was memorize the names of a few good châteaux and vintages.

Everything was selling, the horizons of growing consumption in the United States looked limitless—sober sources predicted a 100-percent rise in wine drinking during the 1970s—and the British were buying more than double what they'd bought only two years before. With inflation approaching double digits in England, investors poured money into "solid" assets like real estate and new wine investment funds. In truth, the prices for the 1970 vintage had allowed château owners to finally catch up after all the losses they'd sustained during the many bad years of the 1960s. But all this was just the beginning. Nineteen seventy-two and the first half of 1973 were accurately described as "a vinous madhouse."

With the ill-starred 1972 vintage, Hugh Johnson wrote, "prices lost touch with reality." The vintage in Bordeaux was a disaster, "cataclysmic," according to Michael Broadbent, head of Christie's wine department. The wines were terrible, thin, and acidic, downright unattractive to taste. But so what? In 1972 a sign in the window of a retail shop on West Seventy-second Street on Manhattan's Upper West Side prophesied, "It's a fact: wine prices always go up." Château owners in Bordeaux admitted that their 1972 wines weren't very good, but a "keeping up with the Joneses" mentality pushed prices higher and higher in spring 1973, when châteaux offered their yet-to-be-bottled wines to the *négociants.* No one wanted to be caught charging less than their neighbors—they would lose face. All it took was the news that the inferior château next door was selling its wines at 24,000 francs to convince the more prestigious neighbor to ask 32,000. And so it went.

Some desperate trade buyers frantically worked the phones to place orders, worried about whether there would be enough wine to supply demand once the wines were bottled. For the average wine drinker it would mean not very good wines at high prices. But the French assumed the Americans would buy anyway. And why not? French wines were the only ones that really counted. (Later, of course, canny American and English importers boycotted the 1972s, causing panic among the *négociants*.) Prices for 1972 Burgundies were up, too, and American demand for wines from the Rhône translated into price increases of 80 to 90 percent.

Because of a radical change in the French tax code due to take effect in 1972, château owners were also selling off their stocks of older vintages, so as to avoid paying taxes on the wines' current value. Fortunately for them, Christie's, the London auction house, had begun a separate wine department in 1966 and held regular sales. Over the next few years business had boomed, so much so that Sotheby's opened a rival department in 1970, and soon London was the site of forty auctions a year. In America, Heublein, which marketed Smirnoff vodka and wanted to cash in on the burgeoning fine wine market, too, started an annual, very well-publicized auction in 1969 of rare old wines that attracted well-heeled new collectors. Auction prices were often astronomical compared with what wines had fetched in the past—at the 1971 Heublein auction a jereboam of 1929 Mouton went for a then astonishing $9,200.

The London auctions had traditionally been dominated by wine trade professionals who had a clear and fairly realistic grasp of current value. But in the mid-1970s a growing number of American collectors flush with their by-then stronger currency haunted these auction rooms, too, intent on creating instant cellars. So, increasingly, did speculators, attracted by the prospect of fine Bordeaux as a sure investment in financially uncertain times. That demand helped push a major upswing in prices. By 1974, the châteaux were dumping huge quantities of wine into the auction market.

But the French seemed blind to what was actually happening in America and the world in 1973 and 1974. Shelves of retail shops in the United States groaned under the weight of unsold, expensive 1970s. Convinced American demand would continue unabated no matter what, some buyers committed to large purchases of the poor-quality 1972s at high prices—what one critic called, "the flotsam and jetsam of a boom gone bust." Then a bumper crop of mediocre and sometimes atrocious 1973s and the abysmal 1974s followed, and still the châteaux' prices went up. Finally the combination of overheated prices, three poor-to-average vintages, war in the Middle East, the fall of the dollar, roaring inflation, and a world economic slump caused by the 1973 oil crisis pushed the French wine market to the brink of an awesome collapse. Bordeaux's "Winegate," the internationally reported scandal involving the well-connected Cruse family, in which wines were fraudulently relabeled as more prestigious ones, didn't help. The American part of the market, which had seemed gullible and insatiable, crashed, and the French suddenly realized they'd killed their golden goose. Undaunted, they counted smugly on the Brits, but alas, the British economy had also taken a major downturn. Wine was an expendable luxury in inflationary times. Other markets—the Belgians and Scandinavians—couldn't make up the difference. Stacks of cases of thin 1973s and very unlovely 1972s sat in château cellars, and were eventually dumped at low prices.

Although unsophisticated consumers in the United States continued to buy the most renowned labels to impress their equally unsophisticated friends, many wine writers were now quick to point out the problems of Bordeaux. When bottles of the 1971s, which Robert Finigan dubbed "deadly dull," finally arrived in 1976, the San Francisco–based wine critic railed in his newsletter against perpetuating the "charade of superiority, with attendant price gouging, simply because a particular Bordeaux château was included in the moldy old 1855 honor roll." That, he wrote, "is patently absurd."

Burgundy, too, was poised for a fall in the mid-1970s. The wine trade excused dismal bottles from noted estates with the pompous assurance that they "were just going through a phase." In fact, overall wine quality in Burgundy was a scandal, insisted Finigan and others. Some attributed the pitiful decline to avaricious growers who now made their wines in a style for early drinking, a move the growers blamed squarely on the desires of consumers but that neatly benefited the vintners' cash flow. Others pointed to overproduction in the vineyards and growers making more wine from a vineyard than regulations allowed, resulting in less concentrated and flavorful wines. And still others even hinted that a few unscrupulous *négociants*, who bottled and shipped wines, were illegally adding in wine from the Rhône Valley to beef up thin, pallid Burgundies, as they had for the English market a century before. Cynical Burgundians just shrugged. What difference did it make? What did the wine-crazy Americans know about what they were buying, anyway?

American consumers may not have known much, but the revolution in wine appreciation and knowledge across the country meant they were learning very fast, and they didn't like to be fooled or taken advantage of.

Parker was one of them.

As an American trade magazine, *Beverage Media*, had warned, "American interest is not just limited to drinking one wine. The American consumer is probably tasting more wine than any European, who confines his preference to one local wine with no curiosity or interest to try any other type." While French wine prices were zooming and quality was spotty, California wine country was in a frenzied period of expansion and experimentation. Their wines were fast improving, providing the new American market with interesting alternatives. In 1973, there were sixty-two wineries in Napa and Sonoma, twenty-one of which were only a year old. In 1974, French wine imports plummeted 42 percent.

California wine was where the real wine excitement lay in America, West Coast wine lovers predicted, even if many of the East Coast's Francophiles still thought the idea of great wines from the state was preposterous. The Paris Tasting in 1976 had helped convince a few that domestic wines might be worth trying, and their lower prices added to their appeal. For the next decade, comparative tastings of France versus California flourished in public tastings and private homes, and every few weeks, it seemed, some new California "boutique" wine triumphed over Château Latour or Lafite. Many Cabernets produced in the mid-1970s were big and gutsy, with bold, fruity flavors; some were macho blockbusters, impressively pumped up with oak (the if-a-little-is-good-a-lot-is-better philosophy), high alcohol, and tooth-staining tannins. California wine lovers were beginning to prefer the big-is-beautiful style. With wines from brand-new unknown wineries constantly hitting the market, consumers began looking for objective evaluations to tell them what to buy.

PARKER AND VICTOR MORGENROTH, THE MAINSTAYS OF THE NEW tasting group, were more than tasting buddies who respected one another's palates. They also loved to discuss absolutely anything that had to do with wine. It was a guy thing; in the same way men who love baseball spend hours rehashing games and trading batting statistics, they talked for hours about the best method of scoring wines and the paucity of wine writing that went beyond regurgitating the old romantic view of wine to comment honestly on the quality of bottles in wine shops. Nobody, they thought, was writing about the unknown wines for $5 that were just as good as the big-name $30 wines. Both had become annoyed, even angry, at what they saw as the wine industry's over-inflated propaganda and the gouging of the American consumer.

Many new wine aficionados who formed the growing number of tasting groups in America were obsessed with lining up a dozen or so bottles at a time and rating and ranking them to determine which were "the very best." To ensure their judgments of a wine's quality wouldn't be influenced by its existing reputation, they tasted the wines blind, disguising their identity by hiding the labels.

Parker and Morgenroth, too, were great believers in blind tastings. Pulling out the tight corks, slipping brown paper bags over the smooth glass bottles, securing them with wide rubber bands at the neck, then mixing the bottles up and writing a number on each one became an enjoyable pre-tasting ritual, a time to chat before the serious, focused business of studiously assessing the wines one by one and figuring out where they fit in the grand scheme of wine quality. They experimented with various scoring systems, including the letter scale A through F and the very popular 20-point system that was used in the oenology department at the University of California, Davis, the most respected center of wine research in the United States. The two weren't fans of the Davis scale, as it was sometimes called. It seemed like a negative system for what was supposed to be a beverage of fun and pleasure. As Parker saw it, taking points away for what a wine didn't have didn't make as much sense as adding points for what the wine did have. One of the men—Parker isn't sure who—came up with the 100-point idea, which was really a 50-to-100-point scale, as they gave every wine 50 points just for "showing up." Still, having 50 points to play with gave them more flexibility than any other system. Parker spent his evenings after work carefully refining and organizing the scale, eventually allotting up to 5 points for color and appearance, 15 for aroma and bouquet, 20 for flavor and finish, and 10 for overall quality level or potential. This system was much less likely to result in the inflated wine ratings they saw all around them, he and Morgenroth thought, and besides, it was dead simple. If you'd attended school in the United States you knew exactly what 97 on a spelling test meant. By 1976, their 100-point scale was firmly established. Neither questioned whether it was fundamentally odd to assign numerical scores to wine.

But the hottest topic was their mutual problem with their wives, who constantly complained about how much money they were spending on the grape. For Parker, the annual total was now between two and four thousand dollars, a disproportionate amount of the income he and Pat were bringing in. "We can't continue spending this much money on wine!" was Pat's angry, adamant daily refrain. The kitchen was a mess, the house needed rewiring—she loved to drink wine, too, but, hey, there were *limits*.

Sometime in 1975 or 1976, it occurred to the two friends that they should write a wine newsletter for consumers so, as Parker put it, they could "keep buying and tasting so much wine without going bankrupt . . . and avoid losing our wives because of our obsessive behavior." Even if they made no money, they could still tell their wives that the wines they were buying were tax deductible. It seemed like a terrific solution to a joint marital problem. They knew that they were tasting more wines, more seriously, than most of the wine columnists they were reading.

By 1977, the two had decided on the perfect subject to launch the first issue—the 1973 Bordeaux. What wine shops like New York's Sherry-Lehmann called pleasant, fruity wines suitable for drinking while waiting for the great vintages to mature, Parker and Morgenroth regarded as thin, diluted, watery wines from a vintage drowned in rain, a conclusion they'd reached after trying every one they could get their hands on. "What a crappy vintage," they told each other, concurring that only a handful of them were any good. Consumers, they believed, deserved to know the truth before they bought.

Parker especially expressed disdain for the current crop of wine books and columns. His legal training had taught him to be skeptical, to ask questions, to carefully examine the underpinning bias, look for conflicts of interest, and make a judgment. Why weren't writers more critical? The

British writers, most of whom worked in the wine trade, knew what they were talking about, Parker decided. But in his view they rarely pointed the finger at properties that weren't measuring up and frequently recommended wines they themselves were importing and selling without letting the reader know. Parker and Morgenroth's tastings convinced them that wines from many of the best-known properties were poorly made and wildly overrated, and that was all there was to it.

To Parker, American magazine and newspaper writers were worse. He could spot when someone had been wined and dined on a free wine junket. As he later told one interviewer, there'd be a "glowing wonderful story" championing "the next great wine from Brazil," and then when you tried it, you could see the recommendation "was pure bullshit."

THE WATERGATE SCANDAL WAS THE POLITICAL BACKDROP OF Parker's last year in law school and his first year as a practicing lawyer, and it profoundly affected his views on the law, the world, and by extension, wine. The illegal break-in to bug the Democratic National Headquarters at the Watergate hotel in June 1972, the subsequent cover-up, and the eventual revelation of a secret Republican fund to finance massive political spying and sabotage against Democrats became big news, with sensational information emerging on a daily basis. Parker, like his fellow law students, followed the unfolding of the story in the *Washington Post* and through the televised hearings of the Senate Watergate Committee formed to investigate what had happened and who was involved. Sam Dash, a professor from whom Parker took a course covering conflict of interest, was its chief counsel. In fact, many of the key players in the growing government crisis were lawyers, and they seemed to belong to one of two distinct groups—heroes or villains.

The good guys included Dash; Archibald Cox, the Harvard professor hired as the special prosecutor; and the attorney general, Elliott Richardson, and his assistant attorney general, William Ruckelshaus, who both resigned from the Justice Department rather than follow Nixon's order to fire Cox, who was pressing the president to release secret tapes of conversations in his office. Among the bad guys were former attorney general John Mitchell, who had controlled the secret fund; White House counsel

John Dean III, who testified that the president had known about the break-in plan; and lawyer and White House aide Charles Colson. Forty government officials, many of them lawyers who had lied, were indicted, which hardly gave the legal profession a good name. Parker regarded them as an embarrassment to anyone calling himself a lawyer. Even the vice president, Spiro Agnew, had to resign over charges of tax evasion and pay-offs, and finally Nixon, a lawyer himself, resigned rather than face impeachment.

One lawyer Parker did admire was consumer advocate Ralph Nader, whose task forces' investigations and in-depth reports were hot copy in the 1970s. A young lawyer with an encyclopedic memory and deep moral convictions, Nader gained national prominence in the 1960s with the publication of *Unsafe at Any Speed*, a damning critique of the auto industry. When the story broke that giant General Motors had hired private detectives to tail and harass Nader, his book catapulted onto the bestseller list. Throughout the 1970s he was the most visible and ardent advocate of the consumer, his goal to establish organizations that could empower "the little guy." The Center for the Study of Responsive Law, which he founded in Washington, D.C., in 1969, became the "in" choice for top law students wanting a summer job—in 1970, one-third of Harvard Law School's student population applied.

Given the atmosphere of protest against the Vietnam War and general disgust over the Watergate scandal, Nader's crusades attracted many idealistic law students and fledgling lawyers. As Nader's Raiders, they took on established powers in business and government, working indefatigably to expose fraud, the cynical manipulation of consumers, and secret conflicts of interest like the luxurious getaways offered to congressmen and senators "for educational purposes" by industry trade associations. Their explosive reports named names, made the nightly TV news, and garnered regular headlines.

Conflicts of interest like these had been the one topic that spoke to Parker when he was in law school, and he saw evidence of it in much of what was written about wine. How could you appraise a wine objectively if you were also selling it or if you were buddies with wine producers, sucking up free wine and letting them pick up the tab for airfare, four-star hotels, and lavish dinners at elegant restaurants? In his view, the real purpose of wine junkets was no different from that of the industry association junkets for congressmen that Nader had exposed for what it was—influence peddling.

Nader was determined to make reliable product information available to consumers, and he believed independent experts should compare and assess product performance. Business and government had to be held accountable. His crusades enraged business leaders but impressed Parker. Nader, he saw, had become the underdog hero taking on those in power, David slaying Goliath, a role that many Americans admired—the little guy with right on his side doing battle with the cynical exploiters.

When he thought about producing a wine newsletter, Parker kept that image of Nader and his philosophy firmly in mind. The newsletter, he insisted to Morgenroth, had to be an independent consumer publication that wasn't at the mercy of the industry, which meant taking no advertising—a revolutionary stance, he thought—and setting strict standards about conflicts of interest. They would pay their own way, buying wines and refusing any perks that were offered.

Parker didn't see writing a newsletter on the side as a conflict with his job as a corporate lawyer, but the Food and Drug Administration, where Morgenroth worked, had complex rules about what employees could do outside their work. The scrupulous Morgenroth would have to obtain the agency's blessing, and they refused. Unfortunately, just as they were ready to make the newsletter a reality in 1977, he had to back out. With his partner gone, a disappointed Parker moped for several months over abandoning his dream. How could he go it alone? Basically an optimist, he decided that if he borrowed $2,000 from his mother and obtained mailing lists from key local retailers and Les Amis du Vin chapters, he could make it work. Pat thought he was crazy. He was a lawyer with a good, secure job. Wine was fine as a hobby, but wasn't this taking it too far?

Parker's indulgent mother, who lived ten minutes away, no doubt found it difficult to deny her only child anything he truly wanted. They had a warm relationship; he had always been the center of her attention, and now all he needed was a couple of thousand dollars for a little project he yearned to start. Years later Parker wryly observed that if she had known this loan would result in his giving up a law career to taste wine, she never would have given him the money.

He knew obtaining mailing lists wouldn't be quite as easy. The best wine shops catered to powerful people in Washington, and they guarded their lists carefully. But without them, he recalled, "the whole thing would fall apart." Parker first approached Addy Bassin, the owner of MacArthur Liquors, whom he respected as a "brilliant, brilliant wine buyer." Though Parker was a regular customer there, he was still nervous about making his pitch. A lot was at stake, because he had told himself that if he couldn't persuade MacArthur's and Les Amis du Vin to part with their lists, that would be the end. Bassin's initial reaction, just as Parker had feared, was to refuse. But Parker had come prepared. On a desk in the store's crowded back room, he spread out a sample page to show Bassin what the newsletter would be like, and words of explanation for his use of the list spilled out rapidly. "Of course I'll pay for it," he told Bassin. "I promise I won't copy it or do anything with it—all I want is to send out a free copy of my newsletter to everyone on the list." He felt he had to warn Bassin, though, that he might be critical of wines the shop carried. That independent stance impressed the shop owner, who agreed to think about the proposal. For Parker, the next few days were tense.

Then Bassin telephoned. "Okay, I'll give it to you," he said. "I think we need a publication like this. I don't agree with everything you're writing. But I like the way you're going to do it." To this day Ed Sands, who owned Woodley Liquors (and still does, having merged it with The Calvert Shop, which he bought in 1982), doesn't know what convinced him to let Parker have his mailing list. He'd never given it to anyone before. Nor had Marvin Stirman, one of the founders of Les Amis du Vin. To keep the organization completely noncommercial he'd always refused to sell his list to importers, distributors, and wineries. But he, too, handed it over.

Parker had that kind of effect on many people when he was starting out. The combination of boyish charm, evident sincerity, earnest self-confidence, and intense passion for wine was powerfully persuasive. Behind the polite, "I'm just a regular guy" demeanor was a forceful and determined personality, and when it came to wine, the zeal of a missionary. Parker was convinced his would be the first real consumer guide to wine. In fact, it wasn't. But that belief was evidence of his single-minded focus and very healthy ego. Other people had their doubts; Parker never did, and he didn't endear himself to other wine writers, later, by repeatedly saying in interviews, "until 1978 most wine critics were essentially on the take."

―――――

THE NEWSLETTER WOULD BE CALLED *THE BALTIMORE-WASHINGTON Wine Advocate,* Parker decided, and to drive home his complete dedication to consumers the logo would be a corkscrew in the shape of a crusader's cross. It was a telling symbol of his deepest feelings about the kind of wine critic he wanted to be. Parker had the earnest zeal of a true convert, a seriousness about the subject that had something in common with that of investigative journalists on the track of political scandal, but with an added touch of dedicated hedonism.

He hired his legal secretary, who lived nearby, to type up the first issue on weekends from his handwritten notes. Deciphering them wasn't easy, but she'd had plenty of practice. They churned out copies on an old-fashioned mimeograph machine, using buff-colored paper and brown ink. One hot summer weekend Parker, Pat, and his mother and father camped on the floor of their living room collating and stapling the pages of some 6,500 newsletters, then folded and addressed each one. The tedious job took a long time, but Parker felt jubilant, filled with the expectation that a good 40 or 50 percent of those who received it would subscribe. It was all consumer-oriented advice, no b.s., and only $10 a year—less than the cost of a good bottle of wine. How could a subscriber lose?

Considering he'd worked at a bank and specialized in commercial law for five years, it seems astonishing that Parker hadn't researched the financial aspects of direct-mail marketing. Yet he claims he was clueless that the normal return rate was 2 to 3 percent. Later he would admit that if he'd known more, he probably would never have taken the risk of producing the newsletter in the first place. He mailed out the issues with a sense of satisfaction, then he and Pat flew to California to spend their vacation doing what he loved: touring vineyards and tasting as many wines as possible.

The first issue of *The Baltimore-Washington Wine Advocate* billed itself as "A Consumer's Bimonthly Guide to Fine Wines." The cover page showed the heavy hand of Parker's legal training; a précis of what the newsletter would offer appeared under the heading "Purpose and Scope" and was written in the royal "we." Parker set forth plainly how the average consumer would benefit, pointing to the fact that wines with famous names were frequently disappointing, while less famous properties now making "strides in quality" were the better buys. For wines with prices of $10 to $25 per bottle, he wrote, "a consumer can ill afford such mistakes." "Subscribe now," he urged, "because you may spend $10 for the wrong bottle of wine."

His cut-to-the-chase language echoed his consumer-oriented stance and strong belief that wine could be evaluated objectively—the word "objective" appeared over and over again. Parker assured readers he was unequivocally committed to honesty and placed himself squarely on the side of the wine consumer as "an ombudsman," who would "expose mediocre and poor wines as well as overpriced wines." He contrasted himself with "most wine authorities," who he claimed "ignored" many good buys and "vastly overrated" famous wines and vintages. Finally, he promised to comment only on wines that he had purchased and assessed himself.

The complimentary initial issue had all the earmarks of an unedited, amateur production—crooked titles, repetitive writing, misspellings of even major wine regions (Pomerol was spelled Pomeral), and a tasting note style marked by a fairly limited vocabulary. Words like "splendid," "pleasant," "velvety," and "rich" were used over and over as positives; "thin," "imbalanced," and "acidic" predominated among the negatives, though Parker didn't hesitate to call wines "annoying," "terrible," or even "atrocious." But the most frequent positive references were to fruit, and a phrase that later became one of Parker's signature terms of praise, "oodles of fruit," made its debut in a description of the 1973 Clos du Val Zinfandel from the Napa Valley.

The writing improved over the first year's issues, but certain things didn't change. Parker's logo, the buff-colored paper, the crowded design crammed with appraisals of wines, the 100-point rating system, and a note in the credits that "news media, wine distributors, and retailers may use the material provided credit is given to *The Wine Advocate*," a brilliant stroke of promotion, were the same then as they are today. A schedule of features for the next six issues showed that Parker had given considerable thought to what might appeal to readers.

Parker carried through on his promises by offering a damning view of the 1973 Bordeaux then widely available in Baltimore and Washington wine shops. He warned his readers that the 1973 from famous first-growth Château Margaux was "a terrible wine . . . very thin and acidic with a dull, dumb bouquet and taste. A poorly made wine that should be avoided," and gave it a failing score, only 55 points out of 100. The Leoville-Poyferré met an even worse fate—this "atrocious wine devoid of any redeeming social value" rated 50, Parker's lowest score. Of the sixty-five Bordeaux reviewed, ten ranked below 60, which meant, according to Parker's definition, they were wines "to be avoided." Only three wines scored over 80 points: Château Ducru-Beaucaillou (83), La Tour Haut-Brion (82), and Château Pétrus, at 87, the best of them all. A two-page spread also recommended good-value wines at three different price levels; the most expensive group included his first 90-point wines: the 1974 Sonoma Vineyards Alexander's Crown Cabernet Sauvignon from California at $7 a bottle, rated 91; the 1973 Clos du Val Zinfandel and the 1971 Ruinart Brut Champagne both rated 90.

The 100-point rating system that was to become so famous was featured on page 2, and the meaning of the numbers was carefully described. (A wine rated below 64 was "to be avoided"; a 60–64 wine had "noticeable flaws"; 65–74 stood for "average"; 75–79 "above average"; and 80–89 "very good." A wine that scored 90–95 was "outstanding." The top rank, 96–100, was reserved for "extraordinary" wines.) Now that so many other publications, as well as retailers, have adopted the same system it's hard to realize how unusual it seemed at the time. Other wine publications in America didn't bother with numbers at all, relying on categories or puffs or stars to convey levels of quality, much the way movie reviews did. Decanter, a British consumer wine magazine that had been launched in 1975, used the traditional 20-point scale, but only for major tastings.

Meanwhile, Robert Finigan's better-known newsletter was going from strength to strength. In 1977 he'd added a national edition, and he considered ditching his restaurant newsletter so he could concentrate on wine. Was Parker aware of Finigan's Private Guide? He says he'd never even heard of Finigan until after he'd published The Wine Advocate and people began comparing his newsletter to Finigan's. But several retailers with whom Parker was friendly certainly read it, and Finigan remembers Parker calling him to discuss his doubts about starting a newsletter, and that he had encouraged Parker to do so. Finigan saw no competition there. He was already a name, and Parker's planned newsletter would be a very local affair.

The two newsletters had many similarities in appearance and content. Both were printed on buff-colored paper, unillustrated, and packed with text that consisted primarily of tasting notes. Finigan, too, placed himself firmly on the side of the consumer. He was also highly critical of the wines on American retail shelves and dismayed by the French, warning readers that he perceived "a tendency to rest on laurels and to maximize profits at the expense of quality . . . France is coasting while California is striving admirably to improve quality with each succeeding vintage." In writing about Burgundy, Finigan had called the 1971 vintage "treacherous investments . . . introduced with hysterical ballyhoo . . . intended to keep prices up," and denounced the "wholesale disinformation" on vintage quality perpetrated by the trade. He had slammed the 1972 Bordeaux as unattractive, thin, bitter, and acidic, and had already shared his assesssment of the 1973 Bordeaux in March 1977, rating it somewhat higher than Parker did. Parker praised Ducru-Beaucaillou, Finigan thought it merely average; Finigan saw Beychevelle and Pichon-Lalande as outstanding and fine values, whereas Parker gave the first a 75 and the second a 71. Château Haut-Batailley, which Finigan called delightful, Parker viewed as a poor, thin wine that "should be avoided." But both were in complete agreement on the Mouton. "What a bore! No fruit, no depth, no charm . . . but the Picasso label is nice," Finigan wrote.

What was different was Parker's 100-point scale, his very direct, heavy-handed appeal to consumers, and an almost strident insistence on telling it the way he alone saw it. His views seemed a little simplistic compared with Finigan's more sophisticated, sometimes intellectual analyses of what was going on in the market and his carefully worded tasting notes. Parker's newsletter certainly wasn't as well written, but his conviction that his judgments were unassailable was abundantly clear.

———

DURING THEIR VACATION IN THE BAY AREA, AS HE AND PAT DROVE from winery to winery on the back roads in Napa and Sonoma, Parker felt charged up. He fantasized a vast pile of checks from a couple of thousand committed subscribers awaiting their return, along with frantic notes from the town's tiny post office asking him to rush over to pick them up as soon as possible because they didn't have sufficient space to hold them all. The hundred or so envelopes that had actually arrived were a devastating disappointment. "I can't believe this," he moaned to Pat. "The timing is right. It's a good idea. With no advertising, people have to be interested." Maybe the whole venture had been a waste, he thought. Maybe he should send back the checks that had arrived. But over the next several weeks a few hundred additional subscriptions trickled in, enough to fund the second issue he was already working on—a comprehensive look at California Cabernet Sauvignon.

For the second issue, the mimeograph had to go. Parker consulted a local printer who explained the basic financials of printing—concentrate on signing up more subscribers and the per-issue cost would go down—but it was still typed up by his legal secretary. The design of the sixteen-page second issue mailed in September was an improvement: the titles throughout were straight, and the text and tasting notes were certainly more grammatical if not more graceful. The explanation of Parker's numerical rating system now took front and center place on the cover page, as though he intuited the fundamental importance of his scores in reeling in subscribers. This time Parker also underlined his independence from the trade, reminding consumers that he had "no interest, direct or indirect, financial or otherwise with the importation of wine, the wholesale distribution of wine, or the retail sale of wine."

Volume II included notes on just over 150 wines, a review of two wine books, and a long, highly flattering profile of MacArthur Liquors and Addy Bassin, who had provided Parker with the mailing list he had considered essential to launch *The Wine Advocate*, though that wasn't mentioned. It evidently hadn't occurred to Parker that such a profile could look like a journalistic big wet kiss.

The main feature, on California Cabernet Sauvignon, included a two-page introduction followed by tasting notes and scores on more than ninety Cabernets and Merlots from the 1973, 1974, and 1975 vintages that Parker had purchased in the few local shops then carrying a wide selection of American wines—Morris Miller Liquors and Harry's Waterside Mall in Washington and Wells Discount Liquors and Perring Liquors in Baltimore. Right off the bat Parker attacked the "overpublicized" transatlantic taste-offs of California Cabernets versus Bordeaux that had become wildly popular after the famous 1976 Paris Tasting. Because the two areas produced wines very different in style, Parker wrote, such tastings have "little value in establishing which is the 'best' wine . . . the better French Bordeaux are elegant, delicate wines that possess incredible subtlety and complexity, whereas the best California Cabernets are massive, powerful, assertive wines often bordering on coarseness . . . Because of this huge fruit and high alcohol content, a wine taster is justifiably impressed." Bordeaux was less attractive in youth, but evolved with age into "a complex beverage of considerable charm and elegance," while most California wines were "immensely enjoyable in youth. Cabernets do not develop complexity and grace, but lose the intensity of their fruit and tannin." Further, he insisted that because of their low acidity, they "can taste flat and boring" after seven to ten years of age. Finigan had frequently expressed much the same view, naturally highly unpopular with California winemakers, starting as early as October 1975.

Years later Parker remembered this issue as a serious attack on California's winemaking philosophy. But though he believed that many California Cabernets were "heavy, thick, one-dimensional and somewhat coarse" and that unlike Bordeaux they didn't develop "complexity and grace," he actually praised twenty-four wines (out of the ninety) as "superlative examples" of California-style Cabernet, describing some as "well structured wines that are simply stunning in terms of concentrated fruit and bigness," others as "top rank Cabernets which . . . create a more balanced, harmonious wine without the overbearing bigness of some of the foregoing." The "superb" 1975 Stag's Leap Wine Cellar Lot #2, he wrote, "has that unmistakeable breed and character; it should make the Bordelais nervous." At the same time Parker found the 1975 Callaway "monsterous [sic] to the point of being grotesque," with a "huge, chewy character" and such high alcohol that it resembled port, and lambasted the 1974 Raymond Vineyards as "overbearing and crude"; he also liked some "big" wines and rewarded them with high scores. Parker offered no notes distinguishing the different vintage conditions of the three years, which seems surprising considering 1974 was widely viewed then as a superlative vintage for Cabernet in northern California, equal to, if not better than, the highly regarded 1970s and 1968s, and that it accounted for half of the wines to which he gave high scores.

Finigan had covered many of the same wines a few months earlier, in his May 1978 issue; he, too, criticized those that were "fearsomely inky," complaining that drinking overprized "monster" wines required "a streak of masochism" he didn't possess.

———

PARKER WROTE IN A TINY SPARE BEDROOM AND USUALLY TASTED alone or with Pat in their now-renovated kitchen. Ten of his friends and acquaintances participated in regular blind tastings for the newsletter, held in the evening at his home, in the dining area of the living room, which had a large stone fireplace. Several doubted Parker could ever make a living from publishing the newsletter, but all happily kicked in $10 to $15 each to help defray his wine expenses—an essential supplement to his limited monthly wine budget. Besides Morgenroth, the group included neighbors, an amateur winemaker who later started one of Maryland's first wineries, the owner of a French restaurant, and a few retailers who became good friends. One was Bob Schindler, who then ran The Wine Merchant, a shop he'd opened in Baltimore in 1977 right after graduating from college. The two met when Parker was chasing down a wine he'd recommended highly for his own cellar—the 1974 Sonoma Vineyards Alexander's Crown Cabernet. Schindler had set aside three cases for his best customers but let Parker have one (at $10 a bottle), and he soon was a tasting regular. Another was Dr. Jay Miller, a fun-loving, chubby-faced family and child psychologist who later became Parker's sidekick on trips to France. A self-confessed wine nut, he moonlighted part time as a salesman at Wells Liquors in Baltimore, where Parker was also a regular customer, to support his wine habit and get his hands on rare bottles.

Parker set a theme for most of the tastings, such as California Cabernets or white Burgundies. The wines were always served blind. In the relaxed, guys-only atmosphere discussions were lively, and Schindler was impressed by Parker's knowledge of French cusswords. Pat sometimes tasted, too, but these events often included people who weren't necessarily friends, and she often dropped out. Parker listened carefully to everyone's opinions, using them as a control on his own taste buds, but in the newsletter he only noted when the panel split over a wine's quality. The tasting notes and scores, which he didn't share with anyone in the group before publication, were very much his own.

Parker's first break came at the end of September, about a week after he'd mailed out the second issue. *The Baltimore-Washington Wine Advocate* received a much-needed boost from William Rice, the executive food editor of the *Washington Post*. Rice had received a free copy and a few calls about Parker from curious readers, and sensing a good story, spoke to him over the phone and then asked him to drop by. The balding, bespectacled Rice had been an important fixture in the Washington food and wine world for more than a decade, and a general reporter before that. He devoted most of his weekly wine column to his interview with Parker, who was anxious for the attention and already projected a dynamic image. "Wow, what ambition for a young lawyer from Baltimore," was Rice's first reaction. "Parker sees himself as . . . a Lone Ranger figure riding through the hail of gobbledygook fired at consumers by retailers and industry-influenced publications to tell the truth," he wrote. Parker described his publication to Rice as "a truth-wine newsletter, a work of passion . . . If I can break even and have a lot of fun doing it, I'll be happy. But I have enough drive to want every wine drinker to subscribe."

The Lone Ranger image captured something essential about Parker. The masked cowboy on a white horse, who first captivated radio audiences in the 1930s, was a very American idea of a hero—a fearless idealist fighting alone to right wrongs and teach villains that greed and prejudice will fall before justice in the end. It was an image Parker cultivated, and maybe even believed in.

Parker wasn't a total innocent when it came to self-marketing. In this first interview he claimed to have one thousand charter subscribers, for example, which made the newsletter sound important and viable, though he didn't have nearly that many. And there was little modesty in the way he characterized himself to Rice. "I'm coming from left field," he said. "People will say 'who the hell is he?' But I came to realize my palate was as good as most people's in the trade I ran into and I was tired of reading books that are so clearly wrong. I may be wrong on some wines, too, but I think I'm well qualified and on most I'm going to be right." Contemptuous of the hallowed ranking of Bordeaux in the 1855 classification, Parker grandiosely announced he was "writing a book to put forward my own classification and only Latour and Pétrus will be first growths."

Rice accurately predicted that the number scores, which Parker "really meant only as an accessory" to the comments, were "likely to become the focus of reader attention." Parker's success or failure, he wrote, "will depend on the curiosity of local wine drinkers and the extent to which they agree with his judgments."

The article drew in more subscribers but didn't make the newsletter an overnight financial success. For the next few years each issue would generate just enough new subscribers for Parker to fund the next one, with nothing left over to pay back his mother, something he finally did ten years later. Even for a workaholic Parker's life was full, an uneasy balance between tasting and writing for his side venture and the demands of his day job. His wine and travel expenses grew—in 1978 he began a pattern of traveling to Bordeaux twice a year, bringing Pat along to translate. He was fond of "snooping around châteaux," where chatting with the blue-smocked *maître de chai* in cold, dark cellars gave him the inside story on what was really going on. It was hard work, Pat remembers. Going from château to château, using your powers of persuasion to elicit tastes of the wine since very few of them knew (or cared) who Parker was, having people spit wine on your feet—and always shivering in cold cellars. Parker's limited vacation time, which he arranged to coincide with Pat's, meant tasting trips were packed; during a March 1979 visit to France, the couple rushed through the Médoc, St.-Emilion, and Pomerol in Bordeaux, tasting the 1977s and the 1978s in barrel, then traveled on to Cahors, Fronsac, Bergerac, and Burgundy. There was always time, though, to dine well, and now they could afford the one-, two-, and even three-star restaurants listed in the fat red Michelin guide. Always ready to drop everything to track down an unknown great wine, Parker jettisoned his schedule on one trip for a two-day detour to visit growers in tiny Cornas in the Rhône Valley; the sommelier at the Beau Rivage restaurant in Condrieu had passionately recommended a 1969 Cornas from August Clape, and it had "shocked and astounded" Parker.

The following June, during Pat's summer vacation, it was back to Bordeaux, where he had appointments at over two dozen châteaux and also tasted a couple dozen barrel samples "procured for me by Christine Delorme and Martin Bamford of Château Loudenne." The genial, charming Bamford, who worked for the English group International Distillers and Vintners, had restored the château and made it the center of IDV's Bordeaux-buying business. A gregarious and generous host, Bamford entertained many English-speaking journalists at Loudenne and delighted in educating them in the ways of Bordeaux. Impressed by Parker's passion, he quickly befriended him.

Because the newsletter listed where recommended wines could be purchased and provided short profiles of good stores, Parker also checked out a wide selection of shops and made it a point to meet the proprietors and key salesmen. Through them he connected to importers and buyers who brought in the best wines. This equivalent of wine networking ended up being extremely important in spreading Parker's name to the people who counted in the trade and those who would make a name for themselves in the next decade and vice versa. Take his early connection with Robert "Bobby" Kacher, another young American who'd been seduced by the romance of wine in France, had come back an instant wine expert, and now imported wines for Continental Wine & Liquor Company. They shared ideas, and at one of Kacher's private tastings he introduced Parker to Jeff Pogash, the U.S. representative for French wines from Alsace. Pogash, too, had had the usual epiphany in France, where he'd attended wine courses at Steven Spurrier's Paris shop while studying at the Sorbonne. He was struck by Parker's intensity about wine, but was even more influenced by Kacher's prediction that "he and his newsletter are going to be prominent."

Anxious to get his hands on anyone who could bring Alsace wine to the attention of consumers, Pogash lost no time in arranging a comprehensive personal tasting for Parker, something that in a few years the country's key importers would be clamoring to do. Pogash drove down to Parkton from his home in New Jersey in an old station wagon, one hundred slim green bottles from twenty different producers chilling in coolers in the back. The two tasted all morning, through lunch, and all afternoon, sharing opinions, memories of Alsace, and laughs. Parker whipped systematically through the wines, Pogash providing detailed information along the way. That night, after a dinner cooked by Pat, served with another barrage of wines, Pogash slept soundly on the couch in the Parkers' living room. When he left, he knew which wines Parker really liked, but not his scores. That had to wait until he published the upbeat overview of "Alsatian wines" in issue 2.

At the end of the newsletter's first year, with issue 6, Parker dropped the

Baltimore-Washington part of the title. He was ambitious for a wider readership, and starting in spring 1980, with issue 10, eliminated the list of shops that carried his recommended wines. But the key connections had been made. The twenty-seven area stores mentioned regularly knew about and (mostly) subscribed to *The Wine Advocate*, and so did many of their customers.

In its first two years *The Wine Advocate* returned frequently to some basic themes Parker would expand on in the coming years. Fulminating against "ridiculous prices" (French wine prices were again "soaring to all-time highs"), he often condemned even good wines as too expensive to recommend, waxed indignant over retailers' "distasteful price gouging," and ranted about the poor wine storage conditions in wine shops. While he criticized many California wines for their high alcohol, excessive taste of oak, and outsized proportions, in his tenth issue (April 1980) he conceded that two of the greatest wines he had ever tasted were California Cabernets—the 1975 Joseph Phelps Eisele Vineyard and the 1975 Caymus Special Selection. He hoped that "competition from California will stimulate some of the lethargic French wine producers."

No country or region escaped Parker's exasperation: in a review of Rhône wines, he found many "either sloppily made or an intentional wine fraud." He came back to Bordeaux again and again, urging his readers to buy the 1975s. He scrupulously described for readers the sometimes elaborate, almost obsessive tasting methods he used when evaluating Bordeaux: always blind, decanted three hours in advance, with small amounts left out and tasted twenty-four and forty-eight hours later, a test he'd devised to predict a wine's longevity.

Parker's tasting notes had evolved and improved, and now included free-wheeling comparisons along with blunt language. One Cabernet, he wrote, "has the finesse of a horny hippopotamus." Another "may be hazardous to your health if drunk. A stinky rotten wine." A 1976 Merlot was "the only red wine of Veedercrest I can swallow without looking for a toilet," while a vegetal-tasting Inglenook Cabernet was "a wine for the jolly green giant."

Among wine writers and the trade at the time there was no accepted format or style for tasting notes. Generally they described a wine's taste, smell, and level of quality in a few sentences. Michael Broadbent had attempted to sketch out some parameters in his book *Wine Tasting*, but Parker hadn't seen it. He had read notes in *Vintage* and *Les Amis du Vin* magazines, as well as in the works of earlier British writers, but didn't find them helpful as models. Comparisons to specific woods, fruits, and flowers and high-flying metaphors were really irrelevant, he thought. He wanted to use plain language just as he had in rewriting bank documents and to describe wine as you could describe a person—introverted or extroverted, small or big, thin or fat. His inspiration for the "keeping it simple" writing philosophy had been his revered undergraduate advisor, Gordon Prange. The chief military historian in the Pacific theater during World War II, Prange was famous for his mesmerizing lectures on the attack on Pearl Harbor and his book *At Dawn We Slept*. Keep your sentences simple and keep them short or you risk losing your reader, he'd told Parker after reading his exam papers, and Parker took his advice to heart.

Though many of Parker's views actually matched Finigan's, from the beginning *The Wine Advocate* established a very different tone and sensibility. To some early readers of both, that was a plus that pushed them toward Parker. Robert Millman, then a salesman at Morrell & Company, a wine shop in New York, felt Finigan seemed like "a dandy and a snot nose," with snobbish "high standards of elegance and refinement," someone who looked down on the wines he wrote about, the kind of person who would belong to an exclusive wine society like the Chevaliers du Tastevin. To him, Finigan's wine complaints sounded querulous. Parker, on the other hand, seemed like a regular guy writing for other regular guys who just happened to be passionate about wine. His language evoked the common man instead of some elitist Ivy Leaguer, and he wrote about a lot of wines anybody could afford. To Millman, Parker's newsletter seemed a breath of fresh air.

———

IN THE WINGS WAS ANOTHER FLEDGLING WINE PUBLICATION WITH a seemingly inauspicious future, a flimsy San Diego–based tabloid newspaper called *The Wine Spectator*, which had been started by Bob Morrisey in April 1976. Published every two weeks, it covered primarily the California wine industry. Its logo was a chubby Bacchus, one hand gripping a glass and an arm embracing a barrel. Former New York real-estate investment banker Marvin Shanken loved it. Already the publisher of *Impact*, a wine trade newsletter, he believed in the future of California wine and spoke to Morrisey often. But the newspaper was woefully undercapitalized. In the early summer of 1979, Morrisey finally told Shanken that if he didn't buy it, *The Wine Spectator* would go under. Shanken took the plunge and paid him $40,000. Now he was the owner of two wine publications. But he had big ambitions for this one.

Like Parker, Shanken didn't grow up in elitist circumstances, nor did he attend Harvard or Yale or any Ivy League university—a fact of which he seems proud. Shanken is a self-made man, with an MBA from American University in Washington, D.C., who gave up the prospect of big bucks on Wall Street to follow a dream. But unlike Parker, he was drawn first to American wine. While Parker was traveling around Europe and falling in love with Bordeaux and Burgundy, Shanken was discovering the pleasures of wine in Napa and Sonoma, where he was orchestrating vineyard deals. Collecting Cabernet became his passion—he didn't have much money, and it was easily affordable at $4 to $5 a bottle. On business trips he usually managed at least one expedition to a winery tasting room to snap up rare bottles unavailable elsewhere, then recorded the details of each purchase in a big ledger on his return to New York. It still pains Shanken that when the prized case of 1968 Louis Martini Special Selection Cabernet he'd purchased rolled off the airport luggage conveyor in New York, the cardboard was dripping with red wine; every precious bottle had broken.

On Saturdays, Shanken combed wine shops just as Parker did, sometimes driving 100 miles to investigate a new store or scoop up a terrific buy. Since few wine lovers on the East Coast had heard of top California wines, proprietors were thrilled to part with cases they'd shoved to the back of the cellar. Often they'd been forced to buy them to obtain the winery's line of jug wines. In California, many of the same wines were one to a customer. In 1970, though, celebrating the success of a business deal at La Crémaillère restaurant in Bedford, New York, Shanken tasted his first great wine, and it was French—1961 Château Pétrus, for $100 a bottle. He didn't make it to France's wine country, however, until a few years later, on a press trip.

Shanken had grand journalistic dreams for *The Wine Spectator*, and that included reviewing wines critically. Consumers wanted to know what to buy, and for *The Wine Spectator* to succeed it would have to tell them. That meant a tasting panel and blind tastings, which were in place by 1980, and a wine rating system. The panel first opted for a "modified 9-point University of California at Davis scale," but soon adopted the 20-point scale also used at Davis and familiar in Europe. Neither Shanken nor his writers had ever heard of Robert Parker and his 100-point system.

FOR PARKER, A SECOND BREAK CAME IN THE SPRING OF 1981. After William Rice left the *Washington Post* to become editor in chief of *Food & Wine* magazine in the fall of 1980, Phyllis Richman, who had been the restaurant critic, became the *Post*'s new food editor. Richman knew she had to do something to fill the now-empty wine slot. Several wine writers had proposed themselves, eager to appear in one of the three most important newspapers in America, but no one stood out. To test their expertise she set up two tastings of ten California Cabernet Sauvignons for the writers. They didn't impress her—none of them recognized that two of the wines in each tasting were duplicates. In fact, they didn't seem any better at tasting than the few amateurs she'd also invited. But Richman was intrigued that all the writers had suggested contributing a profile of Robert M. Parker, Jr. Who was he? She got a copy of his newsletter, found he had an underground reputation among wine professionals around the country, and thought, "Why not get Parker to write the column?"

It had never occurred to Parker to put himself forward as the new *Washington Post* columnist. Consumed with producing his newsletter, he was just learning how to write, and besides, he had a full-time day job. He didn't think of himself as a journalist (like Frank Prial of the *New York Times*) or even as a writer (like Hugh Johnson). Instead he viewed himself as a taster and a critic, producing the vinous equivalent of *Consumer Reports*. When Richman offered him a biweekly Sunday column in the spring of 1981, however, he must have realized immediately how much exposure it would give him and his newsletter, a benefit that would justify his accepting the puny fee: $150. In the May third announcement of this new column of wine notes and briefs, Richman put her impressions of Parker in the form of a tasting note: "young and fresh, clean, a touch of sweetness with a nice balance. Enough backbone so you can expect him to develop well over the years . . ."

Parker's column ranged wide that first year, reflecting his extensive tastings of 1,700 wines plus plenty of samples from barrels on winery visits, but he never included scores for the wines he recommended. He must have realized that those numbers had captured his subscribers' attention and that giving them away in a newspaper column would kill off his newsletter. Some shops that subscribed posted his scores alongside wines Parker had praised, inspiring the curious to plunk down their money to see all his ratings. A few, like MacArthur Liquors, kept their copies of *The Wine Advocate* in a fat three-ring binder for customers to peruse and a supply of Parker's subscription flyers by the cash register.

From the beginning Parker's column went out on the *Post* wire. The exposure gave his subscription base another significant boost. Richman observed that Parker still wrote like a lawyer, but when they went out to lunch or dinner his comments on wine and his precise, seemingly photographic taste memory always astonished her. He was good at connecting with people, too, she noted, getting to know the importers, retailers, and collectors who would help build his career. That may not have been his intention, but the effect was the same. He knew what he wanted and set about getting it.

One of Parker's early columns described a three-evening tasting of nineteen vintages of Château Latour organized for six wine "fanatics" by one of the area's "most knowledgeable private wine connoisseurs," Paul Evans, who became a close friend. Parker's excitement over this "wondrous" event was almost palpable. It was a big deal to him, hardly the common occurrence such a tasting would be in just a few years' time, and he was obsessed with getting his "palate, mind, and body in top shape." "Several weeks prior to the tasting," he wrote, "I became so irrationally fearful of getting the flue [sic] or a miserable head cold that I commenced each day with aspirin and Vitamin C supplements to ward off any evil spirits. I took particular care not to get chilled after my daily exercise routine and lived on an extremely bland diet immediately prior to the tasting so as not to upset either my stomach or my palate."

In spite of invitations to tastings, Parker's budget for wines was mounting still higher, to Pat's dismay; the *Washington Post* didn't pay any expenses, so Parker was footing the bill for most of those 1,700 wines as well as travel. At the time the amount of wine he tasted annually seemed huge, but in less than a decade that number would quintuple.

It was time for Parker to start branching out beyond the Beltway and take on New York. He already made periodic trips to the city, ostensibly, as his boss and bank colleagues thought, to attend legal conferences, but in reality so that he could attend an occasional wine tasting, troll wine shops like 67 Wines & Spirits and Morrell's, introduce himself to importers, and show wine people his newsletter. On one of those trips, in 1981, he stopped in at *Food & Wine* magazine, where Bill Rice was now the editor in chief.

Rice had hired me and my husband, John Frederick Walker, to be the magazine's wine editors the year before. Launched in 1978, the magazine billed itself as a more contemporary, journalistic food magazine than *Gourmet*. Rice was a consummate newsman with his finger on the pulse of what and who were going to be the next big things in the world of food and wine. I'd never met Parker or read his newsletter, but Rice suggested we meet with him to see if there was something he could do for the magazine.

In preparation for the meeting I cleared away a pile of manila file folders, several wine bottles, and a plate of leftover chocolate truffles from *Food & Wine*'s test kitchen from my desk. By New York City magazine standards our office was private and spacious—that is, big enough to hold a wooden desk, shelves of thick wine books, and a long glass-topped table with stacks of cardboard boxes of wine samples shoved underneath. My first impression was that Parker was big, over six feet tall, and as he fumbled with his packages he made the office seem cramped. He didn't seem comfortable in a suit and tie. If he were wearing a red plaid shirt and a pair of heavy work boots, I thought, he could pass for a logger, or maybe an affluent farmer, or at the least someone who used to play football in high school or college and still tossed a ball around with friends on the weekends. His hands were thick, with blunt fingers and a simple gold wedding band; his smile was warm and wide and eager, almost Midwestern, and his receding hairline suggested that one day he might have to start thinking comb-over.

He was a distinct contrast to Hugh Johnson, the urbane British wine writer who had spent an hour in the magazine's dining room a month or so before, entertaining us with his sophisticated Cambridge-accented patter. Very much the great man holding court, Johnson recounted amusing stories of well-irrigated wine weekends at various châteaux in Bordeaux and at the outsized home of a Texas collector.

Parker found some empty floor space for a couple of cardboard wine boxes fitted up with rolled handles that he carried and handed me the fat brown envelope he'd tucked awkwardly under one arm. He started talking in an excited, nonstop style, the words rushing out and spilling rapidly over one another, explaining that the boxes contained half-bottles of an underrated vintage of Château d'Yquem he'd spotted at a bargain price in the wine shop around the corner ("what a coup!"), which was why he was a little late. The envelope was stuffed with the last dozen issues of *The Wine Advocate*. John joined us and we quickly flipped through the issues Parker had brought. I read a few tasting notes, noting the manic enthusiasm and pell-mell prose. His 100-point scale seemed unwieldy; when I questioned him about it, Parker was off, passionately defending his system. We wondered about the fine distinction between say, an 85- and an 86-point wine, but to him the difference was as obvious as short and tall suspects in a police lineup. He walked out with an article assignment that seemed perfectly suited to his energy and passion for categorization: taste through and find the most drinkable jug wines on the current market. The article appeared in March 1982. Though an admitted Francophile, most of his picks were from California because of their consistently fresh, fruity style. It was his first magazine article, and he quickly landed another assignment from us.

In December 1981, Parker and Pat flew to Europe to visit winemakers in the Rhône Valley and Piedmont in northern Italy. It was bitter cold, and the hills of Barolo and Barbaresco were white with snow. Pat shivered in vineyards and cellars, as usual, but Parker was in his element, happily discovering the best of the region's deep, rich reds, especially Angelo Gaja's "stunning" Sori San Lorenzo Barbaresco. In the Rhône, a stop at top grower/shipper Etienne Guigal introduced him to two great wines that he would later make famous, "La Mouline" and "La Landonne." But he also discovered a "disturbing development"—some winemakers were filtering their wines before bottling, which in his view "disemboweled" and "eviscerated" them. This became an idée fixe for Parker, and he would fulminate on the evils of filtration for years to come.

The trip included a stay in Paris. How different from that first visit fourteen years before! Parker and Pat dined happily at the world-famous three-star restaurant Taillevent, where they chatted with proprietor Jean-Claude Vrinat, whom Parker regarded as "the most knowledgeable Frenchman in France concerning the wines of his country." At Tan Dinh, a tiny Vietnamese restaurant on the Left Bank, he drooled over the list of five hundred Bordeaux, especially Pomerol. It was, he wrote, "a place worthy of a pilgrimage."

After they returned, on January fourth, Parker plunged into two weeks of bleary-eyed late nights to put out the first issue of 1982. Raves for one of his new loves, the wines of the Rhône Valley, filled half the newsletter's twenty pages. For the first time he praised several "small, innovative, and high quality" importers, including Francophiles Kermit Lynch and Robert Chadderdon. As a stylist, Parker was beginning to loosen up, using references to his favorite movies and music. "The Good, the Bad, and the Ugly," his subtitle for a section on Sauternes, referred to a movie, a Clint Eastwood spaghetti western.

MUCH HAD CHANGED IN THE WINE WORLD DURING THE 1970S. BY 1982 a combination of political and economic factors meant America was once again clamoring for the best French wines. The weather had produced several excellent vintages in Bordeaux, warehouses were full, and prices tumbled as the dollar continued to increase in value against European currencies. For the first time America imported more Italian than French wines. An exploding number of California boutique wineries continued to reap publicity, giving birth to a new species of consumer, the California wine snob. *Time* magazine called this a "golden age for California wine," but ever-increasing prices threatened to burst the still-small bubble. According to a 1980 study, the American wine boom was still limited—only 4 percent of the adult U.S. population consumed 53 percent of all the wine sold in the country.

Finigan's newsletter remained dominant. At the end of 1981, his California and National editions had subscribers in every state and thirty-three countries, including Kuwait, and he was spending $20,000 a year on wine. *The Wine Spectator* had hired a couple of young journalists, one just out of journalism school, to beef up its commentary, and planned to move to San Francisco to be closer to the heartland of American wine. The importance of the *Washington Post* nationally had brought Parker's name to a wider world. Parker noticed a pattern. A wine shop or person in Detroit or Richmond or Philadelphia would subscribe, then a couple of months later several more subscriptions would flow in from the same place. Zachy's Wine and Liquor shop in Scarsdale, one of the biggest purveyors of Bordeaux futures in America, had even quoted Parker's opinion on the 1981 Bordeaux in one of their 1982 ads in the *New York Times*. But he was hardly famous, and had only just begun to break even on his newsletter expenses.

———

STILL, AS SUBSCRIPTIONS INCREASED, "WHEN CAN I LEAVE THE law?"—the question that was never far from Parker's thoughts—hovered over his and Pat's lives. All along Parker had shared with her his dream of a future in which he could concentrate all his energies on wine. Pat found this a frightening prospect; she was a risk-taker in travel but not necessarily in life. She'd put her husband through law school and their common goal had been his law career. She thought of the law as a sure thing. His job at the bank meant security. Now he wanted to throw over all they'd worked for. Who knew what would happen with a newsletter? But she could see the strain on her husband of working two jobs and she wanted him to be happy.

Parker added up their mortgage payments, the newsletter's operating expenses, and what it cost them to live. The key was ten thousand subscribers, he told Pat. That's all he needed. By the end of 1982 the number had climbed to seven thousand. He had "checked out mentally from the practice of law," as he put it, and knew he would eventually leave. He was thirty-five years old. If only there were a way to speed it all up!

4

THE 1982

BORDEAUX

*F*OR MANY SUCCESSFUL PEOPLE, A SINGULAR event catapults them suddenly into a new future. It's as if while they were driving slowly along a highway to their eventual destination something caused them, almost miraculously, to shoot off onto another, faster route, a shortcut to stardom. The watershed event for Parker's career as a wine critic was the 1982 vintage in Bordeaux. Luck had much to do with what happened. Or as Parker would later put it, "Fate smiled on me."

During his flight back to the United States from France on March 21, 1983, Parker worried obsessively that his plane might crash. Though not usually anxious when flying, he was bursting with his first big wine story. He wanted to lean forward in his seat, willing himself closer to the United States and the next issue of *The Wine Advocate*, in the same way he habitually leans forward over the steering wheel when driving from winery to winery in eager anticipation of the next wine. Convinced that the red wines made in Bordeaux in 1982, which he'd tasted during the previous week, were "very great" and the vintage one of the finest of the twentieth century, he was panicky at the thought that if the plane went down he would miss his chance to write about it. He'd crammed his black notebook with notes on dozens of wines. Superlatives studded every page: "stunning," "blockbuster," "prodigious," "incredible," "fantastic," "heavyweight," and more. Parker was champing at the bit to let his readers in on what he considered the greatest wine buys since he'd started tasting wine in 1968.

———

AT THE TIME THE WORLD'S WINE COUNTRIES STILL FORMED A long-accepted hierarchy of quality, like tiers of angels in medieval theology. France resided at the top, of course, and three of its regions—Bordeaux, Burgundy, and Champagne—stood above all others, not just representing wine at its best, but also bestowing status on their buyers, marking them as people blessed with refined taste—and plenty of money. Among this holy triumverate, Bordeaux held primacy of place in the minds (and cellars) of the world's wine collectors. It produced more fine wine than any other region on earth and it was the most aristocratic. Where else could you find so many grey stone châteaux with fairy-tale turrets (and sometimes moats) surrounded by vineyards? The wines of its most important subregion, the Médoc, had been organized and classified into five categories of status (and price)—first through fifth growths—in the mid-nineteenth century. (Only one wine from a commune outside the Médoc couldn't be ignored; the famous Château Haut-Brion in Graves, praised by Thomas Jefferson, had been included as a first growth.) With few exceptions the original rankings of the sixty-one châteaux in that famous 1855 classification had remained intact over 125 years, and a wine's position in it still determined much of its reputation for quality and its price.

Back then, what the harvest was like for Bordeaux was the general bellwether for every new vintage everywhere. Too bad for Italy's Piemont or Spain's Rioja if they had a great vintage when Bordeaux had a poor one. Most of the world assumed their wines from that year were poor, too. History, mystique, and a strong publicity machine combined to elevate Bordeaux to the forefront of everyone's wine consciousness.

The grand properties in the most important communes, or appellations, of the Médoc's boringly flat landscape on the Left Bank (west) of the Gironde estuary—Margaux, St.-Julien, Pauillac, and St.-Estèphe—and a few historic ones in Graves, south of the city of Bordeaux, were large and appeared cold and intimidating; they gave "an impression of class, stability, reliability, elegance, permanence, tradition, and unimpeachable status," as American importer Kermit Lynch once put it. But behind the staid serene facades lay the snobbery, rivalries, and intrigue typical of a tightly closed, class-ridden society and a greed-driven merchant mentality. Since Parker, like so many Americans from middle-class backgrounds, was a self-proclaimed egalitarian with few aspirations to move up the social ladder, it was particularly ironic that his initial step to star critic status would result from celebrating a great vintage in this, the most aristocratic of wine regions.

In fact, Parker loved the wines from the great châteaux in the Médoc and Graves and found some of the proprietors he'd met charming, congenial company. But with few exceptions he regarded the scores of smaller, more modest châteaux and properties in the less-celebrated Bordeaux communes of St.-Emilion and tiny Pomerol on the Right Bank of the Gironde as far more friendly, casual, and welcoming. Most of the warm, generously rich wines from Pomerol had barely been known in England and America before the 1960s, though they had found a ready market in Belgium. After Graves had established its own, less elaborate system of ranking in 1953, St.-Emilion followed in 1955; Pomerol had no system at all. Perhaps that was one reason many wine lovers considered the elegant wines of the Médoc the true aristocrats of Bordeaux, even if the top St.-Emilion château, Cheval Blanc, and later Château Ausone and Pomerol's star, Château Pétrus, were often spoken of in the same hushed tones as the top wines of the Left Bank. In the Médoc of the 1970s, each château's ranking put its owner into a social pecking order. Alexis Lichine once observed to me that owners of first growths only dined with owners of other first growths, and so on down the ranks; he, an American, prided himself on being invited to dine at them all. The official first growths—Châteaux Margaux, Lafite-Rothschild, Latour, Haut-Brion, and Mouton-Rothschild, which had been elevated to this elite group in 1973—often behaved like an exclusive club, meeting occasionally in Paris to discuss prices, but the rivalry among them, as among the best second growths, remained intense. (Not, however, as bitter as it had once been, when Baron Philippe de Rothschild of Mouton reportedly stooped to such petty acts as serving bottles of his cousin Baron Elie de Rothschild's Lafite with spicy curried rice at an important lunch to make sure it didn't outshine Mouton.) Each first growth wanted his wine to be recognized as the best, and all took their exalted rank in the world of wine with great seriousness.

In the past, Parker had observed firsthand at Château Mouton-Rothschild what he termed the staff's "inexcusably arrogant and condescending" attitude toward foreigners. Of the illustrious proprietors of the Médoc first growths, only one, Laura Mentzelopoulos of Château Margaux, met personally with Parker in March 1983. He had to make do with the *maîtres de chai*, the cellarmasters. But for the most part, he preferred it that way. On the Right Bank, in Pomerol, on the other hand, Christian Moueix welcomed him at Pétrus, and, as he had before, shared his passion for and opinions on wine.

Bordeaux also owed its exalted position to the fact that its wines evolved and developed complexity with bottle age, and could, in great vintages, last for decades. That, plus widespread acceptance of the classification system as a guide to quality and the notion of the "château" as a mark of status, would make the wines blue-chip investments that could be held and traded internationally, creating a virtual stock market of wine. Last but not least, the wines appealed intellectually to wine lovers and connoisseurs. Small differences of soil and climate from appellation to appellation and vineyard to vineyard, the slightly different proportions of related grape varieties in each version of the classic blend, and the effects of ever-changing weather on each vintage translated into subtle and complex variations of flavor and aroma in the wines, creating endless possibilities for fascinating tasting comparisons and discussions—to determine the personality of wines from a particular appellation, like Margaux or St.-Julien, for example, or to understand how the vagaries of weather had marked the different vintages at several châteaux. Bordeaux offered endless scope for quibbling, wrangling, and lively debates among critics, connoisseurs, and collectors; the best wines were not just for drinking but also for thinking and talking about. Bordeaux lovers had their own ideas of which châteaux belonged in which tier of quality, and their choices often related to their own ideas of what the styles of Bordeaux should be. In the past couple of decades the 1855 classification system, perhaps inevitably, had come under attack. Many of the most knowledgeable wine writers in England had weighed in with their own revised classifications, just as Alexis Lichine had, based on their perceptions of contemporary quality, and their demotions and elevations of various châteaux fueled endless debates. But few questioned Bordeaux's stature as the pinnacle of wine quality.

For Parker, Bordeaux had become an ongoing motif, a signature concern that he had raised in his first issue of *The Wine Advocate* as well as in his first interview. Yes, he critiqued wines from all countries and regions, but in the first five years of his newsletter, Parker had returned to criticize Bordeaux again and again and again, as if obsessed with communicating his judgment on which of these grand wines were and were not living up to their reputations. In 1983, it turned out to be the right obsession with the right place at the right time for the American consumer as well as the Bordeaux trade. The way the wines were ranked, made, and sold practically begged for someone like Parker to reassure wine-phobic Americans so the Bordelais could make big inroads into the lucrative growing American market. His rise to power was intimately tied to Bordeaux.

Since 1978, Parker had traveled to Bordeaux twice a year, always spending a week or two there during the annual rite of spring in March when the châteaux welcomed merchants and a handful of journalists into their cellars to taste the wines from the previous fall's harvest in barrel. Basically this was a sales preview for the trade to assess the quality of the vintage and determine demand for the wines, and the first part of the *en primeur*, or futures, campaign. Once the châteaux released their *en primeur* prices, usually starting in April, the merchants would decide what and how much of the new vintage to buy and figure out where they would sell it.

Wandering from château to château tasting great wines sounds glamorous; images of posh hotels, charming old winemakers, and fabulous food and seductive rare wines at long, polished tables in elegant château dining rooms with aristocratic owners in black tie come to mind. But to Parker it was a hectic business trip, and he packed his schedule. His time was all too limited. For convenience he and Pat usually checked into the modern Novotel in Bordeaux Le Lac, a nondescript industrial suburb 9 kilometers north of the city. With its bland but comfortable décor, modern bathrooms, and convenient parking, Novotel (then $38 a night for two) resembled an American-style motel on an interstate highway, surrounded by an anonymous business center and industrial park. From Parker's point of view it offered easy access to the highways circling the city. The exit for the D 2 north to the Médoc was minutes away; Graves was a quick drive south; and conveniently to the east, across the Gironde on the Right Bank, lay St.-Emilion and Pomerol.

That year only a few American journalists besides Parker came to taste the wines in barrel; among them were two major wine critics, Bob Finigan and Terry Robards. A former financial reporter at the *New York Times*, Robards had recently been forced to resign from the *Time*'s wine column over a perceived conflict of interest, but now wrote for the *New York Post*; he would soon be named a columnist for *The Wine Spectator*. For most wine writers there was no point in going. The only consumers who cared about detailed notes on individual wines were those fanatics who intended to purchase the wines while they were still aging in French cellars, as wine futures. A few of the largest big-city retailers in America offered them, but since the crisis of the Bordeaux market in the mid-1970s, demand for futures had never rebounded. Besides, analyzing a young Bordeaux in barrel and envisaging accurately what it would become a year and a half later when bottled—much less a decade hence—was thought to require long experience and a professionally trained palate. A *maître de chai*, for example, barrel-tasted his wine hundreds of times, regularly checking its progress from the moment the grape juice fermented into wine until it was finally bottled. A mouthful of raw, unfinished wine typically tasted mouth-scouring; its unpleasant, harsh tannins dominated, but with guidance a professional learned to look beyond that and assess the wine's underlying balance, concentration, and fruit, and its potential quality.

For consumers, wine futures worked a little like financial futures and could be a risky business. You put your money down (a few retailers charged only 50 percent in advance) to reserve a case of wine at a fixed price. If the wine increased in value between then and when you picked it up two years later, after it had been bottled and shipped, you saved money, sometimes a lot of money. As an investment wine could outperform many stocks. The 1975 Château Lafite had sold for $150 to $160 a case as futures; by the time the bottled wine reached New York, it cost triple that, and by 1983 that same case traded for as much as $1,200. But if prices didn't increase, or the exchange rate wasn't in your favor, tying up your capital made no sense. On the other hand, if you didn't purchase top wines produced in tiny quantities when they first hit the market, you might not be able to get your hands on them at all.

The futures market was part of the unique way Bordeaux wines were sold. The châteaux that produced the wine were the highly visible part of Bordeaux, the one that wine-lovers were aware of. The other, more hidden world, was *la place de Bordeaux*, an almost byzantine commercial network of several hundred *négociants*, or merchants, and *courtiers*, or brokers, who sustained the *en primeur* system. The *en primeur* campaign resembled a game, and it worked like this: Each spring the châteaux sold anywhere from 50 to 90 percent of their wines from the previous year's harvest, which was still in barrel, to *négociants*. Acting as diplomatic intermediaries between the châteaux and the *négociants* were the *courtiers*, who put together the deals by acting as brokers to both. The *négociants* kept a (usually) small percentage for stock and sold the rest to importers, who quickly unloaded it to distributors and retailers, who then offered it to consumers. All this was only on paper—nothing more than a sales slip or a contract—as the wines weren't even bottled. The benefits of the system to the châteaux were obvious—immediate cash flow, worldwide distribution two years before the wines were even shipped, and the guarantee that *négociants* would buy even in not-so-good vintages to make sure they would receive allocations of a certain number of cases in the best years. It was left to the *négociants* to build up markets for the wine.

Over the centuries speculation had thrived (as it still does), and during the eighteenth, nineteenth, and much of the twentieth century *négociants* and *courtiers* grew rich and powerful. During that time *négociants* stored the purchased wine in barrel in their own enormous cellars, looking after its development or *élévage*, until it was ready for bottling. Then they bottled, sold, and shipped it. But their power began to diminish around 1970, when all the important châteaux started bottling their own wines, and even further after the market collapse in the mid-1970s, when many went broke. Where once *négociants* held on to huge stocks of wine, only a few now held any, and most functioned essentially as traders or distributors, often specializing in particular markets—Europe, Asia, and United Kingdom supermarket chains.

In this highly complex system, prices and sales were affected and sometimes distorted by many factors, including the quality of the vintage, the size of the crop, strong demand in one country, the state of the world economy, a particular château's reputation, popularity in a specific market, buzz, and gossip. The prices were set by the château owners, who wanted to make as much money as possible—certainly as much as, if not more than, their neighbors. Setting the correct initial price was extremely important, since each player in the chain had to make money. The *négociants'* main concern was to sell all the wines they'd purchased worldwide and make a profit, and they had a much better grasp of the effect of market factors than the châteaux did.

After the 1982 vintage, a new and powerful factor, Robert Parker, entered the mix through an opening in the system.

Since consumers couldn't taste before they bought futures, how did they know whether and what to buy? In England, wine lovers relied on trusted wine merchants who acted as both importers and retailers and had usually tasted the wines. Their lists and descriptions went out to their best customers after they'd decided what to buy from the *négociants*. The situation in the United States was more complicated, a result of liquor license regulations put into place after Repeal. Importers who had tasted the wines had little connection with the general consumer. Instead in most states they had to sell to distributors, who sold to retailers, who sold to consumers. But unlike England, America didn't have a tradition of trusted wine merchants. Added to that was the fact that few retailers—even among those who knew something about wine—had experience with futures. After some of the terrible wines retailers had pushed in the 1970s, few wine lovers had faith in what retailers said. When in doubt they fell back on the best wines in the 1855 classification. A few serious wine lovers happily turned to independent experts like Finigan and Parker.

The *en primeur* market was moribund. Several of the recent vintages had been very good, but everyone was waiting for a truly great vintage that could excite the fickle but rich Americans and once again convince them to buy futures in quantity. To the Bordelais, the 1982 vintage looked like the answer to a prayer.

————

IN WINE, TIMING IS EVERYTHING. WINE BEGINS AS GRAPES, AN agricultural product heavily dependent on nature's whims. They need the right weather at exactly the right time to make great wine. From June until October grape growers in Bordeaux scan the skies, tune in the forecast, pray for sun, and worry about rain and hail. A comparison with previous growing conditions is never far from their minds; this is the way they predict what the style and quality of the wines might be. In 1982, the weather was ideal at two critical points, and this made all the difference.

The first was "flowering"—the crucial period when vine blossoms produce grape bunches. Sunny skies and warm temperatures quite early in June provided the perfect conditions for a successful and early flowering, which indicated, if all went well, an early harvest. Since rain would arrive eventually in the fall, the grapes were more likely to have been picked before then. With an early flowering, you were already ahead of the game.

By the end of June, the number of healthy-looking bunches of infant grapes predicted a large crop and therefore an immense amount of wine. Endless rain and gray days over the summer and at harvest would mean thin, light wines, while dry, warm weather would mean ripe grapes with dark skins and very, very good wines. But how good? July was very hot, and wine growers softly voiced the opinion that a little rain to cool the vines down wouldn't be so bad. August was cloudy and warm. Expectations rose. The last part of the growing season, as usual, was a time of suspense. In September, during a three-week burst of intense heat and a number of days when the temperature exceeded 100 degrees, the skins of the grapes darkened and their sugar content soared to record levels. High sugar translates into high alcohol in the wine, and in some of the resulting wines it reached an amazing high of 13.5 percent, a level then more typical of warm California than cool, rainy Bordeaux. The sweltering Mediterranean days were alleviated only by a gusty hot wind that dehydrated the grapes, producing extraordinary concentration. This final burst of warm weather turned an excellent year into an extraordinary one. As consulting oenologist Michel Rolland remembers it, "Everything came at the right moment. They were the best grapes I had ever seen in my life."

All red Bordeaux are blends of two or more grapes in the Cabernet family, which includes Cabernet Sauvignon, Merlot, Cabernet Franc, Petit Verdot, and Malbec. Tannic Cabernet Sauvignon dominates blends on the Left Bank, while plump, plush Merlot is key in Pomerol and St.-Emilion, along with fragrant Cabernet Franc. The Merlot, as always, ripened first, which meant that on the Right Bank châteaux began picking on September thirteenth, one of the earliest harvest dates ever. On the Left Bank, Mouton-Rothschild, anxious to get the grapes in quickly before rain fell and diluted their flavor, hired six hundred pickers, who worked feverishly in torrid heat to bring in the crop in a week instead of the usual three. By the time heavy rain fell on October fifth, the grapes at most properties had already been crushed and the sweet red juice was safely hissing and bubbling wildly in fermentation vats. The weather was so hot that some smaller proprietors had trouble keeping the huge amount of fermenting wine from overheating dangerously; many *maîtres de chai*, even at estates with modern equipment, slept fitfully in the cellar so they could monitor vat temperatures and avoid disaster. Some resorted to setting big blocks of ice on top of the fermenters—even, it was said, throwing them right into the vats (illegal, as it amounted to watering down the wine). The bonanza of grapes coming in all at once put many proprietors in a panic. Some didn't have enough vats to ferment their crushed grapes. The new owner of first-growth Château Margaux, Laura Mentzelopoulos, desperate to find additional fermenters to handle the overflow, phoned her neighbor Alexis Lichine to see if he knew of any that could be had.

Fermentation transforms the sweet grape juice into drier, more complex-flavored wine; the natural yeasts on the grapes convert the sugar to alcohol. The riper the grapes, the more powerful the wine they make, and in 1982 everyone agreed they were absolutely perfectly ripe. As long as they didn't get out of control, the high temperatures during fermentation would ensure that more color, flavor, and tannin would be extracted, giving the wine more of everything. Tasting his raw young wine, the owner of Château Monbousquet waxed lyrical. It was fantastic, he told the *New York Times*, "like jam. . . . like mashed fruit in your mouth."

Soon the wine was transferred into small oak barrels, where it would age for twelve to twenty-four months. By the time Parker arrived in March, many proprietors had already performed the *assemblage*, selecting the individual barrels they planned to combine to make the final *grand vin*. Other barrels would go to make a château's second wine, if they had one, or be sold off in bulk to beef up bottles of shipper wine labeled simply Bordeaux.

The word on the presumed stellar quality of the 1982 vintage had already gone out from the château owners and the trade, as if from on high, and a then little-known French wine writer, Michel Bettane, went so far as to call it the finest since 1929. But did it deserve all the shouting? Parker closely followed the publicity and pondered the question in his office, driving to and from work, and while he and Pat ate dinner at home. The "vintage of a century" claim from the French had been repeated so often in the past fifteen years it was like the story of the boy who cried wolf.

For one thing, Parker knew that often in the greatest vintages of the past, like 1961, the harvest of grapes had been small. It was as if there was only so much lusciousness and richness in any one year and with a small crop, all of it would be concentrated, guaranteeing that the flavor of the wine would be powerful and intense and that the wines would last and last. But the yield per acre in 1982 had been immense, usually a sign that the flavor would be diluted. Still, Professor Émile Peynaud, the famous oenologist and then consultant to forty châteaux, had told Terry Robards of the *New York Times* in December that growers had never before seen "such a level of richness and quantity together." Peynaud compared 1982 to the few canonized vintages of the twentieth century 1961, 1959, 1949, and 1947.

Parker admired Peynaud and could hardly wait to taste the wines and decide for himself.

Parker's schedule during his exhausting but exhilarating days in Bordeaux was tight; his first appointments were promptly at 9 A.M. Once worried about his ability to rise early enough to hold a 9-to-5 job, Parker was out of bed at 7:30. He tried to get a solid eight hours' sleep, but sometimes he was bleary-eyed from a restless night in a too-soft hotel bed. He hit the road by 8:30 for eleven- to twelve-hour days of intensive tastings and interviews that rarely ended before 7:30 P.M. On a typical day he visited three châteaux by noon, starting on the hour and allotting forty-five minutes each to taste the 1982, retaste the 1981 and 1980, and ask questions; it was the wine critic's version of the psychoanalyst's fifty-minute hour. Sometimes he lunched at a château where he also tasted, then rushed on to work his way through fifty to one hundred or more samples from a variety of châteaux in a *négociant*'s office, as people in the trade did, or went on to more châteaux. Appointments occasionally continued well into the evening. Always in a hurry, Parker bragged that he drove "at a speed that would make macho French actor Jean-Paul Belmondo proud." Frequently he tumbled into bed after midnight and fell asleep looking forward to the same schedule the following day.

It didn't take Parker long to decide he completely agreed with Peynaud. He'd "never tasted so many super-rich, opulent, dramatic wines that had obviously been produced from very ripe, jammy fruit." Cellars in Bordeaux, as everywhere, are damp and cold. In winter months the awful, penetrating cold seeps into your feet and up through your bones until it sets your teeth chattering. March can be one of those times, a raw month with grey skies and rain. But Parker was in heaven standing by the fat barrels stained with wine, happily inhaling the tantalizing strong, sweet smell of aging wine, and he watched with mounting enthusiasm as *maîtres de chai* dipped their pipettes into the barrels and sucked up the amazingly dark, rich-looking red liquid, then dribbled it into his waiting glass. Parker swirled, sniffed, sucked in, and spit the samples onto the cellar floor as beaming *maîtres de chai* looked on. Down in the small black notebook went his notes, with a score range next to them. These were the most thrilling barrel samples he'd ever tasted. Everywhere Parker inquired about the growing season, the harvest, fermentation, and vinification. Pat tasted, too, pulled her coat tighter, and translated.

In Pomerol, at minuscule Château Pétrus, Parker thought their 1982 "was the most perfect and symmetrical wine" he had ever tasted. He was "still staggered" by its sensational quality an hour later. No wonder tall, thoughtful, intense Christian Moueix, the manager and son of one of the château's two owners, told him that the wine "is my legacy . . . the greatest wine" he had ever made. From the first, Parker thought it would probably become *The Wine Advocate*'s first 100-point wine, his highest accolade. It reminded him of the glorious 1947, the only wine he'd ever rated 99+.

Many welcomed Parker and Pat with open arms, happy to spread the word to America of the greatness of the vintage. Parker told everyone he wanted to learn, and both château owners and *négociants* were eager to help. At Château Ducru-Beaucaillou, a second-growth St.-Julien that Parker had praised many times in print, the modest owner, Jean-Eugène Borie, poured his 1975, '76, '78, and '81 for Parker to compare with the 1982 to illustrate just how special and different the vintage was. At first-growth Château Margaux, the grandest and most magnificent château in the Médoc, with its Doric columns and long avenue lined with trees, the final assemblage had not yet been done. Nevertheless the estate manager, Philippe Barré, and Laura Mentzelopoulos, whose late grocery magnate husband had bought the château in 1977, let him taste from several different barrels of Cabernet Sauvignon and Merlot to form his opinion. Parker had been critical of Château Mouton-Rothschild, another first growth, in recent issues of *The Wine Advocate*, so he was pleased when they brought out a sample. He thought it "legendary . . . a wine that will develop into a Mouton of monumental proportions." Raoul Blondin, the venerable cellarmaster dressed in worker's blues and a beret, ranked it with the 1947 and 1961. But at first growths Château Lafite-Rothschild and Château Latour, Parker was told that only staff were permitted to try the new wine before it was sold. Piqued, he attributed their refusal to "nervousness."

———

ONLY A FEW HOURS AFTER HIS PLANE FROM FRANCE LANDED AT JFK airport, Parker sat in the Pool Room at New York's Four Seasons restaurant at their eighth annual California Barrel Tasting Dinner, surrounded by the crème de la crème of the U.S. wine world: press, winemakers, retailers, and a sprinkling of celebrities. The lavish extravaganza designed to showcase the new vintage of California's state-of-the-art wines, in this case the 1982, had been launched in 1976, and it had become the single most important event on the New York wine calendar. Attending was de rigueur for the city's wine writers and editors; I'd been to most of these events but Parker, rooted in Maryland, never had.

The five-hour dinner required serious gastronomic stamina, never a problem for Parker; six wines, including three barrel samples, accompanied each of the seven courses, and then came dessert. Everyone drank too much; Paul Kovi and Tom Margittai, the owners of the restaurant, frowned on spitting out wine during dinner. Between courses we all circulated to trade wine gossip, which was at least as important as assessing the wines. Word that Parker was just back from tasting the 1982 vintage in Bordeaux rippled around the room, and like many others, John and I were eager to hear his opinion on the wines. How good were they? "Phenomenal! They're great!" Parker enthused, as a few people clustered to hear. I told him we'd just chatted with Robert Finigan, who'd also tasted in Bordeaux and didn't like the wines at all.

Parker had never met Finigan. After we introduced the two of them they chatted politely about the 1982 vintage; Finigan was distinctly cool and aloof. Parker could hardly believe what he was hearing: that Finigan considered the vintage overrated, that he'd tasted dozens of poor wines. The encounter remained vivid in Parker's memory for years. He remembered wondering what Finigan could have tasted, and thinking, "If he really writes that, he's gotta be a fool." The paranoid idea crossed Parker's oversensitive mind that he was the "young guy" and that the more established Finigan was trying to shake his confidence.

Finigan had already committed his thoughts on the 1982s to paper; his March thirtieth newsletter would soon be on its way to subscribers. He doesn't remember meeting Parker that night at all.

The exchange certainly didn't alter Parker's initial view of the 1982s, though he planned to retaste in June just to make sure. Supremely confident of his opinion, he devoted most of his April eleventh issue to a preview report that amounted to a hymn of praise. In the first paragraph he urged his readers to "stock up on some sensational Bordeaux wines." He argued that the buying power of a strong dollar (the exchange rate was a high seven francs to the dollar) and the singular greatness of the vintage—the best since 1961—had created the best opportunity wine consumers had had in the last twenty years. Prices remained reasonable so far because of several good vintages—1978, '79, and '81—still in the pipeline. He put together his analysis of reasons to buy 1982 futures as if he were arguing a major legal case that he knew he couldn't lose.

Compared to average vintages, the 1982s were atypical. But then, most great vintages in Bordeaux were. It was unusual to have both a huge crop and such concentrated wines. Their high alcohol content, especially in Pomerol, and their ripeness and intensity of taste reminded many, including Parker, of big, fruity California Cabernet Sauvignons. Plus they were slightly lower than normal in acidity. Typical "classically structured" young Bordeaux usually tasted harsh, tannic, and even unpleasant in barrel, but the tannin in these wines was covered with so much fruity lushness—Alexis Lichine called it "puppy fat"—that they were attractive to drink right from the start. That raised questions about their ageability—one of the hallmarks of a truly great vintage—as if only Bordeaux mouth-puckeringly awful in youth could amount to anything later. Even some in Bordeaux expressed reservations, but Parker confidently predicted they would age beautifully.

"Certainly your financial condition will dictate what you decide to do," he wrote, but "any serious wine enthusiast will want to latch on to some of these wines now or two years from now." He entertained no doubts.

———

PARKER'S OPINION OF THE 1982 BORDEAUX JUMP-STARTED HIS rise. His rival Robert Finigan's opinion of the same vintage would trigger his decline. Finigan first mentioned the potential of the 1982s very briefly (and positively) in the September 30, 1982 issue of his newsletter, then again in November, when he reported that producers like Jean-Pierre Moueix and Alexis Lichine were "rhapsodic about it." Setting off in early March for a "quick but intensive tasting trip," Finigan was confident that the wines would be great. The only question in his mind was how great.

Although he usually had barrel samples collected and brought to a couple of central locations so he could taste more efficiently, no one had yet released their samples to the trade, so Finigan was forced to slog from château to château, "cajoling" tastes from proprietors and cellarmasters by promising he would not print notes on certain wines as prices had not yet been set. After dutifully swirling, sniffing, sipping, and spitting more than fifty wines, "I was startlingly underwhelmed by most of what I tasted," he wrote in his March 30, 1983 issue. He'd found the wines "disappointing," and even worse, "oafish"; they were too alcoholic, he said, and had little flavor concentration. His assessment: "No—the proverbial thousand times no—it was not the 'vintage of the century.'" While not giving up completely on the wines, he cautioned, "I wouldn't rush out and commit serious dollars in the '82 futures market."

Strong words, and there were more. Finigan called the year "something bizarre," the wines "very odd," and suggested that his readers consider buying instead the "supple and charming" 1980s and "brilliant" 1981s.

When Parker saw a copy of the issue, he thought, "He doesn't know what he's talking about."

Parker arrived in Bordeaux a skeptic and left converted; Finigan came expecting greatness and went away disillusioned. The accounts of the two critics' trips and their reaction to the 1982s revealed their profound differences, as personalities, tasters, and writers.

Parker's prose hummed with energy, excitement, and passion. It was punctuated with exclamation points and grand pronouncements. "I predict 1982 will go down in the annals of Bordeaux wine history," he wrote, adding, "Make no mistake about these 1982s, they are destined to be some of the greatest wines produced in this century," underlining these pronouncements for dramatic emphasis. He described the wines as "profound," "opulent," "fleshy," "chewy," "ripe," "spicy," "rich," "viscous," "big," and "breathtaking," almost tasting them for you, and listed thirty as exceptional or extraordinary. He passed on proprietors' opinions, dropping names. His overwhelming love of these big, rich wines and pleasure in tasting jumped off the page. Never mind that once the wines were in bottle they would taste quite different from these unfinished barrel samples, and that predicting how they would age was chancy at best. Parker threw caution to the winds.

Finigan, on the other hand, came across as a weary judge, intent on impressing his readers with his importance as an insider and his critical powers. Tasting the wines, he wrote, "was no easy task," possible only because of personal contacts "who had come to trust him over the years," as samples were being "carefully guarded." He let his readers know what a tedious chore it was to drive from château to château. Everything was secretive: he gave no names of what he tasted or to whom he talked; some proprietors, he intimated, were "privately less than passionate on behalf of the '82 reds," though he declined to name names. The wines were "diffuse" and "short," hardly what he expected from a great vintage.

Appraising the vintage as a whole and waiting to characterize individual wines much later was the traditional way of writing up the wines of a new vintage, and in this Finigan was following in the footsteps of the

British wine critics, who made frequent trips across the Channel to moni-tor the progress of each new vintage. But in comparison to Parker, he sounded niggling, equivocal, and more than a little pompous.

The debate over the 1982s was on, and would rage for years, with Terry Robards on Finigan's side. Before he'd even tasted the 1982s, Robards had reported that the year was a "heralded vintage" but suggested the 1979s were better buys. On the front page of *The Wine Spectator*, in his very first column in the June first issue on Bordeaux futures as investments, he still advised that "prudent buyers will stay out of the market for now and will focus their buying on . . . '78, '79, and '81."

Such comments did not please the trade. Sam Aaron of Sherry-Lehmann complained that he lost $20,000 in orders for 1982 futures after Robards's report. Aaron was so upset he telephoned Alexis Lichine in Bor-deaux, who quickly denounced the American "press campaign" against the '82 vintage. Finigan heard that a group of château proprietors actually discussed banning Robards and him from tasting at their estates in the fu-ture, perhaps permanently. "You would have thought I had compared the wines to rat poison," fumed an annoyed Finigan, now claiming that nei-ther he nor Robards was "negative on the wines—we just weren't positive enough." He cynically observed that the Bordeaux trade had a stake in all the positive comments because they wanted all the wines sold before the franc dropped any lower.

Lichine, a consummate and theatrical promoter, wasted no time. On a hot summer day in mid-June he hosted a barrel tasting and lunch in the attractive cellar of his property, Prieuré-Lichine, for the owners of six top châteaux, to prove once and for all how good the 1982s really were. He in-vited Jon Winroth of the *International Herald Tribune*, *Wine Spectator* publisher Marvin Shanken, and its newly named European correspon-dent, James Suckling. They were completely seduced.

In fact, the majority of those who'd actually tasted the wines, including British Bordeaux expert David Peppercorn and *New York Times* columnist Frank Prial, also had a favorable view of the 1982s, even if they weren't quite as enthusiastic as Parker. But Parker's detailed over-the-top praise for such a large range of wines virtually ensured he would end up being the vintage's biggest promoter and its most important herald.

If Parker harbored any small doubts about his own judgment, he eliminated them in June, during his regular yearly tasting circuit of France, by comparing hundreds of barrel samples in various *négociant* offices. He had a long discussion with Jean Delmas at Château Haut-Brion, and even perused their archives. There he discovered that people had also had doubts about whether the wines from the famous 1929 vintage would last, and the best were still wonderful more than fifty years later. He spent one of his days on the Right Bank, at a tasting organized by oenologist Michel Rolland, who owned a lab and analyzed wines for over one hundred clients. They'd met the previous summer, and Rolland had offered to arrange a tasting the next time Parker was in Bordeaux. When he told clients an American journalist was coming, all happily contributed samples of their 1982s. Parker tasted all morning and afternoon, and he remained just as excited at the end of the day as he was when he started.

Parker was even more certain of his opinion in his August fourteenth issue. The question is not whether to buy, but when, he wrote, predicting, "I feel absolutely certain that the wines in most demand will be significantly more expensive (perhaps as much as 70 percent more) in two years." For the following year, he would reiterate variations of his basic theme—"buy as much great 1982 as you can afford because there may not be another vintage this great for 50 years." Those who purchased the top wines now would "no doubt have liquid gold sitting in their cellars."

―――――

MARVIN SHANKEN DIDN'T LIKE TO BE WRONG. HE COULD SEE THAT *The Wine Spectator* had not been on top of the breaking story. In the August 1983 issue, as if to catch up, Suckling wrote two articles on the 1982 vintage. One described the Lichine lunch, and it praised the wines. Robards kept the controversy going by reporting his own view; after tasting sixty wines, he insisted some were "downright flabby." He predicted that "the vintage will be charming to drink at a young age" but he had "reservations about its potential for long aging," reservations shared by producers like Peter Allan Sichel of Château d'Angludet in Margaux.

So far *The Wine Spectator* had been a forum for fairly independent freelance columnists, like Robards, weighing in with their own opinions. There was no collective staff voice taking a stand on the vintage as Parker had back in April. Marvin Shanken was convinced that had to change, and soon.

A testy Finigan was not about to back down, and he vehemently defended his "journalistic dissent" in two fall issues. Wines in the Médoc and Graves were "amiable and sometimes impressive" but hardly vintage-of-the-century quality. While proprietors in St.-Emilion and Pomerol compared the style of their wines to the legendary 1947, to Finigan they excelled "in that alcoholic weight and richness which many consumers, especially the less experienced, will adore." Was that an indirect slap at Parker? There was no reason to buy now, Finigan insisted, all but accusing retailers and the Bordeaux establishment of fooling the customers.

Why was Finigan's view so different? Some theorized that he'd lazily relied too much on the views and vintage report of one respected château owner in Margaux, where most of the 1982s weren't nearly as good as elsewhere. Or perhaps the disagreement was more about the style of the wine. Many in England and France were convinced that the character of the 1982s matched an American taste for ripeness and fruit. Americans usually preferred their own country's wines in those popular and ubiquitous California Cabernet versus Bordeaux blind tastings. The 1982s, too, were "wow" wines, showy and flamboyant, with in-your-face sweet fruit. Would they peak early and fade quickly? In a *New York Times* ad, retailer William Sokolin offered customers who'd already bought 1982 futures the chance to exchange them for 1981 Bordeaux, writing, "It is clear that '81 is a finer, more classically stated vintage."

———

MEANWHILE, BACK IN BORDEAUX, THE FIRST-GROWTH CHÂTEAUX, which seemed determined to squeeze as much money as they could from the market, had released the wines as usual in stages, or *tranches*. After the first price, the *prix de sortie*, the cost of the next batch of futures was significantly higher, the next still higher, and so on. Lafite's prices started at 70 percent higher than the year before and quickly escalated. Some of the greedy "super seconds" (the nickname for the top-performing second growths, like Léoville-Las-Cases) waited and waited and waited to see what their neighbors' prices would be, causing havoc. Even lesser châteaux held back a bigger-than-normal percentage of their wine to sell when it had all been bottled. The financial success of this strategy depended heavily on prices rising, which depended on growing demand for the wines, which depended on enthusiasm for the vintage continuing.

And thanks to the promotion of Parker's view in America, it did.

Parker's April issue had an almost immediate impact upon retailers in Washington, D.C., but things really began to heat up in May. Several of the biggest retailers in New York City were already acquainted with his newsletter and since 1981 had occasionally mentioned his recommendations in their newspaper ads, usually alongside those of the more famous Finigan, Robards, and others. But now they jumped on Parker's praise for the 1982 vintage with the speed of a movie publicist capitalizing on a rave review in the *New York Times*. As Gerald Asher put it, "Parker came along in the nick of time. The retailers needed a Simon Says, someone whom the public would believe."

In New York, first came Zachy's, a large Westchester County wine shop that had been offering futures since 1978. Their big, splashy ad for 1982 futures trumpeted Parker's prediction that the quality of the wines "will go down in the annals of Bordeaux wine history with such legendary vintages as 1929, 1945, 1959, and 1961" and placed his notes and scores next to wines like Ducru-Beaucaillou, "a staggering wine which should merit a 95 plus score in 2 years." Morrell's championed Parker as "the only wine writer who has comprehensively tasted the '82s on three separate trips to Bordeaux." Sherry-Lehmann quoted him in full-page ads. But few realized just how useful Parker's reviews could be until consumers started buying big. Acker Merrall put fourteen wines and their scores—Parker had rated all but one 90 or above—under the head "Parker's Picks" in an ad, and within five days all the 90+ futures had sold out.

The Wine Advocate had about eight thousand subscribers at that point, but these ads introduced Parker's name to many thousands more. Wine lovers who had never heard of him were soon asking, "Who's this guy Parker?" And when wine shops realized touting this independent expert's view was extremely beneficial to sales, they used his name even more.

The promotion of the 1982 Bordeaux became a national ad campaign for Robert Parker and *The Wine Advocate*, and it didn't cost him a thing.

The buying frenzy nationwide reached almost hysterical levels. In Baltimore, Parker's friend retailer Bob Schindler bought thousands of cases on the basis of his recommendation and sold them all. There were reports of the Los Angeles doctor, purchasing for a group, who placed an order with a San Francisco importer for $100,000 worth of 1982s and the Texas oil-man who spent $12,000 on them in an Austin shop. A Long Island importer sold more than two thousand cases of Château Gloria in its first twenty-four hours on the market; his average customers, who regularly purchased a few cases each from six to eight châteaux, ordered forty to sixty cases. Parker's close friend Dr. Jay Miller rushed off to the State Employee's Credit Union to borrow almost $10,000 to buy sixty cases of wines Parker had recommended. Clearly at least some of the demand was purely speculative, for investment.

"It's like Las Vegas," marveled New York City importer Barry Bassin. "Everyone has to put a quarter in the slot machine." Ab Simon, who by then headed Château & Estates, the largest U.S. importer of fine Bordeaux, admitted "it was almost a stampede." Demand was "way above my expectations, even for an exceptional vintage." The American market gobbled up futures as if it were starving.

All this eventually made an impact in *négociant* offices in Bordeaux; Christopher Cannan of Europvin remembers massive amounts of orders arriving daily from America, and wines made in tiny quantities, such as Pétrus, were soon virtually unobtainable at any price. "It was the heyday of Bordeaux," said *négociant* Dominique Renard. "The Americans were more active than all the other markets and set a precedent for the futures market. It was the strongest sale ever, and the prices soared. Nineteen eighty-two led a whole new group of buyers and speculators to the market."

Americans bought so much that the Bordelais began to call 1982 the American vintage.

———

IN ENUMERATING JUST WHY HE THOUGHT THE 1982S WERE SO great, Parker seemed to be setting down his criteria for what constituted a great red wine: First, an incredibly dark color, which he associated with "wines of great concentration." Second, "amazing fruitiness and ripeness" and "fat, fleshy flavors that were viscous and mouthfilling." And third, the high alcohol content that came from "grapes that had achieved perfect maturity." Of the 1982 Petit-Village, Parker said, "If wine were a candy, Petit-Village would be a hypothetical blend of a Milky Way and Reese's Peanut Butter Cup." This type of wine, opulent, luscious, and tremendously fruity while still in barrel, became a new prototype and set a new standard. In effect, 1982 became the first modern vintage, the harbinger of a particular style of wine. And it was quite different from what people thought of as typical Bordeaux—wines that started out austere and undrinkable, and were slow to develop.

But 1982 was a milestone for other reasons. It came to symbolize a break with Bordeaux's past. It started a new trend of consumers purchasing fine Bordeaux solely as a financial investment. It ignited a boom in the American market, and it marked the beginning of a new, modern way of marketing wine. One *négociant* observed that after 1982, the Bordeaux market would never be the same. American retailers and the Bordelais were just beginning to see how useful Parker could be.

Parker's call on the 1982 vintage was the first big step in building the Parker legend. The debate over the quality of the wines went on for years, although tastings proved again and again that he had been right in his extravagant praise—at least for the very best wines. The speculators who listened to Parker made money. According to *The Wine Spectator*'s auction correspondent, Peter Meltzer, between 1983 and 2002 the wines rose 2,012 percent in value—far outperforming the 770-percent return of the Dow Jones during the same period.

Within a few years the facts had been simplified, streamlined, and distorted in small ways to fit neatly within Parker's "Lone Ranger" image. The idea persisted that he had been the first journalist to proclaim the greatness of 1982, even though that was not true. Parker himself gave credit to French critic Michel Bettane. But Parker perpetuated the idea that he stood out alone against a "mainstream American press" who had "unequivocally criticized the vintage and the style of wines" with "scathing critiques," as well as extending that criticism to him. In reality, that "mainstream" consisted of three naysayers—in addition to Robards and Finigan, Dan Berger of the *Los Angeles Times* had been lukewarm on the 1982s—and none of them had written anything negative about him personally. In the first five years of *The Wine Advocate*, Parker had frequently sought to characterize himself as quite different from other wine writers, standing alone for the consumer, and he now used his position on the 1982 vintage to distance himself even more. Later he claimed that at the time the "establishment wine press" had regarded him as "a threat to their

THE 1982 BORDEAUX · 111

... influence." But that was hardly the case; in 1983 he was only just beginning to become important.

———

ALL DURING 1983 AND THE BEGINNING OF 1984, PARKER KNEW HE was leaving Farm Credit Banks and the law. After tasting the 1982s it seemed harder than ever to go back to an office and attend boring meetings about contracts. He could barely keep up the necessary effort and attention. Fortunately he had the loyalty of several lawyers he'd hired, who covered for him when he stole off to New York for a tasting. He was looking for every possible way out. Early in 1984, Robards, now entrenched at *The Wine Spectator* and *The New York Post*, heard that Parker was trying to set up a meeting with someone at the *New York Times* to discuss taking on the vacant wine critic slot. But nothing came of it; perhaps Parker realized that the *Times* would never allow him to keep publishing *The Wine Advocate*. In any case, Frank Prial returned to New York after his stint in the Paris Bureau office and went back to writing the "Wine Talk" column.

After the April issue reviews of the 1982 Bordeaux and subsequent publicity in wine shop ads, and the continuing controversy with Finigan, many more subscriptions had flowed in to *The Wine Advocate*. The number had jumped to nearly nine thousand. It was close enough. With the 1982 vintage, Parker was now completely convinced of his ability and knew he was outworking his competitors. Finigan, he was sure, was in decline. Parker had put in eleven years at Farm Credit Banks and his ten-year pension benefits were vested. In March 1984, with Pat's blessing, he finally resigned from the practice of law.

He and Pat threw a "retirement" dinner at their home, inviting a few close friends, and they all "got blitzed" on 1980 Chalone Chardonnay and 1961 Jaboulet Hermitage La Chapelle. Parker felt incredibly happy, as though a weight had been lifted from his life. Two weeks later he and Pat left for Bordeaux to taste the 1983 vintage—and to see how those great 1982s were coming along.

5

TASTING

10,000

WINES A YEAR

*T*HE LIFE AND TIMES OF THE MAN WHO WOULD come to dominate the world of wine began in earnest after the end of his law career. Parker now embraced the grape totally, completely, obsessively. He fell into a pattern of ninety-hour work weeks immersed in wine—researching it, buying it, studying it, writing and talking about it, and above all, tasting it.

On most days, Parker commuted a mere few dozen feet from the breakfast table in the kitchen to the combined office and tasting room he'd added on to the house. Pat had no problems with the expenditure for that, noting that the guest bedroom where he'd worked was so small that his papers had long been creeping into her space. Not to mention the wines, which drifted into just about every room of the house but the bedroom. Function, not fashion, dictated the décor of Parker's new office, though in designing the tasting area he had insisted on an extra-deep sink and had copied something he'd seen at *négociant* Jean-Pierre Moueix's office in Bordeaux: a white tile tasting counter with black tile bullnosing on which you could write.

Within a few years, even that space was too small, and Parker added on two more rooms. No sleek laboratory look for this office and inner tasting sanctum, either. Instead, there was a desk soon strewn with papers, a sink, a rack of clean wineglasses, a messy overlay of boxes and filing cabinets, and shelving stacked with papers and books and a few personal touches, such as the sign propped behind the sink's faucet that read "Spitting on Sidewalks Prohibited, Penalty $5 to $100." There were no fluorescent lights, which could make red wines appear orange, only an incandescent ceiling fixture, a couple of lamps, and natural light from the window. Racks on one wall and under a counter held reds and a massive temperature-controlled unit the chilled whites waiting for their scheduled moment of truth in the glass. Clumps of bottles cluttered countertops and other available surfaces, and sometimes the number standing on the floor threatened to block the path through the room to the outer office where his secretary, Joan Passman, who had given up her own accounting and word processing business to work for Parker full time, fended off the outside world. Despite the seeming disorder, Parker knew exactly what was where. Many more wines rested unseen in his then two underground wine cellars, one of which had been blasted out of the rock under the house in 1977.

Eventually the original office became a storage area; the countertop later held his prized espresso/cappuccino machine. A treadmill stood nearby for rainy-day exercise.

The extra-deep stainless steel sink had been chosen so that Parker could spit during tasting sessions without concern for possible backsplashes— not that he ever dressed for the occasion. His standard working uniform consisted of blue jeans or sweats in winter, shorts and a T-shirt in summer, and tennis shoes. Lifting a glass of wine to one's lips hardly sounds like physical labor, but Parker's marathon tastings each week were nothing short of athletic events. On those mornings, he shunned his beloved espresso and drank only mineral water to ensure an untrammeled palate. Dinners the night before were extremely bland—no spicy food, no strong flavors like chocolate that would linger and interfere with his judgments. Garlic didn't bother him, but watercress, he discovered, was the worst offender.

Since his schedule called for four or five days a week of dictation or writing, Parker crammed his all-important tastings into two days, usually Wednesday and Friday. He looked forward to them; if he didn't feel like tasting, he knew it meant he was too tired and not mentally ready for it. But he was almost always eager.

On a typical tasting day, Parker bounded into his office shortly after 8:30 A.M. He carefully checked his tasting glasses for cleanliness and stray off-odors imparted by improper rinsing. If one seemed suspect he breathed into it, clouding the inside of the glass with moisture, then sniffed again—he later referred to this, half in jest, as "the Parker exhale test"—and rinsed out the glass with bottled water just to be sure. He took his job seriously, and that included using the best equipment. He considered the 8-ounce tulip-shaped ISO (International Standards Organization) and slightly larger INAO (Institut National d'Appellation d'Origine) official tasting glasses excellent for his purposes—they had been developed scientifically over a number of years—but he also leaned toward the expensive handblown crystal "Les Impitoyables" (the pitiless) glasses, designed and made in France, which he ordered direct. The shape intensified, even exaggerated, a wine's aroma. Unfortunately, they didn't fit in the dishwasher. Both would eventually take a back seat to designs by Riedel and Spiegelau, as he continued his experiments with different types of glasses. The temperature of the wine counted too. Parker wanted both reds and whites at 60 to 65 degrees—if too warm, a wine tasted flat and the aroma dissipated; if too cold, he could barely distinguish its flavors.

Dr. Jay Miller, Parker's good friend and since April 1985 his part-time assistant, uncorked anywhere from 80 to 125 bottles of wine on Fridays, hiding their labels with aluminum foil and numbering each one. Parker had selected the themes—a range of good values, or newly released American Cabernets, or perhaps wines of one area of the Rhône—but left it to Miller to set up the tastings, like against like, in flights of at least six wines, but often up to a dozen or more. These tastings, like most, were usually double blind, which meant that Parker knew the type of wine, but neither the producers nor which one was which. Occasionally they were single blind, where he knew which producers were included. Blind tasting, he thought then, was the democratic way to taste.

Mornings he reserved for the reds. Although Parker regarded Burgundy and Pinot Noir as relatively easy to taste, he needed a fresh palate for the more difficult task of detecting the fruit lurking in bigger, more tannic reds. The afternoon was for whites; they were simpler to taste, Parker believed, because there was less to look for.

Miller poured the wines, about an inch-worth in each glass.

A wine had only a minute or so to make its impression. Parker sniffed all the samples first. If the musty, dank smell of a corked wine hit his nose the wine was banished, as were any that smelled foul, weird, or spoiled. He tasted standing up. The method was precise. He picked up a glass, eyed the color, swirled the glass briskly, sniffed the wine, focused on it. The smell was the most important factor, the taste only a confirmation of that. He took a large sip, sucked in air over the liquid through pursed lips, quietly gurgling as he worked it around his tongue, almost chewing it, and held it in his mouth for a few seconds; then he unceremoniously spat it out into the sink. After the briefest of pauses he scribbled down a few notes or recorded his comments on tape. He felt no need to linger. His nose, mouth, and brain went through a systematic series of steps as he registered his impressions. The weight, worth, and character of the wine came to him almost immediately, unbidden; he had trained his mind and palate to methodically analyze the scented liquid, breaking it into odor and flavor components and the constituent factors that gave rise to them—the grape; the kind of oak, if any; the wine's age, and so forth. He thought of the process as a "visual and physical examination," and by now it was automatic.

Parker's secret weapon was his ability to mentally compare the wine in front of him with all the other wines of the same type he'd tasted over the years. He often claimed he could remember every single wine he'd ever tasted and the score he'd given it, something that other wine professionals found hard to believe. Though in the beginning he'd mentally added up points for color, aroma, and so on, now, in a flash of intuition, a number would occur to him, as if rising out of the glass. Sometimes he found himself contemplating a range of numbers, say 87 to 89, but often the precise score would come to him like a vision of the wine's inner meaning, the exact measure of its inherent worth. It was like looking through the lens of a camera and bringing the picture into focus. It required immense concentration, a kind of tunnel vision; all other matters, great and small, faded into peripheral dimness. The ability to focus on each individual wine, Parker said, was the ability that counted most.

Parker was a great believer in the reliability of that first take. "The first impression is critical," he often said. But after finishing each flight, he repeated the process a second time to double-check the accuracy of his initial impressions. That was it. If he had doubts, which wasn't often, he put the wine aside and gave it another chance later in the day. Otherwise Miller dumped the remaining liquid down the drain or down the hill outside and hauled the empty bottles to the garbage bin.

Every couple of hours Parker took a fifteen-minute break, downing glasses of mineral water, often with a slight effervescence, to clear his palate. By lunchtime his teeth and tongue and hands were stained purple and he was ready to move on to the whites. He rarely bothered with lunch; he liked to be a little hungry when tasting. All this went on continuously until 4 P.M. Tasting was serious business, but hardly silent. Miller tasted along with him, and though they had lively discussions they usually agreed on the quality and character of the wines. They considered it a good day when 25 percent of the wines turned out to be recommendable, headed for the pages of *The Wine Advocate*. It was a working relationship that would last until 1998, when Miller entered the wine trade by opening a wine shop on Baltimore's revitalized waterfront. He came to know Parker's palate so well, he says, that if he tasted a wine he could guess what Parker's score probably would be, within one or two points.

Parker didn't worry about palate fatigue. His method of avoiding it was to drink up to 5 liters of water a day. Although many writers and wine professionals felt their powers of discrimination diminished after a couple of dozen wines, Parker insisted he'd experienced a serious bout of palate fatigue only twice—once tasting the controversial 1983 red Burgundies (grapes in some areas were tainted with rot, affecting the wines' flavor) and the second time on a marathon trip to Italy tasting very tannic Barolos from the cask. He had no problem sampling between fifty and 125 wines twice a week. The toughest were wines high in alcohol and tannin, which sometimes made him feel "as if I've taken head shots from a boxer."

But he did worry about the effect of putting well over 10,000 wines a year in his mouth. Was the acidity in the wine destroying the enamel on his teeth? What was happening to the membranes in his nose? What about the effects of abrasive tannins in young wines and barrel samples on his throat? He periodically chewed antacids to counteract the effect of acidity on his stomach (he eventually gave that up) and consulted an ear, nose, and throat specialist every six months, just in case. And to ward off a cold and the stuffy nose that would disrupt his compulsive routine, Parker dosed himself daily with liberal amounts of vitamin C. If he felt a cold coming on, he turned to echinacea.

——————

ALTHOUGH HIS REPUTATION WAS STEADILY GROWING, PARKER continued to lead the same modest, regular-guy life he'd had while a lawyer. Like many other baby boomers he drove a Ford Bronco; followed college basketball on TV and at games, cheering for the University of Maryland team, the Terrapins; jogged (sometimes with Pat) or bicycled in the late afternoon to stay in shape; listened to Neil Young and Bob Dylan on his state-of-the-art sound system; and entertained an inner circle of loyal close friends, who loved wine almost as much as he did and enjoyed the game of making him guess the wine when they invited him to dinner. Parker wasn't above showing off, and remarked often, "I've never heard of anyone who's any better at guessing wines—if not the exact wine, then the region, the grape." Nor was he above trying to fool them by decanting an old Domaine de Vieux Télégraphe Châteauneuf-du-Pape into a Bordeaux bottle and slyly sliding it into a blind tasting of first-growth Bordeaux. His friends admired his wine expertise, the collectors among them happily pulled out their best bottles, and a few of the most wine savvy, like Miller and wine retailer Bob Schindler, seemed to take easily to the role of acolytes, as they usually shared his views on wine.

Parker was living out his fantasy life: his job and his private life were now almost one and the same. When in Parkton, cooking, eating, tasting, entertaining, and writing took up most of his time. He was surrounded by loyal, protective women who doted on him. There was Pat, of course, and his mother now worked part time for him answering *Wine Advocate* correspondence and fed him a tough, critical line when he was in a self-pitying mood. His soft-voiced secretary, Joan Passman, installed in the outer office, deciphered his scrawls and transcribed his dictated notes. She'd learned how to spell the château and winery names but didn't even drink wine—though she admitted to me that she had liked a Château d'Yquem Parker had served her. Passman had started working for him part time in 1982, maintaining his database, answering correspondence, and keeping his increasingly complex accounts; by now he considered her indispensable. During the 1980s demands for his time from importers, wineries, and the media became so pressing he had to stop taking phone calls; she screened those as well as the requests to lead consumer bus tours of wine regions, to lend Parker's name to a line of stemware the way Hugh Johnson had done, and to participate in dozens of other ventures, few of which interested him.

There were always an assortment of dogs underfoot—basset hounds, bulldogs, and border collies, who lay under Parker's desk and snored while he wrote, often with Neil Young's high-pitched voice and twangy guitar in the background. The dogs' flatulence didn't bother Parker unless he was tasting, but Passman and the other part-time secretaries he later hired held their noses and shooed them outside.

The one lack in what Parker and his wife considered an otherwise idyllic existence was children, and after years of undergoing sometimes humiliating fertility tests they decided to adopt through Catholic Charities in Baltimore. When they picked up the baby they named Maia Song Elizabeth Parker at the BWI airport in the fall of 1987 and Parker first held her in his arms, he cried. At the age of forty he'd finally become a father, and he shared the fact with his readers in an emotional, heartfelt postscript to his newsletter.

Pat quit teaching and became a stay-at-home mom, though she continued to edit and proofread *The Wine Advocate*, as she had from its beginnings. In the next few years little changed at home, except for more renovations and Maia's transition from toddler to kindergartner. Despite the fact that Parker's schedule left few openings for vacations, he made time for essential family excursions to Disney World, where he happily wore a Mickey Mouse tie just like any other doting dad. Having Maia along shifted the kinds of vacations the Parkers took; their beach stays in the Grand Caymans, where their daughter could dig in the sand and splash in the waves, became the ultimate vacation for Parker. Pat observed that here he seemed to truly relax, sometimes snorkeling slowly for hours or reading on the beach while watching his daughter. He'd left behind, albeit temporarily, grand cuisine and great wines.

Parker's home base was a comforting space, an oasis of sanity and support as he built his wine career and his power and influence throughout the 1980s. He continued to cultivate the naïve-country-boy-with-a-supernose image, which helped to convince people that he was indeed a regular guy unspoiled by success. "Bob never forgets who he was, he doesn't put on airs; what you see is what you get," his friends said. He put together his own local rat pack, who enjoyed his ribald sense of humor, appreciated his vast collection of jokes, and knew how to party hard. His friends considered him loyal and generous, as he did himself. One of his closest and oldest friends, Maryland attorney Steve Jacoby, once observed that Parker maintained his sense of proportion when people in the wine business fawned over him and that he remained polite even when angry, though he would disparage them later to his friends. "He can divorce himself from the industry," Jacoby was quoted as saying. "Because really, his wife is still his consuming passion." Parker was happy to have more money, which paid for the continuing renovations, but money didn't drive him as long as he had enough to pay the bills and buy the wines he wanted to taste.

———

PARKER'S MANIA FOR DISCIPLINED ROUTINE EXTENDED NOT JUST to his workday and workweek, but even to his yearly travel calendar. The most important stops on his annual global circuit were his fifteen-day trip in late March and early April to taste the new wines from barrel in Bordeaux, followed by another visit in June. After that he drove to Burgundy or the Rhône, often spending an entire month in France. In August he swept through California, if he hadn't zipped through the state the previous January. Then in the fall, it was back to France and southern Europe, although he wasn't a regular visitor to Italy. He encountered the wines of Germany only in bottled form. Each trip was arranged with military precision so he could see as many people and taste as many wines as possible in the shortest time, but it still added up to three or four months of travel a year and thousands more wines tasted. Over the years the dates shifted, but March and June visits to Bordeaux continued to be sacrosanct. Despite a punishing schedule Parker enjoyed visiting producers, and he especially seemed to relish proprietors who were unusual and eccentric. One of his favorite stops in Bordeaux when he was starting on the wine trail was Château Lafleur in Pomerol, a tiny estate run by two diminutive unmarried elderly sisters, Thérèse and Marie Robin, who kept ducks, chickens, and rabbits in their minuscule *chai* alongside the barrels of wine. They called Parker Monsieur "Le Taureau" (bull), which he assumed related to his size compared to theirs. Though the wine was little known in the 1970s and early 1980s (only one thousand cases were produced), Parker and his tasting buddies had often found it just as compelling as Pétrus. The 1947 Lafleur from his birth year, which he shared with Pétrus owner Christian Moueix in 1987, the year he turned forty, was, he says, perhaps the greatest single red wine he had ever tasted, so spectacular "it brought tears to my eyes." Sometimes he dreamed about buying the estate.

One summer, after three weeks in Burgundy, Parker and Pat drove down the Autoroute du Soleil to Provence to visit Domaine Tempier, a small wine domaine whose vineyards lay on terraced slopes high in the rugged rocky hills above the quaint coastal town of Bandol. The sky was blue and cloudless, the Mediterranean Sea a flat expanse of reflected light. It was incredibly hot. They had spent five hours sweating in a car that had no air conditioning and when they arrived late in the afternoon, incredibly thirsty, the owners, Lulu and Lucien Peyraud, greeted them with cool glasses of the deliciously fruity, vibrant rosé that Parker had called the best rosé in France. Parker slaked his thirst by slugging back one glass after another with abandon. Then came a long dinner, more wines, laughter, and crazy conversation with the family, their American importer Kermit Lynch, and a couple of other guests. Parker discovered he was speaking fluent French for the first time. The rest of the evening disappeared in a fog.

He woke in the middle of the night at the bed-and-breakfast in the nearby village where he and Pat were staying. The shutters on the windows were down; he couldn't find the light switch and felt a claustrophobic panic attack coming on. He passed out again. In the morning he found their car parked horizontally across several parking spaces, with the windows rolled down and the radio blaring.

That was the last time he would be able to get drunk in a wine region. Parker's reputation was already beginning to box the couple in; in a few years the intense scrutiny of their movements in Bordeaux would make Pat feel uncomfortable and highly self-conscious. Where they dined, what they ate, whom they saw and talked with, sometimes even what they'd said would be common knowledge, passed on via the region's active gossip chain.

Before Maia arrived, Pat accompanied Parker on most of his trips to Europe, which he scheduled during her school vacations. Her primary role was as translator, but both Parker and the French regarded her as much more important than that. Her charm and warmth soothed many proprietors skeptical of journalists, as well as those annoyed by Parker's sometimes blunt and intrusive questioning and resentful of his growing power, especially in Burgundy. Pat's smile and soft manners balanced his impatience and what was perceived as his arrogance, as well as his tendency to lecture. Several wine producers in Bordeaux and Burgundy told me Parker was softer himself when she was with him, that she had been his one essential asset. And, says Pascal Delbeck of Château Belair in St.-Emilion, "she was a very, very good taster in her own right."

For a few years, Pat had contributed restaurant reviews to *The Wine Advocate*, but she also viewed Parker's wine-tasting trips as a way to keep her French fluent to enhance her teaching. During her sabbatical year, 1986, she spent several weeks in France without him, attending La Varenne cooking school in Paris and working the harvest in Bordeaux, something Parker had never done. He is allergic to bee stings—one of the hazards of picking grapes, as bees are attracted to the sweet juice that quickly covers your hands. For eleven straight days at Château Gruaud-Larose, Pat was in the vineyard until 7 P.M.—"I was the slowest," she confessed; she was proud the regular team of pickers threw her in the grape bin on her last day, since it meant they'd accepted her, that she was "one of the guys."

When Pat stayed home, Jay Miller took over as Parker's travel companion. Their trips over the next decade were buddy movies—two friends on the road in Bordeaux, Burgundy, Alsace, Italy, and California, some twenty trips in all. Their first, to taste the 1985 vintage in Bordeaux in spring 1986, set a routine Parker would follow for the next seventeen years. His exacting two-week itinerary aimed for maximum productivity, and by now he'd learned which châteaux and *négociants* would see him at 9 A.M. (or before) and on the weekend, even on Sunday. His visits were short, forty-five minutes to taste three vintages and ask questions, and he tried to cram six or more into a day. He knew he could do it faster, but felt spending even that brief time was a matter of courtesy. Many of the forty or so proprietors he visited invited him to lunches and dinners, and Parker was happy to have Miller along, someone to joke with later about the notoriously bad "mystery meals" they'd been served. One exception was a dinner at media-savvy Château Mouton, where the marketing manager pulled out a 1947 Mouton, Parker's birth year—the least they could do for the man who'd given their 1982 a perfect score of 100.

A handful of trusted *négociants* set up large half- or full-day comparative tastings with 150 to 200 barrel samples they'd gathered from various châteaux so Parker could try the wines several times in different contexts and see how they stacked up against their peers. Archie Johnston, who had befriended him from the very beginning, had the status of a mentor. Dominique Renard, usually seen with a white silk handkerchief pushed neatly into the breast pocket of his tailored suits, occasionally accompanied him to châteaux. The much less formal Bill Blatch, who specialized in good-value *cru bourgeois* wines for the American audience, dressed casually in shirtsleeves, occupied a warehouse-style office in the midst of an ugly industrial district, wrote one of the region's most respected vintage reports, and relished Parker's tasteless Washington jokes. He had been impressed with Parker's tasting abilities from his first visit. "Parker," he told me, "has got a nose like a dog. He sees wine like other people see color."

Parker whipped his way through the dozens and dozens of wines at the crowded tastings put on for journalists by the Union des Grands Crus, a promotional organization that represented dozens of châteaux, and by the *premiers grands crus* in St.-Emilion. But he was annoyed that the first growths and even some of the super seconds "play the game of not sending samples to a comparative tasting." Some of the châteaux didn't even want to show him the current vintage in the barrel. Reputations were at stake—and egos were involved. Miller had brought along a copy of Parker's recently published Bordeaux book, and wherever they went he asked owners to autograph the review of their wines. At the St.-Emilion tasting, Thierry Manoncourt, owner of Château Figeac, graciously agreed, but then looked more closely at what Parker had written. Annoyed by the description of his quality level as equivalent of a second growth, he crossed out the word "second" and wrote in "first." "Then," says Miller, "he went over to Bob and gave him a piece of his mind."

Every wine region was different, and each was grueling in its own way. In Burgundy, American importers like Robert Haas set up group tastings with their growers for Parker, usually in someone's cellars; at some he sat at a table, and one by one the growers would bring their wines up to him and "do a little talk and dance," as Parker termed it, while he tasted the wine. He and Miller usually spent a day with each importer, and made individual visits to important shippers like Louis Jadot and famous estates like the Domaine de la Romanée-Conti. On one spring trip he came away with notes on nearly one thousand wines. Through the 1980s he was just another journalist to many of the growers, but the importers knew the American market and recognized his growing power.

Parker's punishing routine of travel and tasting, his maniacal focus, and the incredible number of wines he tasted all helped him to gain attention from every part of the wine world during the 1980s. Quite simply, he was tasting more—and perhaps traveling more—than just about any other American writer. Surprisingly, his isolation in Maryland when he wasn't traveling ended up becoming a factor in his success.

In New York City and San Francisco in the mid-1980s, the country's importers, winemakers, and château owners from Europe came to town to woo wine writers at tastings and lunches. New York City was the major European wine entry point to America and the country's most important media market. California winemakers, too, believed they couldn't succeed nationally unless they conquered the city's skeptical Euro-tilted sommeliers and convinced them to put California wines on the best restaurant wine lists.

But unlike ambitious people in many fields who rush to place themselves in the center of the competition, Parker had no intention of moving from the small rural town where he could maintain his low-key life and do things his own way. Though he was now a full-time wine critic, he had no interest in joining the tight group of Big Apple wine writers who met at tastings, traveled together on junkets, and after the 1982 Bordeaux asked one another, "Have you ever met this guy Parker? And where is Monkton,

Maryland?" Parker described the wine-writing community acidly in 1986. "These people meant nothing to me except that they were hacks and charlatans who had lived off and been writing for magazines that paid them because the magazines didn't know the difference between good wines and bad wines. I had no interest in associating with them, being compared with them," he fumed. "These people didn't know anything." In almost every interview he gave during the 1980s he slammed other wine writers, both American and British, contrasting his virtues with what he considered their limitations. They didn't work hard, they weren't doing their job, they were too cozy with the trade, and he'd had to step in to fill the vacuum.

By staying in rural Maryland, Parker set himself apart. Being far from the obvious centers of wine power contributed to an aura of mystery and continued to mark him, as he once described himself, as "well ahead of the pack." It was really a self-indulgent choice rather than a calculated move, but it fed the growing Parker myth: he was different, special, unique. The gossip was that he was a quirky genius who sipped and spat alone, a supertaster in a mountaintop fortress of solitude. In truth, he was on the phone daily with a network of key retailers, importers, and distributors, working the marketplace for up-to-date information, and spoke regularly with me about article ideas for *Food & Wine*. And he did make forays to New York and Washington, D.C., for tastings and private dinners, including one at the Greenwich Village apartment John and I shared in October 1984, where he and Pat joined René and Veronica di Rosa, who owned Winery Lake vineyards in the Napa Valley; Pierre-Gilles Gromand, who owned Château Lamarque in Bordeaux; and Pat Brown, then editor of *Cuisine* magazine. It was a simple meal—mussels and *boeuf en daube*—intended as a backdrop to a barrage of wines, including a "mystery" bottle served blind for fun, a 1970 Biondi-Santi Brunello di Montalcino that no one, including Parker, guessed correctly. But then, no one was expected to; you can discuss a wine intelligently without being able to put an exact name to it.

———

BUT AS HIS NEWSLETTER BECAME MORE IMPORTANT AND INFLUEN-tial with retailers and consumers, people in the wine world increasingly felt they had to court Parker, and that meant making the trip to Parkton.

Even in the mid-1980s the wine world was enormous, with thousands and thousands of wines made every year on every continent around the globe but Antarctica. Fine wines, the ones Parker wrote about, were far fewer in number, but growing. Those from France, Italy, Germany, and elsewhere came to America through a small number of importers, who then sold them to the distributors who had licenses to sell them in various states. The distributors, in turn, sold them to thousands of retailers, who sold them in their shops to consumers, the end point of a long series of sales points, each of which made money. This complicated three-tier system dated from the repeal of Prohibition. In some states, like California, and in the District of Columbia, retailers could also hold a separate import license; but in most, that was difficult or impossible. So to drum up interest in their wines importers hosted portfolio tastings several times a year in major cities for the trade—distributors, retailers, and sommeliers—and gave special lunches and tastings for wine journalists, hoping for positive reviews. In New York City I often attended one or two tastings a day, even more in the busy fall and spring seasons. The biggest importers crisscrossed America to visit important wine shops and tell the story of the producers they represented and their wines, hoping to unload as many bottles as possible.

Parker's initial power began with the retailers' use of his ratings to sell wines. So many new wines were coming in that it was hard for consumers to choose, and the ones to which he gave high scores quickly became the prize wines they asked for and eagerly bought. Even consumers who'd never heard of Parker figured that a wine with a rating was better than one without. Selling these wines required little effort; in this they were like the well-advertised big brands. Retailers didn't even have to taste the wines to know they'd sell, so naturally they were anxious to order Parker's recommendations from distributors. Increasing numbers of them relied on his taste buds. To ensure and even speed up those orders, importers and American wineries saw the virtue in making sure Parker tasted and rated their wines.

In the early years of *The Wine Advocate*, the importers and winemakers who made the trip to the boondocks came to Parker's home, where Pat welcomed them—she saw it as "nourishing" the newsletter—but increasingly it became an intrusion into the couple's privacy. The Milton Inn, a fireplace-laden restaurant in a 1740s fieldstone manor house in rural Sparks, Maryland, an easy drive from Parker's home in Parkton, soon became his other office. Sometimes importers arrived with their entire portfolio of fifty to one hundred or more wines and gladly opened and discussed them. Leonardo LoCascio, the head of the import firm Winebow, was one of them. After Parker had reviewed one of his stellar Italian wines favorably, the two met by chance at the Food & Wine Classic, the magazine's annual wine and food festival in Aspen, Colorado. LoCascio started "a dialogue," as he put it, by sending Parker a large selection of his wines. Parker found much to recommend, and he gave LoCascio the opportunity to stage a portfolio tasting at the Milton Inn the following year. This became an annual event. LoCascio also began introducing selected winemakers to Parker, sometimes over lunch in New York. One of them was Marco de Bartoli, a Sicilian who'd revitalized the production of high-quality Marsala. LoCascio, he told me, felt winemakers had to pass the "Parker test" to get anywhere in America.

Though importers sought the privilege of paying court to the new sovereign of taste, and brought thousands of dollars of wine for him to taste at the Milton Inn, Parker always insisted on paying for lunch. Eventually he was penciling in as many as forty or fifty of these invaluable appointments with importers or producers on his annual calendar. As he built his reputation they gave him the insider's track on the newest, hottest, scarcest, and most interesting wines brought in by dedicated people whose palates he respected to share with his voracious subscribers. Fiona Morrison, who worked for importer Château & Estates, brought Christian Moueix of Château Pétrus when he was ready to launch his California wine, Dominus. After the 11 A.M. tasting with Parker there were, she remembers, "long, soul-searching conversations about tannins" over lunch.

The tastings were firmly under Parker's control. He kept a set of his preferred tasting glasses at the restaurant, decided how long the tasting would last and set its pace, and asked the questions. Importers told the stories of their wines as best they could. Parker tried to connect with them personally, but his focus remained on what was in the glass.

In the 1980s Parker himself sought out importers whose wines he'd purchased and liked. Most were a new breed, innovative one-person enterprises who looked for high-quality wines from tiny European estates

Americans had never heard of. Parker called them the "avant garde." They represented a powerful change in the wine world that both contributed to and reflected his influence.

The first two he'd singled out for praise were pioneers Kermit Lynch, who started the trend in the mid-1970s in California, and Robert Chadderdon, based in New York. Parker took up a third of an issue with tasting notes on the wines in their portfolios. He clearly felt a kinship with the two men, characterizing them as "lone rangers" just as he was. He admired their "guts" and independence, which he saw as mirroring his own. They too had traveled to France, fallen in love with wine and the style of life there, experienced the standard cultural and gastronomic epiphanies of Americans at the time, and returned home determined to bring the delicious wines they'd tasted to an American audience. Of course, this also meant that they could spend significant time in France. Both developed strong points of view on wine, and unlike most national importers, who were interested primarily in wines produced in significant quantities, they happily ferreted out the esoteric and obscure, tracking down "authentic" wines (in contrast to those that were bland and commercial), and were satisfied if they could scoop up a mere thirty cases of something special. They didn't advertise, did all their own tasting and buying, and spent time in the vineyards establishing close relationships with growers, sometimes convincing them to bottle wine they'd previously sold to *négociants*, to be put into village blends. They even ventured to suggest ways to improve the wines. In fact, on Parker's first visit to tiny properties in the Rhône Valley he usually heard that Lynch had already discovered them; the laid-back former hippie with the kindly face and big ears was a legend there, known as "*l'homme de Californie*." After initial forays into Burgundy, Lynch had rapidly expanded his list with unusual wines from Alsace, Provence, the Languedoc, even Corsica, and to enthuse his customers he penned amusing and entertaining short essays about his experiences with eccentric producers for his regular customer newsletter. Long before catalogs like Banana Republic and J. Peterman used stories to sell clothes, Lynch cleverly used engaging tales of his on-the-road adventures to market his wines.

Lynch felt the impact of Parker's extensive review immediately. "He legitimized me," Lynch told me. "I was no longer somebody importing these weird things. All of a sudden I was somebody doing good work. He gave me a national presence." Lynch had survived and even flourished regionally because Berkeley, California, nearby San Francisco, and the Napa Valley to the north were home to a clientele more adventurous about food and drink than many other places in America; Alice Waters of nearby Chez Panisse became a friend and one of his key fans and supporters.

Chadderdon's personality and palate were more in tune with Euro-centered New York; Parker found him snobbish and arrogant, but his wines superb. Chadderdon reached farther afield than Lynch then, bringing in a few Australian and Italian bottlings as well as top Burgundies and a Parker Rhône favorite, Château Rayas. He was noted for the unconventional way he portioned out to his approved list of retailers (and still does) only those wines he felt they deserved. His offerings came minus prices— a take-it-or-leave-it proposition. Like Parker, he was supremely confident of his ability to spot quality wines and had a low opinion of most wine journalists' tasting abilities. He refused to sell Parker his wines, but suggested he drop by his office in Rockefeller Center to get acquainted when he was in New York.

Parker finally did. Chadderdon poured him a couple of unidentified wines, a white and a red. Something to taste while they talked face-to-face, he told Parker, and asked him for his comments. He really wasn't expecting Parker to guess the identity of the wines, but he'd often tested journalists this way (including me) and wanted to hear what Parker would say. Parker took the bait and went so far as to identify the white as a *grand cru* white Burgundy and the red as a *premier cru* Côtes de Nuits red Burgundy. Chadderdon had poured out two Swiss wines, a Chasselas from the Vaudois and a Dole from Valais, and he was not impressed. According to Chadderdon, Parker tried to figure out why he had gone wrong but seemed embarrassed by his performance and eventually left. Parker remembers the encounter differently and says one of the wines was a Loire red that he did guess correctly, but he was offended all the same. He never forgot and never went back. Although he thought Chadderdon a "complete asshole," he praised his wines anyway. Sales of Chadderdon's wines benefited, but unlike others in the wine world, he had little respect for Parker's methods or abilities.

For the world's most interesting fine wines from outside the United

States, the future lay with the small importers who modeled themselves on Lynch and Chadderdon. Marc de Grazia, Robert Kacher, Eric Solomon, Jorge Ordoñez, Joe Dressner, Dan Philips, John Larchet, and many, many more would tap into America's new interest in buying beyond the big brands and famous châteaux. Thierry Theise, an importer of top German and Austrian wines, began staging two-day marathon tastings for Parker in 1988, inviting a couple of other writers and merchants so as not to waste too much of the 220 wines he poured. The wines had so much acidity they could make a taster's ears tingle, and Theise desensitized his teeth by brushing with Sensodyne toothpaste for five days before the tasting. Not so the indestructible Parker, who, despite the volume of wine being tasted, brought special wines for everyone to enjoy with lunch. He'd wrapped them in paper bags so they could be tasted blind. Within a couple of months his comments and ratings on the wines, mostly but not all positive, took up several pages in *The Wine Advocate*.

Being anointed by Parker was the quickest way for a small importer or winery to obtain recognition and convince the serious retailers and distributors there was a market for their wines. Sometimes they lobbied to show him their wines, sometimes he invited them, and most happily journeyed to his Maryland haunts. What he had done for Lynch and Chadderdon demonstrated to Theise and others that they, too, could be successful if Parker sprinkled stardust on them.

———

MORE THAN ANYTHING ELSE, IT WAS PARKER'S 100-POINT SCORING system that made him a force to be reckoned with in the American market and brought attention to him as a wine critic.

Tales of "score-blinded" wine customers abounded. As an experiment, the manager of Wine House, a Los Angeles shop, arranged stacks of two California Chardonnays that Parker had just reviewed. Displayed by each pile was a sign with Parker's score and tasting notes. The Chardonnay rated 92, he told a reporter, "outsold the wine with 84 points by ten to one." When he took away the scores, and left only the tasting note descriptions, sales of the two evened out.

When a new issue of *The Wine Advocate* came out, Sam's Wine Warehouse in Chicago, too, created a display based on his latest ratings. Steve

Henderson, then the food and beverage director at the Omni Inner Harbor Hotel in Baltimore, added Parker's scores to the restaurant's wine list and sales of wine went up 40 percent.

Parker guarded the new scores closely before an issue of *The Wine Advocate* was mailed, careful never to reveal them at a tasting or even to his close friends. They knew not to ask. Parker had realized early on that someone who had the scores ahead of everyone else could use that knowledge to corner the market on a highly rated wine. Now that a high score guaranteed immediate frenzied sales, whether from distributors to retailers or retailers to wine lovers, there was money to be made by buying big before the demand hit and then later jacking up the price. Parker, fiercely determined to hold on to his integrity and worried that someone might try to bribe his printer, insisted that the printing contract include a stiff penalty if the scores were leaked before publication. Rumors persisted for years that Parker made the scores available early to certain people—favored brokers in Bordeaux, certain importers and retailers, and perhaps a few friends. But his enemies' suspicions notwithstanding, no one has ever come up with any convincing evidence of this.

But knowing those scores as soon as possible became so important that around 1988 Bob Schindler, owner of Pinehurst Wine Shop in Baltimore and Parker's friend, started offering (and still does) a "Wine Advocate Bounty" to get a jump on his competition. Whoever brought him the new issue of *The Wine Advocate* before Schindler's copy arrived in the mail could pick out any 90+-point wine in the store, for free. If someone received his copy before 9 A.M. Schindler would send a driver to pick it up. He would quickly flip through the pages, note the five top-scoring wines, and then get on the phone to his distributors. If he had the issue as little as an hour before anyone else, he'd be able to put a lock on the wines and buy all the available cases. If he didn't, he'd probably lose out—someone else would get them.

What *was* it about those numbers? It was partly an emotional thing for Americans, who shared a collective memory of the scale from years of being graded in school. As a result, a score of 98 had an almost visceral impact; it evoked the same feeling of elation as seeing that score on an exam. A 79, by contrast, was synonymous with disappointment, with not making the grade, with your parents chewing you out for not working harder.

The true genius of the 100-point system was that it was universally understood by average Americans and seemed completely straightforward. At the time, seeing a wine slapped with a number was new and startling.

But the system tapped into a deeper fascination with numbers and ratings some thought was peculiarly American. To outsiders, Americans were preoccupied with who or what was *numero uno*—whether it was a baseball team, a rock album on the top-20 chart, or the person with the highest I.Q. Americans rated food products and restaurants, measured job performance and who was watching specific TV shows, and ranked the richest people by income and assets, ballplayers by batting averages, movies by box-office receipts, and high school students by SAT scores. And when it came to wine, they carried around tiny pocket vintage charts and only purchased the vintages considered best, regardless of whether others would be better to drink now. Connoisseurship-by-numbers by meant all you had to do was buy wines Parker rated 90 and above, which is all that Vanna White of television's *Wheel of Fortune* told her Beverly Hills wine merchant she wanted in her wine cellar.

A score of 90 was the minimum for excitement; 95 or higher and a virtual stampede began. So many wine lovers rushed like lemmings to purchase wines rated 90 and above that as early as 1986 Parker felt obliged to plead with his readers in *The Wine Advocate* not to ignore 84- and 85-point wines, pointing out they were often bargains that could be drunk sooner and that he bought them for his own cellar. He professed to be completely surprised by the overwhelming response of consumers to his 100-point system and told more than one interviewer, "I thought they'd read the tasting notes, and the score would be sort of an easy way they could see how scores ranged in a certain peer group of wines. I never realized it would take on the importance it has in the American wine market, and how many other publications would duplicate this kind of scoring system." To stop the practice of wine shops advertising numbers without tasting notes, which took up much less space and seemed to ignite the most sales, and the outright misstating of scores, Parker engaged a New York law firm to send out letters to the biggest offenders. Though it had some effect, he was fighting a losing battle.

Perhaps it was the seeming precision of Parker's scoring that was so reassuring to wine-phobic Americans—and that also added to his mystique.

The system gave the impression that his tongue could register tiny differences in quality in a way that other tongues and palates could not. Other wine writers might group their recommendations in vague categories of quality, but Parker apparently had such a refined palate that he could detect the flavors and aromas that made an 87-point wine ever so slightly better than one to which he'd given 86 points. It implied that every wine had some magic quality number that only a super-taster like Parker could intuit. Like the Hindu idea of a secret cosmic tone for every material in the world, there seemed to be a preordained number for every wine in the world, and whoever recognized that number had to be the person with the greatest palate. Parker was the one who could detect the secret code, which made him seem all the more unusual and special compared to others who judged wines. English journalist Jancis Robinson referred to him wryly as "the oracular wine grader."

Parker may have seen the numbers as a convenient shorthand, but people seemed to prefer them to the accompanying paragraph of adjectives that always required interpretation. The scores fit an impatient, cut-to-the-chase attitude—just give me the best. To English writer Andrew Barr, Parker's scores were "a victory of American pragmatism over French mysticism."

For some the appeal of his scores probably reflected a new interest in the vital statistics of wine. As winemaking relied more on scientific research and technology in the 1980s, a wine's measurable chemical components became part of its description. A new type of consumer, the wine geek, became fixated on the "specs" of wine just as audiophiles obsessed over the "specs" of stereo equipment. American wine writers, especially in California, started to ask questions for which the answers came in numbers: What's the wine's pH? Its total acidity? How many months was it stored in oak? What was the degree of brix when the grapes were picked? What about the degree of alcohol? What was the percentage of Cabernet Sauvignon in a particular Bordeaux château's wine?

The technical wine snobs reveled in these quantitative details. Questions that had never figured in English and European publications now came to the fore in America, especially in California, where the techno-talkers ruled. A love of numbers seemed to go hand in hand with the American love of technology. A tasting group of Los Alamos physicists, for example, were obsessed with verifying wine aeration experiments. One of them: how many minutes should a sturdy Italian Barolo be uncorked in advance of drinking to have a discernible positive impact on flavor?

By the end of the decade, in 1989, Parker had given only nineteen wines—eighteen red and one white—100 points, "perfect" according to his taste buds, and that was out of the 90,000 to 100,000 wines he estimated he'd tasted since the first days of *The Wine Advocate*. All were French, from a handful of estates. The largest number, nine, came from the Rhône Valley, including seven from a single producer, Guigal. Burgundy's famed Domaine de la Romanée-Conti weighed in with two. Bordeaux produced eight, almost all first growths. But what was a perfect wine? "A score of 100 is a magical moment," Parker once said. "You've got to be totally . . . moved . . . when you taste a wine like that. It's a mind-blowing experience to have a 100-point wine. I mean, you immediately know it when you hit one. Virtually all the wines I've given 100 to, I knew it when I smelled them. And I was just praying that they would taste as good as they smelled."

It was a mark of Parker's importance that his first 100-point California wine was considered big enough news to be reported in London's *Financial Times* wine column. Dennis Groth, a former Atari computer executive who'd founded Groth Vineyards in 1982, first heard that Parker considered his 1985 Cabernet Reserve "perfect" in a phone call from someone in the trade who'd just received the February 28, 1990 *Wine Advocate* in the mail. "That sounds wonderful," Groth enthused, though he was somewhat distracted by the day's construction problems at his new winery. His winemaker, Nils Venge, was skeptical: "Well, how can it be a perfect wine? Nobody knows how it will age. A perfect 100 shouldn't be awarded until you know a wine can age!" "Oh Nils, quit arguing," retorted an exasperated Groth, "Parker must believe it can age." It was a moot point, as the wine was already sold out. Distributors had taken all their allocations and sold them at $26 a bottle to their best customers.

So Groth and Venge smiled and raised a glass to Parker and congratulated themselves. They had kept back about three hundred bottles for the winery library, but it was hard to hold on to them. Friends called almost immediately, including Groth's best duck-hunting buddy. Groth regularly traded wines with other top producers, and they were more eager than ever to exchange their best cuvées for Groth's 1985. Still, as usual, Stag's Leap Wine Cellars' Warren Winiarski demanded two bottles for one of his Cask 23 Cabernet on the grounds that his wine cost more. Regular consumers tried all sorts of angles to wheedle a few bottles of the wine. Groth's wife told one insistent caller, "You don't understand. The only wine we have left is for our family."

"Do you have any daughters? What about them? Do they have any allocations of the 1985?" he asked.

"Well, yes, of course," she replied.

"Do you have any that are eligible for marriage?" the caller queried. At this point Groth's wife wasn't sure if it was a joke or not.

The scores became useful tools at all levels: producers used them to sell wine to importers, who used them to convince distributors to stock them, who used them to get retailers to buy, who posted them so customers would buy, who used them to impress their dinner guests. Numbers were the new way to market wines. And Parker was the first generator of the numbers. No wonder that in a 1988 survey of retailers' ranking of the country's ten most important wine writers in *Liquor Store* magazine, Parker rated 96 on a 100-point scale. It was the retailers' highest score, 10 points ahead of their second-ranked writer, Terry Robards of *The Wine Spectator*. Parker's former rival, Robert Finigan, received a 79.

———

THE NEW—AND RAPID—GLOBALIZATION OF TASTE IN THE WINE world during the decade, caused primarily by the spread of improved technology, fit easily with Parker's it's-what's-in-the-glass-that-counts philosophy. It, too, contributed to his power and his increasing ability to make his individual taste and judgment count on every winemaking continent.

The most important person behind that transformation in France was Émile Peynaud, a professor of oenology at the University of Bordeaux who was considered the guru of modern Bordeaux winemaking. He invented what is now termed the international style of wine, especially red wine: ripe, fruity, with soft tannins and the flavor of new wood, wines that tasted delicious when young but could also age. Throughout the 1960s and 1970s he had insisted that grapes must be picked only when they were fully ripe and asserted that ripe tannins were what allowed wines to age. In Bordeaux that was a revolutionary thought. Producers traditionally picked earlier, worried that if they didn't they were taking a big risk. If rain came, the grapes might be attacked by rot; with hail they could lose the crop altogether. Peynaud pushed for temperature-controlled fermenters to preserve freshness in the wine. He preached hygiene in the cellars and scientific vineyard management. He proposed the idea of radical selection—taking only the very best barrels of wine for the *grand vin* and putting the lesser barrels into a second wine. He advocated new oak barrels. During the 1970s many châteaux, who had finally begun to make some money thanks to the American market, began to engage him as a consultant, not just to rescue a wine in trouble but to help make sure it was good from the beginning. In time Peynaud would be advising more than seventy châteaux. The 1982 vintage that wowed Parker and the Americans was the first year when the benefits of his ideas shone through in the wines of Bordeaux. Ripe fruit was the hallmark of the new wines, the ones that Parker liked.

In fact, the 1980s were a heady time for those making wine everywhere, from California to Burgundy. Change was in the air—technology brought improved methods of vinification and communication, and easy travel connected the world's winemakers more than ever before. They tasted one another's wines and learned from each other. California winemakers made pilgrimages to Burgundy to try to discover the secrets of making great Pinot Noir, and Europeans who wanted adventure, to try something new, headed for California and absorbed some of the lessons of American technology. Italians who wanted their wines to have the same international acclaim as French wines did planted the new international grape varieties, Cabernet Sauvignon and Chardonnay, and experimented with French oak barrels. The new generation in traditional winemaking families studied oenology and worked for a few months or a year in other countries and returned with a wider perspective on winemaking. And estates everywhere in the world began to hire consulting oenologists to help make their wine. The one many began to call was Peynaud's pupil, Michel Rolland.

On a July day in 1982, Parker had poked his head in the open door of Rolland's lab in Libourne and introduced himself to the short, stocky, dark-haired man with an open-necked shirt and a welcoming, slightly wicked grin standing over an assortment of test tubes. It was the beginning of probably the most important wine relationship Parker established during the 1980s. It was a meeting of equals, two young professionals born in 1947 on the way up, whose reputations and influence on the wine world were only in the beginning stages. Parker was not yet Parker, the great wine critic, and Rolland was not yet Rolland, the world's most influential wine consultant. Rolland's career had begun first, in 1973, when he established the laboratory to which about 120 properties on Bordeaux's Right Bank brought their wines for analysis.

The day was warm, work at the lab was slow, and Parker was interested in tasting, so Rolland invited him and Pat to his family property in Pomerol, Château Le Bon Pasteur. For three hours they tasted from the barrel and, via Pat's translation, "chewed the oenological fat," talking about wine, about tasting, about everything. The two men found one another *sympathique*. They had much in common. Both were the first university-trained professionals in their families, very ambitious, outwardly modest but with plenty of ego, and both loved wines, especially rich, supple ones based on the Merlot grape. Both had attractive, devoted wives—Rolland's wife, Dany, also an oenologist, worked with him. Rolland, too, was bored with his primary job, lab work, and he was just as interested in putting his views on winemaking into practice in the vineyard as a consultant as Parker was in being a full-time critic. "I had a good moment with him," Rolland recalls. "I found him very interesting and enthusiastic, a sweet and pleasant man."

This became a powerful symbiotic relationship; the two rose in a similar, sometimes intertwined trajectory to become international superstars, the work of each one adding something to the success of the other. Rolland became part of the Parker Bordeaux circuit, driving him around, helping him understand viticulture and oenology, and introducing him to tiny up-and-coming properties where he was a consultant, in Bordeaux and later in far-flung countries around the world. Parker was a lover of the Rolland style of winemaking, acknowledging that one of the things he gave the highest value to was fruit. "If all you are tasting is wood or tannin or acid, that wine is never going to do well in *The Wine Advocate*," he revealed in an interview in 1989.

He applauded most of the wines Rolland made or consulted on, and wrote flatteringly about the "brilliant" and "gifted" winemaker. Parker even recommended him for his first consulting gig in America, at Simi Winery in Sonoma. Rolland's fee was then a mere $200 a day. By the end of the decade he had learned to speak English, consulted on three continents, drove a Mercedes, and wore sleekly tailored suits. He, too, was a workaholic. Still, both men understood professional limitations. For the 1989 vintage Parker rated Rolland's Château Le Bon Pasteur much lower than usual, only 84 to 85. Rolland was stung by his criticism, and when they saw one another later that year, Parker asked him, "You want to speak with me again?"

———

SHERMAN MCCOY, WALL STREET BOND TRADER, "MASTER OF THE Universe," and protagonist of Tom Wolfe's 1987 novel *Bonfire of the Vanities*, symbolized the money men of the 1980s. It was a bull market. Excess, greed, self-indulgence, instant money—all of it played into the new interest in wine. Baby boomers in the thirty-to-forty-four-year-old age group were the wine drinkers, especially well-heeled psychiatrists, lawyers, doctors, bankers, and Wall Street types who'd made it big. They were a new kind of consumer, hungry for experiences and willing to pay for them. Luxury products and hedonistic experiences boomed with the economic success of the "me" generation. In a country with a fluid class system like America's, they helped define who you were. Wine was the perfect hobby: it conferred status but didn't require an elevated social background or much knowledge, and it was fun, too.

Parker's newsletter was the insider's track, a way to instant insight. Like financial newsletters, it purported to immediately put readers in the know. Subscribers included former president Richard Nixon, Malcolm Forbes, senators and congressmen, and even a couple of CIA operatives. At a dinner in New York, fledgling wine enthusiast Bob Lyster noticed that as wines were poured and tasted, the phrases "Parker said," "Parker thinks" peppered everyone's comments. He'd never heard of Parker, but quickly picked up that he was supposed to have. The same thing happened to novelist Jay McInerney, who had recently discovered wine and was hanging around in London in 1985 with a friend, British novelist and wine lover

Julian Barnes, "who kept mentioning Parker." Parker's name was on the lips of all who wanted to be wine insiders, and they became "Big Bob" supporters and subscribers. Lyster, McInerney, and Barnes pored over every issue and bought wines Parker recommended, like Philippe LeClerc's Burgundies. The Parker package exuded a fanatical love of wine, of having a great time. Tom Ryder, then publisher of the American Express magazine division, invented his own must-buy system by working out the price–value ratio of Parker's recommendations: he divided the number of points by the price, ignoring wines scored below 90. He bragged about discovering this code-within-a-code that identified the biggest bang for the buck. By following Parker's numbers and notes it was simple to amass a large wine cellar without having to spend time learning much—all had been vetted by the Great Bob. At some dinner parties, serving a Parker-gave-this-94-points wine had more cachet than offering the old château-bottled wine with a famous name on the label. Even if you had a deep-seated inferiority complex when it came to wine knowledge, with a wine highly rated by Parker you could be smugly superior.

Americans may have a habit of thumbing their noses at authority, but since Parker promoted himself as a rebel against the old authorities it was easy to fall in line. It was like taking advice from an expert buddy.

The most serious collectors, like Texas neurosurgeon Marvin C. Overton III, southern California particle physicist Bipin Desai, and German businessman Hardy Rodenstock, became famous during the decade for their marathon two- and three-day tastings of dozens of rare wines. These tastings were the epitome of conspicuous consumption—sixty different wines from the great 1961 vintage in Florida, fifty vintages of Château Haut-Brion in the Texas hill country—and often seemed like potlatches among a tiny international wine-collecting circuit, more about ego than connoisseurship. Even château owners invited to attend had never tasted so many of their own wines at one sitting. Michael Broadbent, who'd sold some of those wines to the collectors in the first place, was a frequent guest, as was the odd journalist (myself on several occasions), as most collectors were hoping for a write-up. They quickly added Parker to their list of desirable media, anxious for him to bestow his presence on their latest events. When he couldn't attend Desai's 1988 Los Angeles Château Margaux tasting, the collector lured Parker to a sampling of its highlights with

a tale of the sensational old vintages he'd missed, especially the extraordinary 1900. What clinched Parker's attendance in Paris at Taillevent in November was Desai's message that according to Margaux's winemaker, the
"vintages 1982 and 1900 were almost identical." A privately printed book
for tasting notes commemorated the occasion. And they drank two bottles
of the priceless 1900. Parker sent "a gracious note" of thanks and gave the
wine 100 points in his newsletter. Nineteen eighty-nine brought an extraordinary tasting of fifty-five vintages of Château La Mission Haut-
Brion, a wine from the Graves region in Bordeaux that Parker loved, and
another that included one hundred Bordeaux from the 1982 vintage; the
following year featured a tasting of nearly all the vintages of Grange Hermitage, the greatest wine in Australia.

But by the end of 1990, Parker had become disenchanted with these
massive tastings, pronouncing them "ostentatious . . . excess to its extreme." He had become close friends with several serious collectors, but
his idea of wine fun was the more gourmandish Les Oenarchs, a group of
eight "local forks"—wine-loving Baltimoreans—that he founded in 1991,
modeled after the group of the same name in Bordeaux. Members included Jay Miller; Parker's neighbor Bert Basignani, who owned a nearby
winery; and orthopedist Hank Dudley, who also belonged, with Parker, to
an amusingly named group, the Wine and Spine Society. Each member
hosted a dinner with a wine theme and provided "good stuff" from his
own cellar—twenty-four California Cabs from the 1970s, for example, or
a series of vintages of a Parker favorite from the Rhône Valley, Hermitage
La Chapelle—for the group to taste blind with dinner.

━━━━

JUST ABOUT EVERY ACTION PARKER TOOK SEEMED TO CONTRIBUTE
to his influence and power in the marketplace. Outworking all his competition kept his name and ratings out in the world, but he also turned out to
have an exceptional talent for self-promotion. Take insuring his nose and
palate. To protect Pat now that he was self-employed, Parker had consulted his insurance advisor on what to do. He recommended a disability
policy, outlining the benefits if Parker found himself unable to work due
to accident or illness. Parker countered that it wasn't really worth it unless
the policy covered loss of his sense of smell or taste, now the center of his

livelihood. That was the disability that counted, and the idea of it had secretly worried him ever since he'd read that English wine expert Harry Waugh had completely lost his sense of smell after hitting his head against the dashboard in a car accident. "I've never had a client who asked for that," his advisor confessed, "but I'll ask." Lloyds of London refused, but Tennessee-based Provident Insurance agreed to coverage of $1 million. Ten years later Parker tried to double the amount, but the company refused.

When news of his million-dollar nose and palate appeared in early articles and interviews, it underscored the already widespread belief that Parker's sense of taste and smell must be extremely special—in fact, unique. No other American wine writers were insuring their noses. This put him on a par with celebrities, among whom insuring body parts was nothing new. Rock guitarist Keith Richards had insured his hands; why not Parker's nose?

A book on Bordeaux helped solidify Parker's reputation in America after the 1982 vintage. Previous books on this famous wine region had come from the British, but since the first days of *The Wine Advocate*, Parker had been dreaming and talking about writing a book in which he could provide his own answer to the 1855 classification. Once again, the connections he'd made in the wine community were crucial. When he confided his dream to Kermit Lynch, the maverick Californian urged him to call New York literary agent Robert Lescher, whom he'd just seen in New York and who, he told Parker, happened to be a subscriber and a Parker fan.

Lescher and his wife, Susan, also a literary agent, were on vacation in southern Vermont at their favorite inn when they received the message that a Robert Parker had called. Susan, convinced it was Robert B. Parker, the best-selling mystery writer with a strong fan following, was elated. Bob Lescher, who represented many food and wine writers—including Robert Finigan—hoped it was the new star of wine journalism.

The two met in Lescher's New York office. Parker was impressed. Lescher speaks in orotund tones about literature and has that disheveled literary style of appearance cultivated by many academics, suggesting the life of books and the mind was far more important than mere fashion or worldly endeavor. But it was the cases of wine from the Rhône Valley on his shelves that convinced Parker the two were meant for each other. Both, it turned out, loved Château Rayas, a red wine from Châteauneuf-du-Pape whose reclusive owner, the eccentric Jacques Reynaud, had refused to see Parker for three years. Once he hid in a ditch when Parker and Pat drove down the bumpy dirt road to his very modest château, where chaos and "a pack of undisciplined dogs" greeted them.

Lescher knew just whom to approach with a book by Parker—another wine lover, Dan Green, the associate publisher of Simon & Schuster, which had made a bundle from their list of contemporary wine books, most by well-known British authors such as Hugh Johnson and Michael Broadbent. Parker, Green, and Lescher met for lunch on New York's Upper East Side to discuss ideas. Green had checked out Parker's draw with his eldest son, who was working in the wine business; he told his father, "Sign him. He's the best." But Green quickly dismissed the initial proposal of a book on the red wines of the world. "That's the wrong book," Green insisted, shaking his head. The 1982 Bordeaux was front-page news, and Parker's name was attached to it. Green had already decided the right book was one on Bordeaux timed to come out just as the bottled 1982s arrived in the stores in 1985. Now *that* would have built-in publicity.

Parker, who'd come prepared to be as persuasive as possible, and was probably rehearsing in his head all the reasons why he was the right person to write any wine book at all, instead found himself being wooed by a publisher to write the book that he would have written for nothing. The lunch had been a triumph, he thought, marveling at his good luck on the train ride home; as he told one reporter years later, "I felt like Sylvester Stallone in *Rocky*." For the next year he poured his time and energy into the book, drawing on his notes and his newsletter. In fact, the book began Parker's systematic reuse of those newsletters, a recycling scheme that any author would envy and one that added significantly to his income. To complete the perfect sales circle, a paragraph telling readers how to subscribe to *The Wine Advocate* appeared on the book's last page, and subscribers to *The Wine Advocate* could purchase the book, autographed, through the newsletter, something many did. Editorially the only problem was length—a fight ensued because the manuscript was longer than expected, but Parker was anxious not to cut anything.

The hefty 542-page *Bordeaux: The Definitive Guide for the Wines Pro-*

duced Since 1961 was published in November 1985, with praise from British wine heavyweights Hugh Johnson and Michael Broadbent on the back. The *New York Times* reviewed it twice.

At a celebratory lunch, Parker presented Green with a double magnum of a 1982 Bordeaux *cru classé*. And the book started selling immediately. The marketing key was not the bookstores, but America's serious wine shops, who subscribed to *The Wine Advocate* and used Parker's notes and scores to sell the wines they carried. In most states (but not New York, where the liquor lobby prevented it) these shops displayed books, wine openers, and much else besides wine. Green had been counting on them to sell the book, but even he was surprised when Washington D.C.'s MacArthur Liquors ordered some 2,500 copies. New York City retailer Acker Merrall & Condit held a book signing for Parker by partnering with the closest bookstore, Murder Ink., which specialized in mysteries, to handle the actual sales. The line of their customers waiting for Parker's autograph snaked out the door and around the block.

Thanks to the promotion of wine shops the initial printing sold out quickly, and by the end of January the book was in its third printing, had been chosen as a selection of the Book-of-the-Month Club, and had sold 65,000 copies. A British edition appeared in September; rights were sold to France and Germany and eventually to Japan, Sweden, and Russia. The French edition became a selection of the month in France's Book-of-the-Month Club in 1991.

Meanwhile, Green had suggested Parker compile each year's tasting notes in an annual guide, yet another way to recycle them. And what about books on other wine regions? Burgundy was the obvious choice, but Parker argued his beloved Rhône should be next. At this point no one disagreed. A handful of people rushed to hammer out a multi-book deal in the publisher's small boardroom, before Green, who was leaving to work at another company, could snatch Parker away. The advance Lescher asked for was enormously high for wine books, but when publisher Jack McKeown took out his calculator and figured how many copies of *Bordeaux* had been printed and sold, the answer was yes.

A few years later, while on a tour of Bordeaux with a friend, French author Dominique Lapierre, Dan Green noticed copies of Parker's *Bordeaux* book displayed in every château he visited. Yet all, he said, "pretended tremendous dismissiveness of the book and Bob . . . they just couldn't admit to being interested."

After John and I had left behind our editorial duties at *Food & Wine* in the mid-1980s we continued to write a column together for the magazine that alternated with Parker's, but our paths crossed less frequently. Many times I would be somewhere that Parker had just been, and no one who wrote about wine could miss his impact or the controversy he inspired.

In journalistic circles, when newspapers take over a story by concentrating their reporters and resources so that no aspect of it escapes them, it's called "flooding the zone." From 1985 to 1993 Parker did just that to the wine world. In those key eight years, in addition to his initial book on Bordeaux, he published one on the Rhône, one on Burgundy, a revised edition of *Bordeaux*, and three *Wine Buyer's Guides*, the last over 1,100 pages. While publishing seven books in eight years he maintained his punishing tasting and travel schedule and continued to publish his bimonthly newsletter, which by 1993 was forty-eight pages an issue with tasting notes for as many as 764 wines, regular columns for *Food & Wine*, and for part of that time, columns in *Connoisseur* and *The Wine Enthusiast*.

Even if Parker was a master at reworking his material, other writers wondered, when did he sleep? His presence was starting to be felt everywhere, and his impact was often presumed to be equally pervasive. Frequently, in a winery cellar somewhere in the world, a journalist would bring up Parker's opinions, as one American writer did on a press trip I was on in Cornas, in the tiny, dirt-floored cellar of August Clape, a Parker favorite. Clape, clearly uncomfortable, listened, then said in a flat, hard voice, with a brisk tap on his cigarette so that a long ash dropped directly in front of the journalist, that Parker had nothing to do with how they made their wine—*comprenez-vous?*

———

IN 1986 PARKER'S NEWSLETTER COST $28 PER YEAR, HAD 17,800 subscribers in thirty-seven countries and a renewal rate of 70 to 80 percent, and grossed about $500,000 a year; Parker spent $60,000 to $74,000 on wines, which accounted, he claimed, for about 75 percent of all the wine he tasted in bottle. Still, all this had major paybacks—directly in the form of tax deductions, as well as indirectly by enhancing the image he wanted to project to the consumer of the writer who pays his own way. The other 25 percent of his wines were free, unsolicited samples of not-yet-released wines shipped to him (as they were to many wine writers) or the lines of wines importers brought to the Milton Inn for him to taste. Three years later the number of newsletter subscriptions hit 28,000, it was forty pages long, and the price had gone up a mere $2, but Parker now made serious money, grossing about a million dollars a year including the income from his books.

The publication of every book was yet another chance for reviews and interviews, many of which included information on how to subscribe to *The Wine Advocate*. For the most part the reviews were glowing, but, more important, all spread Parker's name and story, which he polished with each retelling. His was an age-old story that played well with Americans: the tale of coming out of nowhere to succeed, of the hero who "goes where no man has gone before," the honest, plain-speaking rebel challenging gentlemen thieves. And Parker didn't hesitate to pat himself on the back. Non-wine journalists at national magazines—*People, Time, Newsweek*—ate it up. Just being who and what he was turned out to be the best story of all.

In 1988, when Parker became the paid wine expert for the Prodigy Interactive Personal Service system, one of the first nationwide computer networks, he brought wine discussion into a new medium and helped usher in online wine geekdom. Prodigy, a joint venture of IBM and Sears, Roebuck, hired a small staff to edit content as well as experts in various fields, like Parker, to write weekly columns, respond to e-mail questions, and participate in discussions on the active bulletin boards. And they marketed their experts aggressively. So Parker became the first online wine guru, and his scoring system found a ready acceptance among many of the techies who joined the wine bulletin board for online chatter. "He attracted the geekiest of the geeks," says Mark Squires, who became the board leader. "Everyone wanted to talk to Parker—or yell at him." It became a tight-knit community, which even met offline for parties. The majority were Parker supporters, who ended up forming the base of a sort of Parker fan club. They stood behind him and defended him from all criticism, spreading his name and ideas about wine on other bulletin boards. Others ranted and raged, calling him an idiot and even comparing him to Hitler. Heated online discussions were often virulently for or against; egos were at stake. A surprising number of wine lovers spent so many hours on the bulletin board that the online staff called them the "get-a-life club." As Prodigy started to decline and lose money and cut the experts' fees, Parker opted out. His connection lasted only a few years, but many on the bulletin board hung together as friends, who would regroup later when Parker began his own website.

Parker picked the venues for his public appearances carefully, turning down lucrative TV and radio offers—he relished his privacy—but welcoming opportunities to press the flesh of potential subscribers among the wine-converted at important shops like Zachy's and at organized tastings as a guest speaker. As a wine celebrity he generated so much excitement that being near him affected some very strangely. One New York City retailer who picked up Parker at the airport was so thrilled to meet him that as they drove into Manhattan he kept his eye constantly on his passenger instead of the road. He drove erratically while chattering about wine, and as Parker watched with horror, kept smacking into the concrete-and-metal barriers at the side of the highway, damaging one side of his car.

One of Parker's first public tasting appearances, in 1984, a Bordeaux seminar put on by Howard Kaplan and Robert Millman's Executive Wine Seminars in New York City, was typical of many to follow for the next decade. Eighty people, mostly men, paid $28 for the privilege of a two-hour blind tasting of a dozen wines from the 1978 and 1979 vintages with Parker. It was held in a private banquet room at the Doral Inn on Lexington Avenue. Parker arranged copies of *The Wine Advocate* in a neat pile on a white tablecloth on the front table and wore a suit. Many in the audience had never heard of him—the wines were the draw, not Parker—and the tasting sheet didn't even list his name. What won them over, say those who were there, was his sincerity and passion. He seemed a down-home kind of guy, who described wines as "cunningly made," not too technical or pompous, in fact the kind who wouldn't look down on you if you were a New Jersey podiatrist with a hairpiece and a dozen cases of wine in the basement. He received a loud ovation.

Kaplan and Millman, wine salesmen at Morrell's, were already Parker fans, and they aimed their specialized seminars at executives anxious to become wine savvy—exactly those people Parker was trying to reach. They paid Parker no fee; as he so often did back then, he bartered for publicity. They included a *Wine Advocate* flyer with their regular mailing as well as a plug, something they never did: "In our judgment if you are going to subscribe to one wine publication, it should be *The Wine Advocate.*" Thus began a long relationship with Kaplan and Millman; Parker, intensely loyal to all who supported him when he was a nobody, continued to appear annually (though at an ever-increasing fee rate), and still does. Within a few years, Mr. P., as they referred to him, drew double the crowd; they moved to the Doral's ballroom and sold out a couple of months in advance. They were getting calls from corporations offering a four-figure finder's fee to persuade Parker to give a talk for their executives. And by 1990 the suit and tie were long gone; Parker dressed as he pleased.

Everyone wanted Parker. He spoke at the International Wine Center, packed tastings in London, headlined a Pinot Noir conference in Oregon, led tastings at charity events in Washington, D.C., and even taught a course in wine appreciation at Goucher College. In 1993, he was leading a tasting of the 1982 Bordeaux at a restaurant in Detroit. A few years before at a similar function in the city, he had been just another guy who had happened to praise a vintage. Now titans of Detroit's auto industry and Midwestern "wine wonks" paid $275 to as one newspaper put it, "be in the presence of the crown prince of wine."

———

PARKER'S FIRST COMPETITOR, ROBERT FINIGAN, WAS STEADILY losing ground. Because of a legal tangle involving an investor going bankrupt, he was forced to suspend publication of his newsletter for more than a year, just as Parker's star was rising and bottles of the 1982 Bordeaux had begun to arrive. When he started up again, he'd lost momentum to Parker and never really recaptured it despite a newsletter redesign. In fact, in October 1987, because of subscribers' letters asking for more precise wine ratings, Finigan had to succumb to the inevitable. He didn't "believe that sensory experiences of any sort can be reduced to the sort of scoring appropriate for a math quiz," but gave in to using the 20-point scale. He thought Parker's 100-point system was ridiculous.

The up-and-coming competition was Marvin Shanken's *Wine Spectator*. In late summer 1985, Shanken flew from New York to San Francisco for one of the two-day freewheeling creative meetings he convened regularly in *The Wine Spectator*'s office to energize his young staff and work on improving the magazine. Cigar clamped in his mouth, jacket on his chairback, clad in shirt and his trademark suspenders, Shanken leaned forward at the conference table and brought up the issue of Parker.

Two years before, only a few months after Parker had championed the 1982 vintage, Shanken and *The Wine Spectator* staff did an about-face and published an issue devoted to Bordeaux, which also praised the 1982s. Then, to boost their credibility with consumers and show that *The Wine Spectator* could be just as critical of wines as Parker was, they added a new section to their wine reviews in which they listed wines they'd found too poorly made to recommend. Advertisers who found their wines on the list were angry and let Shanken know it.

"We've got this competitor, Robert Parker," Shanken began, his eyes narrowing as he looked around the table at balding Harvey Steiman, Jim Laube with his trademark black mustache, and newly hired Jim Gordon, who would become the managing editor. "What could we do about this? Could we adopt the 100-point system? Let's talk about that." It was phrased as a question, but according to Gordon everyone on the staff guessed that Shanken had already decided but wanted them to agree. It was clear that Shanken and *The Wine Spectator* were competing with Parker for influence. He was looking for share of mind rather than market—*The Wine Advocate* didn't go after advertisers, so there was no competition there. After two days of on-and-off discussions, everyone was on board. Parker had test-marketed the idea, proving people could understand the system and that retailers loved it. Laube voiced the opinion that by using it they would be taken more seriously as wine critics. Shanken decreed that only staff writers could use it, to ensure that there would be no conflicting scores in the magazine. Basically leery of independent wine stars out for themselves, he wanted to build a team that collectively made the judgment calls, held the power, and exerted the influence. His new policy of no press junkets for the staff looked very much like a response to Parker's no-conflicts-of-interest stance. While it welded Shanken's team together even more, it separated them from their wine-writing colleagues.

In the best co-opting tradition, Shanken called Parker about writing for the *Spectator* (Parker said no). He even expressed interest in purchasing *The Wine Advocate* (another no). But rivalry and even hostility thrived. In one apocryphal story that won't die, Shanken supposedly invited Parker to dinner in California, at which the first words out of his mouth, before his guest even sat down, were, "What do you think of me stealing your 100-point system?" Parker's retort was, supposedly, "What difference does it make. You give everything 90 points anyway." Shanken shot back, "We're going to put you out of business."

Shanken did attempt to enforce a rule that the *Spectator*'s advertisers couldn't use Parker's scores in their ads, but it didn't work. In fact, in 1990 Zachy's, one of the country's top wine shops, took out a one-page ad headlined "Zachy's Presents Parker's Hall of Fame 95–100." It listed forty-three wines Parker had rated 95 and above, including his score and tasting notes for each one. Their full-page ad in 1991, however, included scores from both Parker and the *Wine Spectator*.

The media adulation of Parker, which elevated him into the world's greatest wine critic, rankled not just Shanken but everyone else on the staff. Each time another article appeared, Jim Gordon remembers, they would all roll their eyes and groan, "Oh, no, not another story about Parker and how he grew up drinking Coca-Cola!" They were sick of hearing Parker's story, felt slighted, and decided that while they would refer to *The Wine Advocate* if a news story about wine futures required it, they weren't going to make a big point of writing about him. After all, did *Newsweek* write about *Time*? There was no a hard-and-fast rule, but Gordon knew that if Parker was mentioned in a story, Shanken wanted to see it before it went to press. Occasionally Gordon fought to keep a reference to Parker in, but when it didn't seem important, Shanken would tell him, "Cut Parker out." Naturally, though, they read every issue of *The Wine Advocate*, passing it around and initialing it, and they reviewed Parker's books, usually favorably.

Shanken, too, was a driven workaholic, but his taste in music tilted towards James Brown, godfather of soul. Much more of a conscious businessman than Parker, he continually explored ways to expand his growing empire. While Parker relied primarily on himself and word of mouth to spread his name, Shanken undertook more typical business moves to increase *The Wine Spectator*'s power and influence. First he changed it from a newspaper into a four-color magazine and worked on getting it sold in wine shops. Selling anything but wine in New York stores was illegal, so he hired lobbyists to dog the New York legislature to exempt magazines. By mid-1991 some 1,500 stores across the country carried his publication, with the intended result. Retailers testified that customers bought the *Spectator*'s recommendations. That was the kind of influence Parker had, and Shanken was aiming for it, too. He wanted *The Wine Spectator*'s scores right next to Parker's: RP 90, WS 91.

Another business success was Shanken's annual Wine Experience event, a three-day weekend extravaganza at which the owners of the world's top wine estates paid to pour their wines for free at consumer sit-down tastings and seminars. Shanken thought of it as the Super Bowl for fine wine producers. One formal photo taken at the first event reveals just how Shanken positioned himself in the wine world. He stands dead center in a crowd of vintners from the world's most illustrious estates. Though plenty of people in Bordeaux and Burgundy disliked him personally, regarding him back then as another example of "The Ugly American" abroad, he managed to coax producers to attend. In the process, the globalization of the wine world got a boost. Robert Drouhin, the famous Burgundy shipper, introduced himself to Christian Moueix of Bordeaux's famed Château Pétrus. They had never met. Winemakers from Italy had a chance to meet those from France and taste their wines for the first time.

But it was also a way to emphasize to the Europeans the importance of American consumers and marketing. Angelo Gaja was surprised that so many wine lovers were interested in lining up to shake his hand. Shanken was anointing the most important people in the wine business. If you were invited, and you didn't accept, you wouldn't be invited again, unless of course you were one of the very top properties. Shanken wanted whoever was important in the wine world to be part of the event, and in 1988 he even persuaded Parker to headline a tasting of Rhône Valley wines, providing excellent publicity for his recently published book on the subject.

By 1993, Shanken had moved the magazine to New York and vastly expanded the numbers of wines it rated. In terms of capturing American media attention, though, *The Wine Spectator* had a fatal flaw: it was a collection of voices. American heroes rarely came in bunches. Like Parker, they didn't belong to a club. Parker's story was the kind Americans liked, the story of an individual—one person, one palate.

Other American publications also felt the pressure to adopt the 100-

point rating system if they wanted their recommendations on wines to count. One was *The Wine Enthusiast*, an outgrowth of a catalog business of the same name that sold wine gear and home-sized temperature-controlled units to store fine wine, which had been started by a gangly wine salesman-turned-entrepreneur, Adam Strum. Thanks to Parker's promotion of the 1982 Bordeaux, people who bought dozens of cases of wine scrambled to purchase these temperature-controlled units in which to store them, and the company had made a lot of money. The magazine agreed to insert a *Wine Advocate* ad in their mailings; Parker penned a few articles for them drawn from his newsletter, without, of course, his precious scores. He also granted the magazine several exclusive interviews, which their readers devoured, good publicity to drum up more subscribers for *The Wine Advocate*. *The Wine Enthusiast* had started out rating wines with a star system, but by 1993, Strum insisted on numbers. Once again, says former editor W. R. Tish, "it was a marketing decision, to get market share."

But in the battle of the numbers, at the end of the decade, Parker's still counted most.

THE OUTSIDE WORLD TOOK NOTE OF PARKER'S INFLUENCE. SOME had noticed signs of it even before the 1982 vintage. In the beginning of July of that year, a young *courtier* named Erik Samazeuilh had been tracking the wine orders coming in to the firm Tastet & Lawton, on the curving Quai des Chartrons in the city of Bordeaux, the traditional location of the city's *négociants*. It faces the Gironde estuary, and on sunny afternoons the light hitting the eighteenth-century stone buildings colors the ornate facades a soft gold. Orders for bottles from the best-known châteaux arrived with the steady regularity of a heartbeat. But late that day a flurry of calls for what was an obscure fifth growth, Château du Tertre, created a sudden aberrant blip. A mystified Samazeuilh asked a colleague, "What's going on with Château du Tertre? I've just taken twenty-five calls and now we're sold out."

"Oh, some American named Robert Parker just wrote something good about it," was the reply.

In Parker's June review of the 1979 vintage in Bordeaux, he'd devoted four lines to Château du Tertre. The restored and revitalized property was "beginning to turn out first class, deeply flavored classic wines." His 85+ score and "highly recommended" notation promoted it quickly to the level of wines that cost quite a bit more.

The dramatic impact on prices when, in a later issue, Parker bestowed 100 points on the 1982 Château Mouton, gave some indication of what was in store. The Mouton had come on the market at 170 francs in 1983 and was up to 330 francs by September 1984, but immediately after Parker gave the wine 100 points in February 1985, "at that moment the prices went to 450, 500, we rode it up to 750 francs," recalled *négociant* Bill Blatch, showing me the appropriate pages in the oversized notebooks he keeps meticulously to track wines and prices. "To quadruple the price within three years was unheard of. That would never have happened without Parker." In 1986, Parker's influence extended beyond the retail level, leading growers to put up prices or hold back their stock. And even in Paris, at his Caves de la Madeleine wine shop, Steven Spurrier could tell exactly what wines rated highest in Parker's most recent newsletter by the sales in his shop.

By 1986, château owners, *négociants*, and brokers in London and elsewhere in the international wine market had their eye on Parker and watched his influence and power climb steadily. But Blatch and many in Bordeaux believe his assessments of the 1989 and 1990 vintages in Bordeaux were when his impact suddenly skyrocketed. Parker initially went overboard in his praise of the 1989 vintage, Blatch is convinced, but the effect was that "the châteaux finally understood that they could make a lot of money." It was the most expensive vintage Bordeaux had ever seen. "There were prices," *négociant* Dominique Renard told me, "that we couldn't believe."

———

AS HIS HIGH SCORES FOR WINES TRANSLATED INTO IMMEDIATE sales, the reactions of some of the proprietors Parker visited and the importers he tasted with changed. Many genuinely seemed to like him—he was, in fact, a very likeable guy, warm, friendly, and enthusiastic, though people had begun to remark on his well-developed ego and his dogmatic certainty about his opinions on the quality of a wine and, increasingly, the correct winemaking methods—but all understood what was at stake in remaining on his good side. As with a newly crowned prince, they offered Parker gifts—plane tickets to wine regions and stays at winery and château guest houses, invitations to lunch and dinner, even valuable music boxes and, once, a Porsche. After a tasting at the Milton Inn one importer urged him, "Take my beach house in Barbados for a week," mentioning another wine-writer friend who frequently took advantage of the offer. Parker interpreted the sly way a couple of *vignerons* in France introduced their daughters as some kind of indecent proposition, sex in exchange for a good score for their wine. He had declared from the beginning that he would pay his own way, so he refused these offers, bribes, and propositions. Wine people were often stunned when Parker picked up the check in a restaurant, although he did accept lunches and dinners in people's homes. First-growth châteaux didn't try to ship an annual case of wine to him, as one English wine writer told Parker they always did for him.

In Italy, Angelo Gaja told me with disgust that European journalists often left the trunks of their cars open during visits, expecting to find them filled with a couple of cases before they drove off. That would never happen with Parker, people said, and he used this reputation to set himself apart, to reinforce how different he was and emphasize that he was uninfluenced by the need to win advertisers or pay back favors. The implication was that his judgments were completely reliable. It was a familiar stance to wine lovers who regularly consulted *Consumer Reports* when they bought a dishwasher or a car, and they trusted it.

But drawing the line on potential conflicts of interest proved more complicated than it originally seemed from Parker's morally rigid viewpoint. Parker, too, accepted hundreds (eventually thousands) of unsolicited sample bottles for his regular blind tastings, as most wine writers did. When Alexis Lichine invited him to dinner at Château Prieuré-Lichine he went and appreciatively drank unobtainable older vintages, as he did over lunch with Christian Moueix at Pétrus. If he dined in a restaurant with any château owner or grower, he drank the wines they brought, even if he paid for the meal. Like all wine journalists, Parker depended on wine people to let him taste from the barrel in their cellars—there is no way to pay for that.

Neither was there a way to attend the Hospices de Beaune, an invitation-only charity auction held annually in Burgundy since 1851, unless you were the guest of a *négociant* firm who paid for your auction ticket. When Parker turned up as the guest of Maison Louis Jadot in 1988, Bob Foster, a Prodigy subscriber and criminal prosecutor in the California Department of Justice who spent much of his time sorting out conflicts of interest, raised this on the Bulletin Board as an ethical issue. After all, hadn't Parker—as was customary—hired Jadot to finish and bottle the barrel of wine Parker had bid on and won at the auction, then a few days later tasted Jadot's wines at their offices and given them high scores? Wasn't that a conflict of interest? Parker, however, didn't see it that way. He'd paid for the barrel of wine, for his own plane ticket, and hotel room.

And what about the inevitable gestures of professional friendship? After Parker and Pat adopted their daughter Maia in 1987, Angelo Gaja in Italy and Dominique Lafon in Burgundy, both of whose wines Parker regularly reviewed in glowing terms, each sent him a case of wine from her birth year. Returning them, Parker decided, would be too insulting, so his solution was to donate the dollar value as charitable contributions in their names.

Very early Parker's power began to distort even simple interactions in ways that he probably barely realized. While he often attributed low motives to his journalistic rivals, he frequently took at face value the compliments and offers of friendship of many importers and producers who sought him out. "I call wines the way I see them," he was fond of saying, but to do that he had to taste them, and the only way he might get to try the less available, less known wines was if someone recommended he taste them or introduced them to him in a lineup of samples. Did he realize how many times yet-to-be-discovered wines were reserved for him, knowing he'd be excited to be the first critic to taste them? Possibilities for opportunism and exploitation existed in every wine region. When someone has growing power in an industry, it acts as a magnet. Everyone who has something to gain wants to know him.

One of the Bordelais who was eager to meet Parker was Dr. Alain Raynaud, a handsome bon vivant society doctor whose family had owned a little-heard-of Pomerol château, La Croix-de-Gay, since the fifteenth century. They met through Michel Rolland. Raynaud's wine, even the 1982, did not impress Parker; he called it "a picnic Pomerol," with none of the intense flavor he'd read that it had showed in the 1940s and 1950s.

When Raynaud found he couldn't sell his wine after a Union des Grands Crus tasting in the United States, he lost no time in sending Parker a letter. As he described it to me over an aperitif in his garden, with *négociant* Jeffrey Davies and winemaker Jean-Luc Thunevin, "I wrote, okay, Mr. Parker. After two weeks with my other friends and comparing my wine with the others, I must agree that you were quite right. So let me prove to you that I have understood what you mean, and let me prove to you I can produce something that could be a good wine." After their meeting, as Raynaud tells it, "I started to understand what he liked. I understood that I must reduce my yields and improve the winemaking." From Raynaud's view, it may have seemed like the beginning of a beautiful friendship.

Parker must have been flattered—he told Raynaud that he was the only Bordelais who had ever reacted this way, that proprietors usually took his remarks personally, responding with insulting letters and calls. A few years later when Raynaud asked him to be his new baby's godfather, Parker told me he didn't see how he could refuse. But it was a relationship that, in time, would bring trouble.

─────

PARKER'S TRAVELS HAD BECOME A KIND OF ROYAL PROGRESS, AND Parker began to notice a shift in how he was received in wine country, especially Bordeaux. The region is famous for its formal hospitality to those it considers "a potential ally," as British wine writer Stephen Brook observed.

When Parker rolled up to Château Brane-Cantenac for his short tasting stop expecting only the friendly and charming Lucien Lurton, he found the entire family, including many of Lurton's ten children, there to greet him. Laid out on a big table for him were two dozen vintages back to 1961. The family members ranged themselves along the wall of the tasting room and watched, in total silence, as he made his way through wines he considered mediocre at best, and dirty and sloppily made at worst. Fortunately they didn't ask him directly what he thought of the wines. As he drove on to the next stop he thought, "It doesn't get any worse than this."

The welcome at first-growth Château Margaux in 1986 was clearly for visiting wine royalty, as Jancis Robinson described it in a column for the London *Sunday Times*. The large circular table in the immaculate new tasting room was covered with a thick white linen cloth so the wine king could judge the color of the wine. Elegant, perfectly polished long-stemmed crystal glasses stood at each place, accompanied by large, attractive saucers of carefully arranged sawdust for spitting. Displayed alone on the sideboard was Parker's recently published book on Bordeaux. The château director, Paul Pontallier, *maître de chai* Jean Grangerou, and the *régisseur* (manager) Philippe Barré were all in attendance. Since the final blend hadn't been assembled, they poured samples from six different vats and inquired as to Parker's preference. He allowed as to how he favored the last, a "galumphing blockbuster," while Pontallier favored the second-to-last glass, a more aromatic, classically Margaux sample. Grangerou and Barré, Robinson noted, "immediately pointed out that they, too, actually, had always rather liked the last sample."

But to Parker the first indication of a seismic shift occurred in 1987, when, for the first time, the owner of famed Château Lafite-Rothschild, the tall, suave Eric de Rothschild, flew down from his family's bank in Paris to be there when Parker arrived for his forty-five-minute appointment. He greeted Parker with an urbane smile and didn't keep his right hand neatly in his jacket, Napoleon-style, the way he was often photographed, but thrust it forward to give Parker a warm handshake. Now even Lafite permitted Parker to taste barrel samples in March, even though they hadn't yet made the final blend.

Les Oenarques, a tasting club composed of eight of the most important names in Bordeaux, including *négociants* Jean-François Moueix, also co-owner of Château Pétrus, and Pierre Castéja and Jean-Bernard Delmas, the administrator and winemaker of Haut-Brion, now invited Parker to join their intimate tasting dinners once a year. It was a club with a long history, whose members wore matching red-and-green plaid jackets and deep red bow ties. They tested each other's palates by tasting the wines at each course blind; those who guessed the commune and château correctly were rewarded with, Bravo!

In Burgundy and the Rhône, Parker's arrival was not yet considered a command performance except by importers. Busy Rhône producer Emile Champet, out looking after his other farming interests, skipped two rendezvous. But in Alsace, Madame Faller, the owner of Domaine Weinbach, an estate that Parker had put on the American map a few years earlier, laid on a lunch so extraordinary, with every specialty of the region, that Jay Miller told Parker afterward he'd had enough and simply couldn't go on to the next stop. He had to go back to the hotel to recover. Parker left his weak sidekick behind and pressed on to Josmeyer, where a tasting of Alsace wine and cheese awaited him.

———

PARKER'S FIRST APOTHEOSIS AS A WINE CRITIC OCCURRED ON JAN-uary 30, 1993, when François Mitterrand, the president of France, be-stowed on him one of France's top awards, *La Croix de Chevalier de l'Ordre National du Mérite*. Few Americans had received it, and none from America's wine world. The ceremony took place in Libourne, on the Right Bank in Bordeaux. Parker's nomination, he knows, originated there, in Pomerol, though he claims he had no idea who began the process. In his speech Mitterrand pronounced him the "worldwide reference on the wines of France" who had had "a singular impact on the increase in qual-ity of French wine." Considering Parker was an American, that was quite a statement. But after all, he had written four books on French wine in the past few years. Parker gave, in the best French he could muster in front of five hundred people, an impassioned and emotional speech, dedicating the award to his wife. She listened with tears in her eyes.

———

BY THE END OF 1993, THE NUMBERS LOOKED GOOD FOR PARKER. His newsletter, now fifteen years old and thriving, had 29,000 subscribers. He'd tasted approximately 130,000 wines since he began it and had writ-ten seven books. He'd been honored by the French, gained 70 pounds from eating all that rich food in their great restaurants, and was a celebrity in wine circles. Resting on the couch in his living room was a pillow, sent by his literary agent, embroidered with the words "When Parker spits, the world listens."

But it was only a dress rehearsal for what he would accomplish next.

6

PERILS

OF THE

TASTING ROUTE

*J*ACQUES HÉBRARD, THE MANAGER OF CHÂTEAU Cheval Blanc in St.-Emilion, was very, very angry at Robert Parker. In an issue of *The Wine Advocate*, Parker had described barrel samples of the château's 1981 vintage as "disappointing" and "mediocre" and characterized it as no more than average. Hébrard believed Parker was wrong. He got so steamed up about it that one evening he telephoned the hotel where he knew Parker was staying to tell him so and demand that he retaste it. Parker, finished with the day's visits, was sitting quietly in his room, reviewing and amplifying his notes as he usually did on a trip, mulling over what he'd just tasted, and doubtless sipping mineral water. The phone call was an unwelcome surprise. Yes, he understood that Monsieur Hébrard was upset. The following day's schedule was tight, but yes, he could come to the château in the evening after his last appointment. They agreed on a time.

Château Cheval Blanc, ranked as one of Bordeaux's eight greatest wines and one of the two in St.-Emilion rated *premiers grands crus*, uses more Cabernet Franc in its blend (the remainder is Merlot) than any other Bordeaux wine, over 50 percent, giving it a silky texture and distinctive character. Plump, almost sweet, and deceptively easy to drink when young, the wine lacks the powerful tannins that give wines made from primarily Cabernet Sauvignon, like Château Mouton-Rothschild and many other wines in the Médoc, backbone and a dense structure. But even so, in good vintages Cheval Blanc ages brilliantly. Parker in fact had praised the famous 1947 Cheval Blanc as one of the greatest wines he'd ever tasted. Hébrard thought it was probably the greatest wine in the world, but then his wife's family had owned the château for the last 150 years. The 1947 was the vintage that gave the château its modern reputation and convinced English connoisseurs that wines from St.-Emilion could be as great as the aristocratic Médocs.

The following evening Parker pulled up at the modest two-story château with white shutters that sat in a pocket of green, with a tiny chapel and lovely orangery on one side of it and the *chai* on the other. He knocked on the door. Hébrard opened it, and out tore a small dog, a miniature schnauzer, which made straight for Parker's calf and bit down ferociously. Hébrard, a large man with greying hair, a couple of inches taller than Parker, stood impassively by the door, watching the astonished wine critic shaking his leg vigorously to get the dog to let go. It took several attempts, but finally the dog ran off, growling.

A shaken Parker accompanied the silent Hébrard to the château's office and realized that not only was his pant leg ripped, but his leg was bleeding. Instead of getting the bandage Parker requested, Hébrard rummaged around for his copy of *The Wine Advocate*, which he angrily threw on the desk, saying, "This is what you wrote about my wine!"

Parker tried to pacify him by reminding Hébrard that he'd come to retaste it. With a stony face, Hébrard suddenly said, "Well, I'm not going to let you taste it."

By now Parker had had enough. He could feel blood oozing down his leg. He told Hébrard, "You wanted me to retaste your wine. I'm here. If I was wrong, I'll say so." (Hébrard later claimed there was no blood.)

Hébrard left the room. Minutes went by. Just as Parker was about to give up and leave, back came Hébrard, who said gruffly, "Let's go and taste."

In the tasting room, Parker tried the wine twice. Hébrard might be right, he thought. Now that it had been bottled it seemed like a different wine, spicy and plummy, much better than he had first thought. In his final report on the vintage, a year later, he promoted it to one of the top thirteen wines of the vintage, and eventually in his book *Bordeaux*, he blessed it with a score of 90. Nonetheless, Hébrard barred Parker from tasting the 1983 from the barrel.

Eventually their relations became more cordial, but a few years later, when his friend Bob Schindler mentioned he was visiting Cheval Blanc on a coming Bordeaux trip, Parker jokingly asked him, "Kick the dog for me, will you?" When the dog finally died, Hébrard let Parker know.

———

HÉBRARD WAS NOT THE ONLY WINE PRODUCER WHO WAS OUT-raged when Parker criticized his wines with poor scores and sometimes savage tasting notes in his newsletter. As he busily and brilliantly built his reputation in the United States throughout the 1980s and early 1990s, Parker left a number of grumbling proprietors in his wake wherever he traveled. Few attacked him directly, the way Hébrard (or rather, his dog) did—no one wanted to risk Parker's ill will—but privately they complained. The more influence Parker wielded over which wines American merchants and consumers purchased, the more what he said about individual wines and wineries mattered to their business. Which meant money.

To retaliate, some Bordelais, like Hébrard, simply refused to permit him to taste wine from the barrel (although Hébrard relented after a year) and didn't send samples to the *négociants* who set up his tastings. In 1987, Lucien Lurton wouldn't allow Parker to taste at his three châteaux in Margaux; Château Rauzan-Ségla followed suit. But by 1990, denying Parker access to your wines in barrel was a chancy business for the Bordeaux châteaux, who relied on the futures market. His scores had become essential to selling wines as futures and many buyers relied upon them. In fact, in 1990, several "super seconds" had taken the obvious step of not releasing their prices to the *négociants* until after Parker's scores of their wines had been published. Those with high scores knew they could ask much higher prices as a result, and ended up reaping extraordinary financial rewards by waiting. Nonetheless, importer Robert Chadderdon discouraged estates he worked with in France and Italy from inviting Parker to taste. He didn't need Parker's imprimatur to sell their wines in America, he told them. And generally speaking, benign neglect was better than a low score.

One obscure Beaujolais producer sued Parker for *not* writing about his wines. Parker received a letter saying, in essence, Monsieur Parker, we'd like to know why you don't review our wines. Parker had never heard of the producer, sent him a note asking where he could obtain the man's wines in the United States, and subsequently found himself with a lawsuit on his hands. The producer's charge—that he couldn't sell his wines if Parker didn't review them—was thrown out of court. Some producers wrote letters of complaint, and then complained he never replied. Parker tried to be deferential and polite in person, and retasted under pressure, but his standard reply when producers tried to argue with his scores was, "My first duty is to my subscribers. I call it the way I see it." (In truth, he sometimes later upped a score, but for the producer it was too late.)

People who make wine and tend vineyards are an opinionated bunch. They often disagreed with Parker's assessment of their wines and sometimes objected to his ideas about how great wines should be made. In the Rhône, at Domaine de Mont Redon, a heated exchange with Jean Abeille turned into a shouting match. And in the Napa Valley, Randy Dunn, who'd made a name for himself when he was winemaker for Caymus Vineyards and now ran his own eponymous winery on Howell Mountain, told me he'd warned Parker to be careful who he accused of adding acid to their wine. When one Napa winemaker, Robert Sinskey, wrote to Parker denying that his wine had been acidified as Parker claimed, Parker didn't correct the record—and Sinskey's wines were given lower scores the next time they were reviewed. I heard dozens of similar stories.

In his newsletter, Parker took every opportunity to let his readers know when furious producers wouldn't let him taste or even banned him from their estates. It underscored his reputation for honesty in bringing them the unvarnished truth about a wine's quality.

The French, in particular, were not used to this kind of wine writer, one who had so much influence. It had been a long time since the powerful Raymond Baudoin of the *Revue du Vin de France* had pointed an accusing finger at underperforming producers and struck fear into the hearts of

French châteaux and domaines because of his highly critical view of their wines back in the 1930s. They were in a bind. On the one hand Parker was a French wine enthusiast who had helped them expand their sales in America during the 1980s by virtually creating demand for Bordeaux futures and championing the neglected wines of the Rhône Valley, but on the other, even though the dollar/franc exchange rate was not nearly so favorable as it had been when the 1982s arrived, his scores had begun to affect the Bordeaux châteaux' opening prices in the Place de Bordeaux. That was frightening—and serious—even if they were making more money.

The buzz on Parker turned up in unlikely places, such as ordinary dinner parties in Paris, where guests regularly discussed and denounced this American who seemed dangerously close to controlling their wine market. "America is trying to become the center of the world," they complained in dismay. "But they should leave wine to us." In terms of image, prices, and general wine culture, French wines had basked in high regard for centuries, largely for historic reasons, and the French wanted to keep it that way.

Even some whose wines Parker's palate had anointed were becoming uneasy about his power to move markets, and many in the wine trade, as well as other wine writers, started sounding the themes of criticism that would be leveled against Parker from now on. His take on certain wines and wine regions was called into question, and his scoring system, his tasting methods, and his personal taste preferences all came under attack. Not to mention his certainty that he was right.

In fact, as early as 1987 Parker was becoming a lightning rod for complaints about what was happening in the wine industry, much of which was not caused by him. Ironically, he himself had already begun to worry about "a certain backlash" and confessed to a reporter, "As you're being discovered, they all rave about you. All of a sudden you get to the top, you're influential and they say he's too powerful and the scoring is a gimmick." Mark Squires, his colleague on the Prodigy Wine Board, observed, "Being number one meant, as the Japanese say, getting the most wind."

———

DURING THE 1980S AND EARLY 1990S THE WINE INDUSTRY GREW worldwide, climbing to an $8 billion-a-year industry by 1993. For most of that time France was happy: during the 1980s Bordeaux had had six astonishingly good vintages, Burgundy four, and Champagne five. For areas that were usually lucky to get two to four, this was blissful. But by the end of the decade the value of the dollar had dropped 50 percent, there was a glut of good French wine already in the U.S. and U.K. pipelines, and the number of high-quality California wines was growing, capturing more and more of America's fine wine market. The fine wine business in the United States was fast becoming a public relations and press-driven industry, but many of the French, as well as the rest of Europe, hadn't yet caught on to how to play the game in America. California vintners, who were anxious to make a splash in the world, had a strong grasp of what it took to capture public opinion and create excitement for their wines. They understood bold, aggressive marketing. For that, publicity and visibility were key. Challenging French wines in comparative tastings staged in cities across the country, the way William Hill Winery and Mondavi did, was one way. A good note and high score from Parker could guarantee both with less effort and expense.

Robert Mondavi had taken upon himself the role of senior American wine diplomat and seemed to be everywhere, his familiar hoarse, honking voice preaching the future of California wine and the Napa Valley in particular to whoever would listen. In the 1980s it became a joke that wherever you went in France, Robert Mondavi had either just been there or was arriving next week. When dining at the Dubern or the St. James or Clavel in Bordeaux, visiting American wine journalists learned not to be surprised when his voice, which never seemed to stop for breath, emanated from an adjoining room. He praised the French as his model, reminding them that Americans were working as hard as possible to catch up.

In contrast, public relations in most wine regions of Europe seemed underdeveloped, offering either repetitious puffery ("it's the vintage of the century") or doom-and-gloom reports, such as for the horrific freeze in Bordeaux in 1991 that damaged the vines, and using both to justify high prices. In the case of Bordeaux, it was a matter of owners entertaining a select few lavishly at their châteaux, not reaching new consumers by opening tasting rooms and giving tours of the cellar to the hoi polloi as wineries did in America. This lack of promotion was a far cry from the sophisticated publicity efforts and shrewd marketing strategies at the end of the nineteenth century for Champagne, which succeeded in establishing it as the ultimate luxury drink, "one of the glories of the French nation," as Kolleen Guy chronicles in her book *When Champagne Became French*.

Often Mondavi traveled with Harvey Posert, an industry veteran who was in charge of his public relations. "The French just didn't understand public relations in the way we knew it in the United States," Posert explained. "The world had come to them. They were the pinnacle and believed it. Writers came to worship the gravel. The Bordelais and Burgundians were private and reserved and insular and had relied on *négociants*, shippers, and importers to sell their wines abroad. Why should they go anywhere themselves to promote their wines?" But they had become obsessed with Robert Parker. Wherever Posert and Mondavi went, they heard "Parker was here" or "Parker is coming next month." And the question coming from the lips of many French producers, especially in Bordeaux, was a hand-wringing, "What should we do about Robert Parker?"

Behind that question was the specter of American dominance over the wine industry. America already dominated the globe militarily, economically, and culturally. Now the power and preferences of one American were threatening to hold sway over wine, too.

———

AMONG PARKER'S MOST VOCIFEROUS CRITICS WERE OTHER WINE writers in the United States, France, and especially England, which for decades had been the kingdom of wine writing, the last word on French wine in the English-speaking world. Books by many of them, especially Hugh Johnson, had made an impact in America throughout the 1970s and well into the 1980s. Parker was the Johnny-come-lately as far as many English wine writers were concerned, and while he'd praised their books, he'd also criticized them for their blurry alliances with the wine trade.

Even before the publication of Parker's first book, reigning British wine writers were taking him to task. His name came up over prime steaks and a few bottles of Bordeaux at Sparks Steakhouse in New York in 1985, when my husband and I dined with three of them, Serena Sutcliffe, her husband, David Peppercorn, and Michael Broadbent. Sutcliffe, with her swinging blonde hair held in place with a velvet band, wondered aloud how Parker could presume to pronounce so definitively on French wines when he visited a region only a few times a year. "It's astonishing," she said archly, "that he feels confident enough to do that. David and I pop over there all the time, so we can see how the wines are coming along."

Sutcliffe's strong, square jawline and aristocratic nose gave rise to her being dubbed the Vanessa Redgrave of the wine world. Noted for her fashionable attire and being the second woman to become a Master of Wine, she and Peppercorn imported wines; the grey-haired Broadbent, head of the wine department at Christie's, always had his nose out for a great cellar the house might auction off. A renowned taster, especially of old wines, he had been carefully recording his impressions in little red notebooks of every wine that had passed his lips since September 1952, though his wine descriptions were quite different from Parker's and he favored stars rather than scores. As he talked, Broadbent frequently waved his hand backwards, as if to dismiss most of what he heard, and occasionally let his opinion of American vulgarity surface in print, as in "If I have any criticism at all of the otherwise engagingly warm hospitality of our American friends, it is that their largesse is sometimes a little too large: a dinner party or tasting is planned as though it is the last they or anyone else will ever attend, often with rather too many courses, too many wines."

Sutcliffe, Peppercorn, Broadbent, and others didn't have the same preoccupation with conflicts of interest that Parker did and didn't feel their commercial involvements compromised what they wrote. The newspaper columnist Edmund Penning-Rowsell, who wrote weekly for the *Financial Times*, was also president of the Wine Society, Britain's first mail-order wine club, though he was scrupulous about any overlap. And the affable Hugh Johnson, with his bushy eyebrows and banker's suits, who wrote with rare style and grace on wine, was involved with the Sunday Times Wine Club and was a director of Château Latour. Nonetheless, both Broadbent and Johnson had written glowing blurbs that appeared on the jacket of Parker's book *Bordeaux*. The word was that Johnson had thought Parker's scores were printer's marks, and that if he'd realized what they were he wouldn't have endorsed the book. But the first to officially introduce Parker to the British public in a newspaper interview was one of the new generation of British wine writers, Jancis Robinson, whose story appeared in the *London Sunday Times* in 1986. In it she noted that his impact was beginning to hit Britain—even the auction house Sotheby's now used his notes in their wine catalogs—and referred to his influence as "the

Parker phenomenon," a phrase that would enter British wine conscious-ness.

Parker was the super-taster with a self-educated palate; many of those who wrote about wine in England had undergone rigorous training by comparison, and some took an almost scholarly, historical approach to the subject. The Institute of the Masters of Wine, started in the 1950s to ensure a greater level of knowledge and professionalism among the British trade, gave a comprehensive professional course—including tast-ing exams—that entitled its graduates to use the letters M.W. after their names just as professors used Ph.D. Robinson became the first English journalist outside the trade to pass.

To Americans this hardly mattered. As Matt Kramer argued in *The Wine Spectator*, "One of the compelling attractions of American culture is its very freedom from self-appointed gatekeepers." Still, some said that if there were more degreed professionals in the American wine industry, there might be less reliance on simplistic number ratings for wine prom-ulgated by self-trained experts like Parker.

The British attack began after *Bordeaux*'s publication in England in the fall of 1986 and Parker's subsequent trip to London to promote it, during which he spoke at several public tastings arranged by U.K. wine merchant John Armit. Armit, then a consultant to Corney & Barrow, had put Pomerol on the map in the United Kingdom, and was the founder of a wine investment fund that catered to a group of wealthy private cus-tomers. Though many English writers found Parker—to their surprise—likeable, down-to-earth, a very good taster, and even somewhat modest in person, many of them denounced his scoring system as bunk, his sam-pling methods as inadequate, and his taste preferences as skewed. But worst of all, in their eyes, were his blithe arrogance and unshakable cer-tainty about his judgments in print.

Peppercorn's measured but quite positive review of *Bordeaux* in the En-glish wine magazine *Decanter* zeroed in on those issues. "For the English reader, there is a degree of culture shock in the very title of this first book from his pen, *Bordeaux, The Definitive Guide* . . . this impression is con-tinued in the body of the book by Parker's extremely decisive judgments, which to European ears, may seem over-dogmatic, especially in his assess-ments of the relative merits of individual châteaux and the presentation of his own classification." The use of the word "definitive" rankled. Few English writers were willing to give Parker the last word on the subject, especially not those who, like Peppercorn (an M.W.), had written their own books on Bordeaux. He found the scoring system flawed, unhelpful, and highly misleading.

On my travels through London I met up with Bill Baker, a partner in the firm of Reid Wines, near Bristol, whose large, jolly face and ample girth marked him as much of a hedonist as Parker. His blistering review of *Bordeaux* for an obscure regional newsletter had been much, much more critical, in fact polemical in the extreme, and his views on Parker didn't seem to have changed. As he read the book, he wrote, he became "more and more angry," voicing the litany of complaints that would dog Parker from now on. Baker saw Parker's claims that he was the first person outside the trade to write on Bordeaux and that he knew "no one who tastes these wines as frequently as I do" as not just incredibly boastful and self-aggrandizing, but—considering all those with a long history in the wine trade, like Harry Waugh—also untrue. He slammed Parker's methodology of sampling: Parker tasted too early and didn't "get in the car and visit every serious château in each commune." Of his massive tastings at *négociants*, he wrote, "I only hope he doesn't mix the glasses up." He found Parker's judgments "full of over-assessment, under-assessment, and just plain hype of certain properties which he clearly considers the best in Bordeaux at the moment." Then there was what Baker saw as his obvious taste preferences for hugely fruity, immediately appealing, concentrated wines. He wrote, "It is this power and richness that characterise Parker's 100/100 wines. . . . Another key to his likes and dislikes can be seen in his treatment of wines that are full of fruit but which have the sort of austere classiness that I look for in good mature claret." He worried about the amount of power in Parker's hands, having heard that Americans were so swayed by his ratings that they purchased only wines scored 90 and above. "Should such a thing happen here," he wrote, "I feel I should be forced to go into selling motor cars or chest freezers."

Not all of this criticism was accurate, of course; for example, Parker did regularly visit châteaux in addition to tasting dozens and dozens of samples in *négociant* offices. But Parker's April 1, 1987 letter and five-page point-by-point rebuttal to Baker showed just how thin his skin and sensitivity to criticism could be. After all, Baker's review had only appeared in a periodical with a tiny readership. Who but a handful of Brits had even seen it? Did it really matter enough to merit response? Clearly Parker thought it did. The letter verged, as Parker's responses to criticism sometimes did, on the paranoid and even the vindictive. Parker accused Baker of jealousy—"your untrue letter is that of a small, jealous person." He faulted Baker for failing "to grasp the one reason for my success" which was because he "works harder" and has "been more reliable than anyone else in my field." Parker asserted that he was "simply content to judge a wine by what is in the bottle, paying my own expenses, without favor, without influence all along the way. No British writer who specializes in the subject of Bordeaux can make that statement." Parker trotted out a point-by-point defense of each of his highly rated wines on which Baker's opinion differed and accused Baker of being snide and contemptuous, insisting that Baker's main complaint was that Parker, an upstart and an American, was so widely respected.

Parker's reaction seemed over-the-top to Baker's friend Clive Coates, who published a monthly newsletter, *The Vine*. He reproduced these juicy lines from the letter in "Baker vs. Parker" in his July issue and commented acidly that "instead of having the honesty and humility to accept that he himself might be in error, or at the very least not be 100 percent infallible, Parker accuses Baker of bias, prejudice, vindictiveness and malice." He went on, "Mr. Parker may be treated as God in certain parts of the USA . . . He may even assume that the deference with which he is treated by the *négociants* and proprietors in Bordeaux—or by some of them—is his just due. But could it be fear for the power of *The Wine Advocate*? Is it merely sycophancy?" Finally, he warned: "A little humility is now in order, Mr. Parker . . . You are not infallible and those who disagree with you are neither fools nor knaves."

Clive Coates, M.W., had been a wine merchant for twenty years when, inspired by the success of *The Wine Advocate*, he "retired" and launched *The Vine* in 1985. He sometimes rated with points, too, but used the more familiar 20-point system, and his tasting notes were short, almost clipped, and avoided the strings of exuberant adjectives in *The Wine Advocate*. He didn't taste blind. Coates thought of himself as anti-bullshit, against the "precious elitist approach of yesteryear," but spoke out equally against the

100-point scale and rating wine by numbers in general. Parker had given *The Vine* a rave, writing that it was the one wine publication he looked forward to reading. Coates felt Parker was "pretty good on the wines of the Rhône," but like many other writers often defined himself by the ways he differed from Parker, especially on Burgundy. Entertaining but pompous, with distinctive black eyebrows, a white beard, and a penchant for bow ties, Coates regularly led Burgundy tastings in the United States. Under the usual hotel ballroom chandeliers, behind rows of glasses on white-clothed tables, Coates usually began with an exhortation that Burgundy was "a wine of delicacy and nuance," a fact missed by "those who would judge Burgundy through Cabernet spectacles." Every wine lover in the audience knew that was a not-so-subtle swipe at Robert Parker.

The Letters to the Editor column in *Decanter*, England's foremost wine magazine, which had been founded in 1975, quickly became a highly entertaining forum for both Parker bashers and defenders. In just about every other issue it carried passionate missives about Parker and his scoring system, both for and against, though mostly the latter, and still does. They had headlines like "Parker and his points: world-wide impact" or "Wine and numbers" or "Parker's points" or "The number's up for wine reviews." Occasional replies, like "A Plea for Parker," were more positive; some defended him as "one of the brave few who will actually give their opinion of individual wines, whether they think them good or bad," and praised his palate and honesty, but more typical was one that referred to him as the "100-point pulverizer."

Parker often wrote in to hotly defend himself.

Hugh Johnson had once been the most important and well-known wine writer in America, but by the end of the 1980s Parker was taking over that spot. As far as influence among the American wine trade went, the British wine writers were losing market share to Parker and *The Wine Spectator*. A band of younger British writers entered the arena—Robinson became a columnist for *The Wine Spectator* for a short while in addition to her outlets in England—but though they influenced English consumers, they, like Clive Coates, didn't have the clout among the French, the power to move markets the way Parker did. On the world stage of wine writing, the English influence seemed to be declining at the same time the California wine industry was exploding.

ON BOTH SIDES OF THE ATLANTIC THE MOST HEATED CRITICISM was heaped on Parker's 100-point scoring system, which the wine world was rapidly embracing. And that focused more attention squarely on Robert Parker as the person who first used it, even though other American publications, including *The Wine Spectator*, had adopted it too. Of course, there is nothing better than a noisy polarizing controversy to spread someone's name and add to his reputation.

Robert Finigan called his scoring system a "gimmick." Hugh Johnson disliked all numerical ratings because they gave "the impression that all good wine is a competition or a race. . . . Each good wine is trying to express something different. And that is not the same thing as competing." Tim Atkin, then editor of *Harpers Wine & Spirit Weekly* in England, felt scoring was "treating wine like a basketball game where there has to be a winner." It was reductionistic, making wines nothing more than winners and also-rans. Wasn't a wine more than that? What about the human dimension, the background story, the cultural context? Sooner or later everybody trotted out the aesthetic argument: imagine going to an art museum and assigning points to this Picasso and that Matisse. How useful would that be, and how accurate? In brief, wine had been enjoyed for thousands of years without any numbers attached. Why bother?

Those objections, which started in the mid-1980s, fueled an ongoing debate among writers, wine lovers, importers, retailers, wine producers, and marketing and public relations executives that approached the obsessive and continues today. You heard them at dinner parties, from professionals leaning into microphones on panels at industry events, in cellars in every wine region in the world, and read opinions ad nauseum in wine publications and later on wine bulletin boards on the Internet. And basically they boiled down to these:

Tasting wine was a subjective experience. Wines could seem different to the same person under different conditions and at different times. The order in which they were tasted, the types of glasses used, the temperature of the wines, the mood of the taster—all affected perception of quality, and wines themselves changed and evolved even after they were in the bottle. No matter how objective you were, a score was only a snapshot, and a blurry one at best. Though Parker's methods aimed at precision, and he seemed to be remarkably consistent, he wasn't a specially calibrated machine into which one could pour an ounce of wine, wait a few seconds, then take a digital readout that would give the same score time after time. There was, as Michael Broadbent pointed out to me, no such thing as a "94 wine."

Another problem was the fact that the 100-point scale was misleading. You could measure chemical aspects of wine—level of alcohol and acidity, for example—but not quality. People who tasted professionally knew it just wasn't possible to make absolute judgments with scientific precision. Statistically, a 92- and 93-point wine were just not that different; the notion that that one was better was absurd. The whole game was pseudoscientific, part of what Theodore Roszak called in his book *The Making of a Counter Culture* "the myth of objective consciousness." The problem was that the number became an abbreviation, carved in stone and therefore "absolute" for all time.

Then there was the question of palate preferences. Within any group of similar wines, one professional might give the edge to big, fruity wines that packed a wallop; another might give higher scores to the lighter, subtler, more elegant ones. When Parker slammed the 1988 Chateau Potelle Chardonnay with a 67 and *The Wine Spectator* praised it with an 88, the California winery's owners printed both scores in an ad suggesting wine lovers make their own decision. In short, there was no universal set of objective criteria.

Scores were limited. They didn't tell you if the wine was delicious, great with a particular dish, or a good value for the price. Whole categories of wines never received a score above 90. But shouldn't a perfect Beaujolais get 100 points too? Wasn't perfection, well, perfection?

And last but hardly least, ratings distorted prices, which of course is what vintners with high scores were hoping for. The point chasers, gripped by 90+ fever, stalked the shelves for high-scoring wines, and with skyrocketing demand, prices followed suit. Parker, who frequently castigated producers for high prices, was contributing to the very thing he railed against.

Parker has made the case for the defense to nearly every interviewer who questioned the 100-point scale: Wine was no different from other consumer products, and the scores were merely a shorthand accompanying his extensive tasting notes. As for preferring certain categories of wines over others, he loved wines from Alsace made without oak and, while he didn't like vegetal character or high acidity in wines, he was a big fan of elegant Bordeaux like Haut-Brion that had both fruit and finesse.

But in any case, was the power of his scores Parker's fault? Even many of his critics said no, blaming the wine marketers and consumers determined to fill their cellars with nothing less than the magical 90-point wines. Except in certain quarters, that phenomenon flourished far more in the United States, the British noted smugly. They would probably have agreed with another Maryland sage, satirist H. L. Mencken, who said that Americans generally seek answers that are quick, simple, and usually wrong.

In America, wine shops were coming to both love and hate the ratings system. Yes, the wines flew out the door when Parker bestowed them with a 90, but 70- and 80-point wines were increasingly difficult to sell. Wine-knowledgeable retailers and importers started to feel that their own expertise was going begging; their customers didn't want to know what they thought about a wine, only what rating Parker had given it. Bordeaux importers ceased sending out their staff notes from the *en primeur* tastings to retailers, who were now only interested in what scores Parker had given them.

By 1993, the 100-point scoring system had become one of the most contentious issues in the wine industry. A Wine America trade show that year in New York devoted two seminars to the topic. And the debate was far from over.

Parker's practice of massive tastings of barrel samples came under attack from no less than the great professor Émile Peynaud, who was by now officially retired but still consulted at Bordeaux's greatest châteaux. "In order to have scoops," he said in a 1987 interview in the journal *Amateur de Bordeaux*, "journalists classify a wine from a vintage far too early." It was, as British journalist Joanna Simon noted in *The Sunday Times* in London, "a brave stand, but it is by no means certain that it is sufficient to kill off the Parker phenomenon."

Many winemakers, such as California's Richard Arrowood, the star of Chateau St. Jean who went on to found his own eponymous winery, readily admitted it was hard to be accurate about barrel samples. Wines often changed radically while they were in barrel, so while it was possible to get a rough idea of the vintage, it was difficult to make a definitive judgment on an individual wine's character.

That was exactly what the British had been saying all along, but it didn't seem to make any difference.

———

WHEN ATTACKED, PARKER OFTEN OVERREACTED. HE JUST COULDN'T let the criticism go; he had to try to correct the record, insist he was right. He seemed to have an almost paranoid concern about his image—but then, his reputation was intimately tied to his ability to taste and judge wines by what was in the bottle. It's easy to understand his frustration with much of the criticism. After all, he was no megalomaniac who had dreamed up the 100-point scale as part of a plot to take over the world of wine. Although he had expected to be successful, and ambitiously promoted himself, he had never sought the kind of power he now held and felt there wasn't much he could do to temper it. His intentions, he felt, were good. He was gratified that many acknowledged he was a catalyst for change and improvement. He reminded himself that wherever he went he exhorted wine producers to do better and praised them when they did, and that his integrity was as unimpeachable as anyone's could be. He believed he was a decent person who tried to do the right thing and who gave time, money, and wine to charity. Hadn't he raised $40,000 at the Seattle Children's Hospital auction by donating a dinner at the Milton Inn with himself and Pat? He was sure he worked harder than everyone else in the field, yet he still gave generous reviews (and jacket blurbs) to books written by his competitors. He was convinced he was a consistent and talented taster and that he deserved his success. Parker was grateful for his new life and fortune. In every December issue he thanked his subscribers profusely. What was the problem?

But in his reaction to criticism, Parker showed a blind spot. Frequently, as he did with Baker, he questioned the motives of his critics, voicing his conviction that they must be envious of his success or have "an agenda."

Even when Parker didn't reply, the slightest questioning of anything he did or wrote was interpreted by many of his loyal followers as evidence of bias and jealousy. When Harvey Steiman of *The Wine Spectator* reviewed Parker's book on Burgundy, for example, pointing out its difficult-to-follow organization and suggesting that some of the tasting notes weren't up to date, Howard Kaplan of Executive Wine Seminars wrote a letter claiming that "Steiman, like all other Parker bashers within the wine industry, is motivated by jealousy, pure and simple. They simply will not admit that Parker got to where he is today with talent, dedication and hard work. Yet Parker bashers ignore his immense contributions to the wine industry, choosing instead to discredit him." Never mind that in the very same issue Marvin Overton, who had doubtless drunk more great Burgundy than most wine writers, chimed in to agree with Steiman's assessment. The war between the Parker bashers and the Parkerites, later dubbed "The Human Shield" and sometimes "Parker sheep," had only just begun.

———

PARKER'S HONESTY AND INTEGRITY WERE WIDELY ACKNOWLEDGED, even by those who disagreed with both his rating system and his tasting judgments. Still, his continued insistence that he was one of the few wine writers, if not the only one, who scrupulously avoided conflicts of interest was almost guaranteed to result in some kind of backlash. When he invested in a vineyard, other critics leapt on it.

In 1986, while on vacation in Oregon, Parker's brother-in-law Michael Etzel discovered an 88-acre pig and dairy farm that had once been a vineyard listed among FHA foreclosure sales. Its location, in Chelahem Valley near Newberg, was ideal. Soon he was on the phone to Parker. Would he loan Etzel the money to purchase it?

Initially Pat was dubious. Mike Etzel thought his sister was probably worried there would be family squabbles and she would end up in the middle, as peacemaker, with her brother on one side and her husband on the other. But Pat was also concerned about whether this would seem to the outside world like a conflict of interest, and cause problems for her husband in the future. In the end, after much discussion between them, Parker said yes to the loan, if the price was right. Two days later Etzel called. He'd purchased the farm for $120,000; no one else turned up at the sale. Etzel moved his family from Colorado to Oregon, intent on making great Pinot Noir. They named the winery-to-be Beaux Frères, French for brothers-in-law. Etzel planted a 16-acre vineyard (later expanded to 34 acres) in 1988, with the rows of vines tightly spaced as was common in Burgundy, and for the next few years apprenticed himself to Dick Ponzi of nearby Ponzi Vineyards to learn how to make wine.

The vineyards and winery rapidly turned into a black hole for funds, so Etzel and Parker sought and took on a third financial partner, a Quebecois businessman, Robert Roy, in 1991, the same year Etzel produced their first wines—seventy-five cases. But their first real vintage was 1992—a grand total of one thousand cases made in Ponzi's cellars with his mentor looking over his shoulder. He sent unfinished samples to Parker regularly for his expert evaluation and advice, and Parker flew out to participate in the final blending. In keeping with his wine philosophy, yields were low, grapes harvested at optimum ripeness, and the wine handled as little as possible before bottling, which meant no fining or filtration to clarify it.

As wine ages in the barrel, the dregs—dead yeast cells, grape seeds, fragments of skin, bacteria, and more—eventually settle at the bottom. Wineries often hastened the process by stirring in a fining agent such as egg whites that absorbed and drew these elements to the bottom. And many filtered the wine before bottling, a practice that was becoming controversial when it came to fine wine. Parker, following Kermit Lynch's lead, was convinced that filtration especially was detrimental to the highest-quality red wines.

Etzel and Parker aimed for a style with intense fruit, a rich, supple texture, and that haunting, seductive Pinot fragrance characteristic of the best red Burgundies. But for Parker the results would prove frustrating. The wines were good, he felt, but they just didn't have the complexity and greatness of Burgundy. A few critics, as well as Oregon pioneers whose bottlings stressed delicacy instead of richness, faulted early bottlings of the wine for being too oaky—but many praised it.

From the first, Parker informed his readers that he wouldn't write or speak about or rate Beaux Frères's wines, and he even refused to comment on Ponzi's twentieth-anniversary cuvée of 1990 Pinot Noir because half the grapes used to make it came from the Beaux Frères vineyard. Yet as soon as the first wine was released at the beginning of 1994, at the then high price of $34 a bottle, more questions and criticism surfaced. Maybe

Parker wouldn't discuss or rate Beaux Frères, but everyone in the wine world knew that he had a financial interest in the wine. Wouldn't his reputation and power ensure sales and positive notes? Then there was the fact that Parker served Beaux Frères to importers who came to the Milton Inn and put them on the spot by soliciting their opinions. If they didn't like it, could they afford to tell him what they really thought?

Although Etzel planned to sell five hundred cases of Beaux Frères exclusively through a mailing list, the remaining five hundred would be distributed by importer Bobby Kacher, a friend of Parker's. One retailer suggested to me that this was a gift—Kacher could make a lot of money from the Beaux Frères without having to make much effort. Besides, this gave Parker a powerful tie to someone in the trade, whose wines he reviewed regularly. Wasn't that a conflict of interest?

Dan Berger, then wine writer for the *Los Angeles Times*, pointed out another: "If you own a winery and agree not to write about your own wines, can you be 100 percent objective when you're writing about your competition?" *The Wine Spectator* jumped in to report on the Beaux Frères winery, and Marvin Shanken told the *Baltimore Sun* in 1993, "It's silly and pompous to be critical of other people when [Parker] himself is violating a principal ethic, which is conflict of interest . . . He ought to decide what he wants to be: a wine writer or winemaker." Still, *The Wine Spectator* judged the wine on its own merits, giving the 1992 a 91 on its release.

Malcolm Gluck, who wrote a witty wine column, "Superplonk," for the *Guardian*, in defending his fellow Brits from Parker's accusations of their conflicts of interest, asked, "Can a man who has received from President Mitterrand the gift of the *La Croix de Chevalier de l'Ordre National de Mérite* write impartially about French wine? Could the part-owner of a vineyard in Oregon be considered a paragon of disinterested virtue when it comes to considering wines from this region?" This was just what Pat had feared.

———

ONE INDIVIDUAL IN THE WINE BUSINESS WHO WAS UNHAPPY WITH Parker went far beyond complaining and accusing. In 1990, Parker was preparing to hit the road for a book tour to promote *Burgundy: A Comprehensive Guide to the Producers, Appellations and Wines*, with a schedule that included stops and signings at important wine shops across the country. But one morning in early October, as his secretary, Joan Passman, listened to the messages on the answering machine upon her arrival at the office, she heard a disturbing threat. Shaken, she found Parker and said, "I think you should listen to this." The voice on the tape warned Parker "you will be killed." The message had been left sometime after midnight, setting a pattern for the nine threats that would follow; all would be left between 11 P.M. and 2 A.M.

Parker immediately phoned the local police and the FBI, who started tapping his phone. He'd already done some signings locally, and Simon & Schuster agreed to provide protection at future events. Parker continued with the tour, but the calls persisted. What was most worrisome was that the caller seemed familiar with Parker's schedule. Simon & Schuster had placed ads in selected wine publications to let the faithful know places and times. "If you show up at Draper & Esquin on Saturday, you will be killed," he intoned in one message. Draper & Esquin, in San Francisco, was one of the shops on the West Coast leg of Parker's tour, which also included other shops in San Francisco, in Menlo Park, and at two locations in and near Los Angeles. At that point his publisher decided they couldn't properly protect Parker. It was too expensive. They advised him to cancel the rest of the tour, and he did.

With the phone tap the FBI was eventually able to determine that the threatening calls were coming from a New York City retailer. Parker informed the local authorities.

" 'You know, you should hire a private investigator, somebody who is a former policeman, and have him go and confront the guy who's leaving these messages,' " Parker told me they advised. " 'That might stop it right there. If you want to prosecute, go forward. But you probably won't have much luck because you don't have any bullet holes or knife wounds.' " Making death threats was considered a felony in both New York and Maryland, but the local police made the point that unless the person said "I will kill you," there wasn't much that could be done. Just warning someone he would be killed or shot wasn't enough.

A worried and angry Parker bought a gun and installed a security system in his house. When packages arrived, Pat was afraid to open them for fear they might contain a bomb. Parker hired a former lieutenant in the

New York City Police force as a private investigator, who soon discovered that the retailer was snorting cocaine at night after he closed the store and making similar threatening phone calls to other people. He had apparently threatened to embarrass another wine writer by outing him as a "fag," even though the writer in question wasn't a homosexual, and in fact was known as something of a ladies' man. After the ex-cop confronted the merchant at the store, telling him, "We have the tapes. You are in serious trouble," the calls stopped, just as the police had predicted.

Parker, seeking justice, went up to the mid-Manhattan precinct with his private investigator. Passing police and people in handcuffs, they headed straight to the back of the station, where they handed over copies of the tapes to the appropriate unit and told them the retailer's name. Parker wanted him prosecuted and spent an hour filling out the requisite papers. A month later he received a letter from the New York authorities saying they declined to pursue the case, so a determined Parker took it to the Maryland authorities. But they, too, refused to prosecute. The problem was semantic: the caller didn't say "I will kill you," only that "You will be killed."

Parker still gets worked up about it.

———

THE CRITICISM AND PRIVATE GRUMBLING SURFACED MOSTLY IN newspaper and magazine articles, less often to Parker's face, and in letters to *The Wine Advocate*, which his mother handled for years. Once, he says, he asked her, "Doesn't anyone send me anything but complaints?" She replied, "Those are the only letters I let you see," a story she confirmed when I met her in his office, a stack of envelopes in her hand.

The unrelenting criticism was the unavoidable counterweight to the growing adulation, making Parker's home even more a psychic refuge to the untrustworthy wine world. It was a place of retreat and rest from scrutiny, where he could taste in private and live like an ordinary person. Pat, though she no longer traveled with him, was (and still is) his most important sounding board and confidante when it came to *The Wine Advocate* and his experiences and role in the wine world, as well as chief sympathizer when he was attacked. Parker relied heavily on her advice. The negative things people said about her husband often made her furious. Sometimes she would ask him, "Do you want me to beat them up?" But what galled her most was when anyone peddled what she knew were lies about his honesty and integrity. "He makes an absolute point of not accepting anything," she insisted fiercely, "and if someone says otherwise I would like to wring his neck."

Both could understand a proprietor objecting to a bad review. "It's like telling a mother her child is ugly or stupid," Parker admits. Pat eventually came to view the resentment of her husband as part of the life of a critic. "If my livelihood were in the hands of somebody, I'd probably resent them too," she told me.

While Parker threw himself into travel and tastings, Pat now immersed herself in taking care of Maia, the house, and her expanding garden. She had always been a passionate gardener, but full-time teaching and traveling had limited how much she could accomplish. Now she plunged in, joining the local garden club and eventually becoming its president as well as a champion daffodil grower. It was not an interest that Parker shared, not the way he enjoyed trips with his tight-knit family, especially to Disney World, which they visited seven times from just before Maia turned four until her eighth birthday.

———

PARKER'S BIGGEST PROBLEMS ON THE TASTING ROUTE AROSE IN Burgundy, where the personalities and structure of the wine business as well as the wines themselves had an entirely different character than in Bordeaux, California, or Italy. Compared with the formality and grandeur of Bordeaux, Burgundy was rustic, a place of independent-minded growers and farmers who tended the tiny vineyards of their family holdings and were often out on their tractors when Parker arrived for an appointment, as well as more worldly *négociants* like Maison Louis Jadot, who made and bottled wines from dozens of different vineyards. Compared to California—where the emphasis was on individual wineries and their star winemakers, where wines were conveniently labeled with the grape variety, and where most people had an international outlook—Burgundy in the 1970s and 1980s seemed wildly complicated, downright confusing, and most *vignerons* highly insular, focused more on their land and family traditions than on marketing. It was a place where voices raised against Parker would turn into a chorus.

What boosted Parker's power in America when it came to Bordeaux was their system of selling futures that consumers couldn't sample. With Burgundy, the situation was different. There were so many names to know: dozens of villages, hundreds of individual vineyards with many different owners, and little-known small growers who now bottled their own wines. Quality could be widely variable, yet poor wines sold at the same incredibly high prices as the good ones. For American consumers anxious to get their money's worth, the region was often referred to as a "minefield." Wine lovers had been burned again and again. What they needed was a knowledgeable guide to sort out the labels to avoid and the hot wines to buy. It seemed a region ripe for Parker's rule, but his influence proved to matter less in Burgundy than in Bordeaux because the wines were made in such minute quantities that many domaines could sell all they made whether Parker liked them or not. What often mattered more to them was his assessment of a vintage as a whole. If he panned it, many Americans wouldn't buy. When he took on Burgundy, Parker found himself enmeshed in a complex web of history, tradition, and regional sensitivities.

The region encompasses five different areas, but the Côte d'Or, comprised of the Côte de Nuits and the Côte de Beaune, is the source of the region's greatest wines, made from two grapes: Pinot Noir for reds; Chardonnay for whites. On this narrow 31-mile-long strip of land, an escarpment of limestone covered with vines and stone walls, lie some of the world's most famous vineyards. Once they'd belonged to a few aristocrats and the monasteries whose wines gained fame in the Middle Ages, but after the French Revolution all the estates were broken up. The Code Napoléon, which abolished primogeniture, hastened fragmentation of property; the resulting patchwork ownership of land and complex web of vineyard leasing arrangements ensure that an enormous variety of individual vineyard plots, winemaking styles, and techniques are crammed into a very small region. A string of twenty-five villages, including Volnay, Nuits-St.-Georges, and Chassagne-Montrachet, give their names to the basic wines; within the villages are 476 individual vineyards classified as *premiers crus* and 32 as *grands crus*. There are more than five hundred appellations. What adds to the confusion is that dozens of growers may each own a few rows of vines in a single great vineyard; the 124-acre *grand cru* Clos de Vougeot, for example, has some eighty owners, about fifty of which make a wine labeled Clos de Vougeot. The finest wines are produced in such tiny quantities that it is easy for demand to outstrip supply.

Up until the 1930s almost all small growers sold their grapes or wine to large *négociants* (among them Latour, Drouhin, Rodet, Faiveley, Bouchard, and Jadot), who cellared and bottled and marketed the wines, though some also owned extensive vineyards themselves. Raymond Baudoin had been the first to urge growers to vinify and bottle their own wines; Frank Schoonmaker and Alexis Lichine scooped them up for the American market, and they were followed by early specialist importers such as Robert Haas, Kermit Lynch, and Robert Chadderdon, as well as Becky Wasserman, an American who had come to live in Burgundy with her artist husband in 1968. She formed a partnership with Christopher Cannan in Bordeaux and became a broker for—and close friend to—many small domaines, selling their wines to importers and distributors in the United States. Parker had praised many of them. At the same time the French themselves began buying direct, as did many English specialist importers.

All this helped accelerate the practice of domaine bottling at the best small estates during the 1980s. Though it was risky for small domaines to bottle their own wine, they made more money this way, and by 1992 about half of all growers, and a high proportion of the very best estates, were bottling their wines. Some growers even became "micro-*négociants*," buying grapes or wine from one or more neighboring small growers and bottling and selling these, too, to specialist importers. Their relationships with importers were intensely personal and emotional, and for the importers, obtaining a few dozen cases often required persistence and delicate negotiation, often over a hearty dinner of boeuf bourguignon and shared bottles of a special barrel. Trust was paramount—could they count on the importer to pay? Many produced so little wine that two or three transactions represented the family's annual income for a year of work.

Burgundy lies at the northern limit for grape growing, which means that every year is a gamble against possible failure—that spring frost will damage the buds, the grapes won't ripen, or hail will wipe out a crop. About six years out of ten the grapes don't fully ripen and the wine must be chaptalized, or dosed with sugar to achieve sufficiently high alcohol. Yet

Robert M. Parker, Jr., and Patricia Etzel, his future wife, at Maxim's restaurant in Paris, 1968.

Parker in Washington, D.C., at the launch of *The Wine Advocate*, August 1978.

Parker at work in his office.

The wall of press articles in Parker's office, 2002.

Parker at home with his beloved dogs Hoover and the late George.

Parker in
the cellar at
Château Rayas,
September 2003.

E. JUNQUENET

Parker with Michel Chapoutier in Hermitage, September 2003.

Hanna Agostini with Parker at her home in Bordeaux in the late 1990s.

Parker and Michel Rolland at VinExpo in New York, October 2002.

Parker tasting at Shafer Vineyards, Napa Valley, California, with winemaker Elias Fernandez, May 2002.

Cartoon from *Harpers Wine and Spirit Weekly*, April 2, 2002.

Château Soutard's pun-filled poster applying Parker scores and tasting notes to screws of various sizes. Translation: "For your celebrations at the end of the year 2001. Parker Screws. Selection of hand-picked samples (a vertical presentation)."

Château Bouscaut's advertisement after Parker had rated their 1999 wine 79-82 on his 100-point scale. Translation: ". . . . a good *terroir* wine? . . . with much *typicité?* . . . without hesitation, a bad score from Parker!"

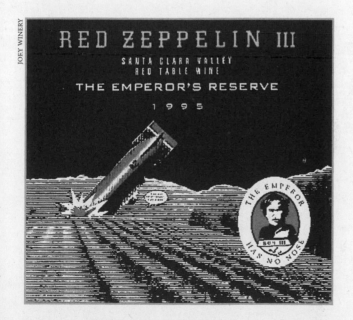

Jory Winery label for The Emperor's Reserve wine, Emperor Bob III, 1995.

Left to right, Maia, Parker, Pat, and Jacques Chirac, President of France, at the Elysée Palace reception after Parker was given the Legion of Honor. June 1999.

One of Parker's favorite pastimes: eating dim sum and tasting wine with friends at Mark's Duck House in Falls Church, Virginia, March 2003.

occasionally the temperature climbs alarmingly. When I followed the Parker trail to Burgundy in 2002, the temperature in the region's main city, Beaune, was 92 degrees Fahrenheit—a heat wave—far different from the rain and even hail I'd experienced on my last visit. Wine quality here can be more variable than almost anywhere else. The thin-skinned Pinot Noir is notoriously difficult to grow; red Burgundy is perhaps the most difficult wine in the world to make.

Yet for all this, Burgundy holds a special appeal for wine lovers. The complex whites are the models for great Chardonnays, and the reds can be the most scented, silky-textured, seductive wines of all, albeit lighter-colored and more delicate than the deep, tannic reds of Bordeaux on which Parker had trained his palate and the rich Rhône wines he had come to love. More than any other wines, they seem to capture the essence of place the French call *terroir*, a complex notion that encompasses geology, soil, microclimate, exposure to light, altitude, and intangibles of the soul and spirit that the Burgundians have elevated to a theology.

By the time Parker started writing about Burgundy, changes that would lead to better-quality wine were under way. A new generation, with professional training in oenology, was beginning to take over their family holdings and looking to science instead of following their fathers' practices in the cellar. Burgundians were beginning to travel, to taste wines from the New World and gain a wider perspective, and most of all were initiating a sweeping viticultural revolution to reverse the widespread use of chemical fertilizers and high-yield grape clones. Kermit Lynch in particular had begun convincing growers not to overfilter their wines and shipped them to America in refrigerated containers, nicknamed reefers, so they wouldn't cook in the heat en route and arrive as damaged goods.

———

ONE OF MY STOPS WAS BECKY WASSERMAN'S FIFTEENTH-CENTURY farmhouse in Bouilland, a tiny village a thirty-minute drive from Beaune. It's the kind of bucolic setting Americans dream of when they fantasize about moving to a wine region: chickens pecking near the barn, a farmhouse with exposed beams and an enormous stone fireplace, shelves sagging under the weight of well-worn books, and a table on a stone terrace outside under fruit trees where you sit to sip some delicious, little-known wine, as I did with Wasserman, her husband, Russell Hone, and her son, Peter Wasserman, who was visiting from New York.

When she first met Parker in the early 1980s, Wasserman, a short, plump, curly-headed expat American, had been living in Burgundy for about fifteen years and knew just about everybody in the region. She started in the wine business by exporting barrels to Americans, then gradually began exporting the wines of small growers, especially to the United States in partnership with an Englishman in Bordeaux, Christopher Cannan. For many pioneering American importers, like Kermit Lynch, Wasserman was the entry point, the go-between they needed to meet the most interesting winemakers.

What first struck Wasserman about Parker was his utter delight and enthusiasm in discovering wines. "You could see it on his face," she said, sucking on a cigarette. She thinks of that time as the age of marketing innocence in Burgundy, before it was "big biz" and producers understood how vital American media approval could be to sales. The important thing was for Americans to learn about Burgundy, she felt. Parker was the new kid on the block, and she happily put together a tasting of wines from her growers for him. In fact, he lauded the "genius of Cannan and Wasserman" in an issue of *The Wine Advocate* in the early 1980s.

In June 1985, Parker turned up at her farmhouse at 9 A.M., sped through about 130 wines, stayed for lunch, tasted fifty afterward, and then went on to tastings at Domaine des Comtes Lafon and Domaine de la Pousse d'Or. Wasserman remembers vividly how all the growers tasted the wines after he left and later compared their notes with his when they were published. To them, his methods were a shock; they were upset by Parker's extraordinary pace, a wine a minute—*and he didn't retaste*. They told her it seemed awfully fast for a judgment that would become a published note. When the reviews appeared, Wasserman and many of her growers agreed that Parker clearly seemed to have definite stylistic preferences, plus a strong bias against filtration. She began to worry that if one of her growers admitted that he filtered "when necessary," it would be interpreted as "bad" and affect Parker's judgment no matter how the wine tasted.

Kermit Lynch had pushed his growers not to filter their wines before bottling because excessive filtration stripped out flavor and color, especially with Pinot Noir. It was an important stance. Many of the best growers fined and filtered as little as possible, but shipping totally unfiltered wines to America seemed chancy. If certain elements in a wine hadn't settled out naturally, they might cause problems in the bottle unless shipped in reefers. Lynch had to educate his dubious American customers that a little sediment in the bottom of a bottle of wine wasn't necessarily a bad thing—it showed the wine was natural.

Wasserman saw a wider implication. "As a term, 'unfiltered' probably had good connotations for Americans," she observed. "Like getting your news 'unfiltered,' it suggests you're getting the complete truth, the real thing." *No filtration* became one of Parker's mantras for great Burgundy; he closely quizzed growers, sometimes even lectured them, and publicized the idea with his readers, regularly mentioning in his newsletter which domaines did and did not filter their wines. Many new American importers, such as Kacher, concurred with Parker and Lynch, and they could also see that unfiltered was the watchword if you wanted a high score. To many winemakers, though, including good ones, light filtration and fining before bottling, done the right way, was necessary and, they believed, undetectable, especially after the wine had been in bottle for several months.

Early along, growers had welcomed Parker—and especially Pat—into their cellars, but now many chafed at submitting samples for him to taste at a central location or in his room at the Novotel outside Beaune. They wanted him to visit, talk, hear their story. And by now some started to resent being told how to make their wines. Wasserman knew for a fact that some growers were beginning to hide their filtering equipment before Parker visited and simply lied.

Gerard Potel, who managed the Domaine de la Pousse d'Or in the village of Volnay, which Wasserman handled, didn't lie. A proud man (he died in 1998) who mentored a generation of Burgundian winemakers, he liked Parker, who had purchased Pousse d'Or wines for his own cellar. But after Potel told him he would filter if he needed to, Parker wrote that he had begun to filter his wine—a misrepresentation of what he had meant—and the domaine lost a lot of their American market as a result. Potel was deeply stung.

On Parker's next visit to the domaine, in 1986, which coincided with that of a German journalist, Potel served them two 1985 wines, one filtered, one unfiltered, just to see what Parker would say. Potel later claimed that Parker not only couldn't identify which was which, he preferred the filtered wine. "My father made a mistake embarrassing Parker in front of another journalist," Potel's son, Nicolas, a *négociant*, now says. Parker, who had called the 1985 one of the best wines of the vintage, never came back; he gave subsequent vintages of the wines much lower scores, and within several years he stopped writing about the wines altogether. "It was a little boys' fight," shrugs Nicolas. "Neither was listening to the other." But it affected Potel until his death. Nicolas Potel believes Parker's insistence on no filtration has resulted in producers adding genetically modified enzymes that don't even come from Burgundy, which he thinks is a much worse solution than light filtration.

Parker claims this tasting never happened. Instead, he says he thought Potel's wines started to decline with the 1986 vintage and looked for reasons. And then he heard that Clive Coates was saying, "Parker never tasted the Pousse d'Or wines he wrote up." Parker sent Potel a letter to ask if he had told Coates that. He never got a response. "What an asshole," Parker thought, and stopped tasting the wines.

But because Parker was important in America, their main market, Wasserman and Hone dutifully continued to put together tastings. In 1988, they pre-tasted the difficult 1986s with producers Dominique Lafon and Daniel Rion, removing in advance the wines they thought Parker would hate. In 1993, the farmhouse was his last stop on a Saturday night and he showed up two and a half hours late. Insulted, the growers had left, and the wines had been opened for hours and didn't show well. "My importers in the United States kept telling me, 'You must show your wines to Bob,'" Wasserman said. "But I kept thinking, Why must I submit everything I do to one person to be judged? Should I show all the wines or just those I know he would like, even if Burgundy is diversity and individuality?"

Parker, like other journalists, visited when it was convenient for him regardless of whether that was the optimum moment for tasting from the barrel. "Drinking wine is a live performance," Wasserman pointed out. The wines could be in the middle of their second, or malolactic, fermentation and taste off. Barometric pressure seemed to affect the taste of wines in barrel. For many of Wasserman's growers, like Michel Lafarge in Volnay, it didn't really matter what Parker said, anyway. His wines sold out.

Parker says he became disenchanted with Wasserman, whom he calls "the diva of Burgundy"; they were philosophically at odds over what they wanted from Burgundy. She wouldn't require her growers to stop filtering altogether, nor all her buyers to use temperature-controlled shipping containers. He says, "I told her point blank: you're the big defender of Burgundy and yet you're not encouraging these people to bottle the essence of their vineyard and you're not taking any responsibility for how it's taken care of once it leaves the cellars. I think that is very irresponsible."

For her part, Wasserman says of Parker, "He believes he is 100 percent right. Saying I don't like it, therefore it is no good; I do like it, therefore it is good, is very American. With great power comes great responsibility."

———

WASSERMAN WASN'T THE ONLY ONE IN BURGUNDY WHO BEGAN TO resist Parker and chafe at his stern opinions. Though Parker singled out dozens of properties for praise and high scores, he wrote frequently that growers and *négociants* were "too greedy and sloppy" and the wines in Burgundy were "perfectly suited for masochists . . . you must suffer a great deal of abuse and pain to find pleasure." Burgundians thought he had it in for the region. In 1992 Parker announced that he would no longer rate the wines he tasted in barrel because "there is little correlation between what many growers present from cask and what ultimately is put in the bottle. This lamentable situation exists nowhere else in France. In Burgundy, growers have a tendency to show only their best lots to visiting wine writers." In one interview he suggested that perhaps the Burgundians were using him to "salt the market." As British Burgundy expert Anthony Hanson saw it, Parker's view seemed to be that the region was the country of wine sins—overproduction, overfiltration, bad treatment of wine, never shipping what was in the cask. Burgundy was guilty before being proven innocent. None of this went down well with growers, *négociants*, importers, and sympathetic writers. They felt it was proof that Parker didn't understand the region, its culture, its wines, the situation of the small grower, the mercurial, delicate nature of Pinot Noir, and the overarching importance of *terroir*. Even people whose wines he praised, like Dominique Lafon of Domaine des Comtes Lafon and Aubert de Villaine of Romanée-Conti, objected to his way of judging wines and his seeming preference for bigness over finesse. Parker, however, called his views and tasting notes "unfiltered."

"He had enormous misconceptions about how the wines aged and he misjudged vintages like 1993," says Hanson, who lived in Burgundy for three years while writing a book on the region and importing wines to the United Kingdom for the firm he once co-owned. "Parker sometimes writes about Burgundy as if he thinks it should be sent to prison," he wrote. "The lawyer's adversarial standpoint is an original one, and it can make for startling diatribes, as he goes for a conviction and heavy sentence, or seeks the acquittal of a plausible felon who has pulled the wool over his eyes in some dingy underground cell."

Still, some producers managed to forge a good relationship with Parker. André Gagey of Maison Louis Jadot was one of the first to welcome him as a young journalist, and Burgundians were convinced that ever after Parker always had a soft spot for the *négociant's* wines. He was one of forty-five guests, including France's top chefs and restaurateurs, invited to the April 6, 1992 celebration for Gagey's retirement after forty-five years in the wine business. Parker flew in on the Concorde (he paid his own way) for the tasting of a dozen wines, almost all from the nineteenth century, held in vaulted cellars beneath the old Couvent des Jacobins, followed by lunch and speeches. And many small growers like Bertrand Ambroise, whose domaine Parker "discovered" in 1987 and continued to praise, were grateful. Ambroise, who met Parker through importer Bobby Kacher and felt a "current of sympathy between us," attributes his success to Parker. Ambroise made 20,000 bottles in 1987; by 2002 he was producing 150,000 bottles a year, most of which went to the United States.

That didn't keep other Burgundians from calling him "The Wine Dictator." "You're not supposed to like anything he doesn't, and you're patently wrong if you do," said one. Parker's skepticism of the mystical importance of *terroir* was tantamount to heresy in Burgundy. He saw it as only one element of making a great wine. "*C'est l'homme qui fait la différence,*" he often said—"It's the man that makes the difference," a sentiment from Parker's lips that others felt smacked of the New World's elevation of the winemaker over the vineyard. "He is . . . imposing on the region a celebrity system, a star is born Hollywood mentality, to which he no doubt believes Americans will easily be able to relate," Hanson wrote. "His approach constantly lionizes certain individuals, making them particularly suitable for a certain type of American consumption. This may be a path suitable for some Americans, but others, and consumers in other countries, may ask themselves if they want Burgundies to be hijacked into an American mould."

Many in Burgundy's younger generation resist that. As Nicolas Potel—in tan shorts, sneakers, and a white T-shirt—led me from barrel to barrel in his *négociant* cellar in Nuits-St.-Georges, he shared his own vision of making wine: "We have to be very careful. The grapes are just a memory, a fingerprint whose variations come from the earth, the wind, the stars. Where the vineyard is located there is force or not that gives that memory to the grapes. With old vines there is more *terroir* definition." To him it was a simple way of thinking, "that there is a connection between everything above and below in the earth." It was a vision that you hear over and over again in Burgundy, and one not easily captured in Parker's 100-point scale.

═══════

L'AFFAIRE FAIVELEY, AS IT CAME TO BE CALLED, WAS A WATERSHED for many Burgundians. Its origin can be traced to two sentences in the third edition of *Parker's Wine Buyer's Guide*, published in fall 1993, at the end of a four-and-a-half-page highly positive review of Faiveley's wines, under the heading Other Thoughts: "On the dark side, reports continue to circulate that Faiveley's wines tasted abroad are less rich than those tasted in the cellars—something I have noticed as well. Ummm . . . !"

These remarks inspired François Faiveley, one of the most beloved and powerful figures in Burgundy, to sue Parker; his publisher, Simon & Schuster; its then chairman, Richard Snyder; and several bookstores in Paris, including Brentano's and W. H. Smith, for libel in February 1994. Faiveley filed the suit in France, under a provision of the Napoleonic Code, claiming that he'd been dishonored because Parker implied he was engaged in fraud.

The Wine Spectator ran a half-page story with a photo of a pensive Faiveley. "Sources close to Faiveley," it read, claimed "Faiveley was upset because he felt the comment implied he provided Parker, and customers tasting in his cellars, with higher quality wines than those that were finally shipped." Faiveley told the *Spectator* that his distributors around the world did not interpret Parker's statement as referring to storage or shipping problems.

Eventually an out-of-court settlement was reached, but in essence, Parker lost. The case and its fallout affected all of Burgundy and, in some ways, still does.

Now, almost a decade later, I sat in François Faiveley's office on a back street in Nuits-St.-Georges to hear the whole story. The headquarters of Maison Joseph Faiveley, founded in 1825 by Pierre Faiveley, is an unaesthetic concrete building landscaped with neat rows of flowers and clipped evergreens that sits above cavernous historic cellars. François Faiveley is the sixth generation to run the company, one of the largest domaines in Burgundy—they own some 275 acres of vineyards divided between Mercurey and the Côtes de Nuits—and also a respected *négociant*. The offices seemed fairly modest for one of the reputedly richest men in France. The contemporary Faiveley fortune derives from Faiveley Transport, a Paris-based company founded in 1919 that invents and sells railway equipment for the TGV in France and around the world. Today, Faiveley S.A. also has a plastics division, offices and factories in some ten countries, and an annual turnover of about 415 million Euros. Faiveley owns most of it, which provides more than enough money for him to indulge one of his great passions, sailing, and to pour his love for wine into making the best.

Faiveley, casually dressed in a blue chambray short-sleeved shirt and dark blue pants, sat behind his desk. There is a bald center in his untamed, wildly curly hair now, but the hair and his horn-rimmed glasses still give him the studious appearance of an intense French intellectual, one of those contemporary theorists who argue politics late into the evening at a great restaurant. In fact, he is a great lover of music, especially Bach, and named two of his sailboats *Glenn I* and *Glenn II*, after Glenn Gould, the great Canadian pianist noted for his interpretations of Bach's *Partitas* and the *Goldberg Variations*.

Parker and Faiveley first met in the mid-1980s, and up through 1993 Faiveley regularly invited the critic to lunch at his home when Parker sped through Burgundy on his tasting trips. His years running the domaine paralleled Parker's years running *The Wine Advocate*; they were only a few years apart in age and were, Faiveley thought, friends. He found Parker "*quelqu'un de sympathique.*" Parker thought Faiveley was "a good guy" and admired his wines, listing him under Best Performances by Wine Personalities/Winemakers of the Year in 1988. In his book on Burgundy, Parker gave Faiveley's firm a five-star rating, his highest. Faiveley, he wrote, is ". . . one of the most impeccably run firms in all of Burgundy, and under the inspired leadership of the young François Faiveley the wines now rival the best made in Burgundy. . . . Since François took over the winemaking in 1978, their quality has soared to the very top."

One night in mid-January 1994, after spending the entire day in Paris at a board of directors' meeting, Faiveley arrived back in Nuits-St.-Georges at 10 P.M. and stopped briefly at his office to check his messages. There was a puzzling personal fax from his distributor in Brazil, Ciro Campos Lilla, who had just returned from a vacation in California with his wife, where he had purchased a copy of the third edition of *Parker's Wine Buyer's Guide*. According to Faiveley the fax said, "I don't know if you have had a chance to read or see it, but it's an absolute disaster because the way he wrote the book he clearly implies that the wines he is given to taste are not the wines you ship to your customers." Faiveley didn't know what to think. He didn't know the man personally, and thought perhaps his English was not very good. So Faiveley called his office the next day and asked him to fax a copy of the page.

When he received it, "I read it. And I reread it," Faiveley said, seeming upset all over again at the recollection. "My father Guy was still alive and I gave a call to him and I said, 'Papa, I'm going to read you something in English. Could you give me your feeling about it?"

"It's difficult to give you my opinion like that," his father replied after hearing the lines. "I'm an old man. Send me a copy. I need to see it printed, and I will give you my feelings about the whole thing."

When his father finally called back, he said, "François, it's very obvious. Parker considers that Domaine Faiveley is run by a bunch of liars. . . . Parker doesn't give you any choice. You've got to do something. *Voilà, c'est tout.*"

Meanwhile, Faiveley also called his American broker, Henry Cavalier, who told him, "I have known for a week or ten days, but I didn't dare call you. I didn't know what to tell you." Two distributors in the United States had already called Cavalier to express concern. Others, including one in

Scotland, called Faiveley directly to inquire outright if the wines they had in stock were the same wines they'd tasted at the Faiveley cellars. Faiveley came to the conclusion that if he didn't do something urgently, the image of his family company would be damaged "forever." During the next month he became extremely depressed and, exhausted, resorted to sleeping pills to help him get through the night.

The previous September, in 1993, Parker and Jay Miller had lunched with Faiveley and his wife, Anne, at their home. "The atmosphere was as friendly as ever," recalled Faiveley. They discussed music, and Faiveley opened a bottle of very old Bordeaux from his personal cellar. He walked them to their car. Parker must have submitted his manuscript to Simon & Schuster long before the lunch, Faiveley observed. "I still say to myself, Why did he accept an invitation to eat *chez moi*, while writing this?" He looked upset at the thought.

To Faiveley, it seemed obvious that "if you eat at someone's table, *en famille*, with his wife, you simply don't write things like that. If you write things like that, you don't eat with them." He viewed Parker's action as "a personal betrayal" of their friendship.

The wine world in Burgundy is small, and word spread quickly. People called Faiveley, offering support, and he consulted friends in the region, including Hubert de Montille, a lawyer in Dijon and proprietor of a well-known small domaine in Volnay, who had acted for him on some legal matters; American broker Becky Wasserman, whom he'd known since 1979; and Dominique Lafon. Within three weeks he'd hired two lawyers in Paris, French-American specialists in defamation cases, and filed an action. A French journalist as well as some others warned him that Parker himself was a lawyer, that the law in America was complicated, that a lawsuit would go on for a year and would cost a colossal fortune, and anyway, it was very difficult to obtain justice against a journalist in the United States. But he believed the honor of the Faiveley name was at stake, and pushed ahead.

The lawsuit seemed to draw in all of Burgundy and provoked deep, deep conflicts. At meetings of growers and *négociants*, individuals heatedly weighed in with their opinions. It was like a referendum on Parker, whom many considered too powerful. Now was the time to "clip his wings." "If you lose," some warned, "All Burgundy will lose." But, Faiveley said, "I told them Charles de Gaulle identified himself with France, but I'm not Burgundy. I'm doing this only for my name. . . . if I go to court and lose the case, François Faiveley only will lose."

―――――

"IF IT HAD BEEN ANYONE BUT FAIVELEY . . . ," SIGHED PIERRE Meurgey of *négociant* house Maison Champy, recalling the situation over lunch. "Everyone had great respect for him; people supported him. He was one of the few *négociants* to choose quality as the key without compromise. It was as if Burgundy itself had been criticized not on quality, but on honesty. It wasn't theater for Faiveley. He was authentically hurt." Faiveley, Wasserman told me, is the closest thing Burgundy has to a saint.

Within this tight wine world, there were two reactions.

Many producers who were Faiveley's friends, such as de Montille, felt almost as insulted as Faiveley and simply announced they would no longer receive Parker. Becky Wasserman felt nauseated. Her growers were shocked; everyone talked about "that terrible *Ummm!*" Growers phoned to say "François, we like you a lot." At dinners and tastings people sidled up to Faiveley and said, "I'm on your side." Small domaines declared that they would not let Parker taste their wines, either.

Others, including many *négociants*, took a cooler view and tried to get Faiveley to reconsider, says Meurgey, because they were afraid of the consequences. They believed Parker would consider "Faiveley's reaction a regional reaction."

At one meeting, Louis-Fabrice Latour, the head of large *négociant* Maison Latour, voiced his view that Faiveley should drop the suit because "it is not worth it to have an enemy of Burgundy." Latour felt Faiveley was acting childish ("someone dares criticize me!"). And he pointed out to me that "the way Maison Latour was treated in the early 1980s was ten times worse than what Parker wrote about Faiveley. Parker was a good promoter of Burgundy in the U.S. There was no reason to have an argument." Latour was one of the few who held this position and, he said, "not because I'm in love with Parker." But in 1993 and 1994, business was tough. He felt they should look at Parker from a strategic point of view. Many of the growers criticizing Parker, he said, didn't even sell their wines in America, but still "they were saying how can *The Wine Advocate* be a good newsletter, if we won't see him and let him taste." They believed that if they cut Parker off, they could punish him; he could not offer his subscribers comprehensive coverage of Burgundy, and his newsletter would suffer.

Bertrand Devillard, head of the *négociant* house Antonin Rodet and at the time president of the Syndicat des Négociants, had heard that Parker was under the impression that the whole of Burgundy was fighting him and that he was the victim of an organized reaction on the part of the Syndicat. He telephoned Parker to promise him that the Syndicat was holding no official discussions on the case. But he says he also told Parker that the accusation against Faiveley was "too rude to put freely on the table" unless he provided evidence. "Parker did not give the same significance to what he wrote that Faiveley and the Burgundians did. He said he didn't mean to say that Faiveley was cheating. But that is the way it was perceived. It was his responsibility to put it in the right words . . . it was only three lines in three pages, but it was *le coup pied de l'ane*—the donkey's kick."

Aubert de Villaine, of the Domaine de la Romanée-Conti, used his influence to personally persuade Faiveley to settle out of court.

———

FOR HIS PART, PARKER TOLD ME AFTER A LUNCH AT HIS HOME featuring his famous crab cakes, "I was blindsided by the lawsuit." The first he knew of it was when he was served with papers at his home in February. As a lawyer he knew firsthand what happened once lawyers became entrenched with a case, so he tried to phone Faiveley to head that off. But he says Faiveley wouldn't take his calls.

Parker had always given Faiveley's wines great reviews, but says that for three years he'd been telling him that when he'd purchased the same wines in the United States they often seemed to be totally different from those he'd tasted in Faiveley's cellars. Were the wines damaged in transit? Parker suspected the problem was the way the wines were handled by Faiveley's American importer, who, Parker says, didn't seem to care about how they were shipped and stored. The largest wine distributor in Texas, who handled a huge number of Faiveley's wines, had served some that tasted terrible at a lunch the distributor threw for Parker and thirty-five fans during his 1990 book tour. A dismayed Parker thought it would be too awkward to say anything at the lunch, but when the distributor showed him the wine inventory afterward, it was clear to Parker what the problem was. The wines were housed in a two-story-high warehouse that seemed the size of three football fields, with a corrugated tin roof and no air conditioning. Worse, all the great wines, including Faiveley's, were stored at the top! Parker thought, if the temperature down here is 90 degrees, what could it be up there? The distributor kept them up there to prevent theft— it took a forklift to get the wines down.

"You should look into this," Parker had told Faiveley on more than one occasion, but it seemed to Parker that he did not want to hear it.

Still, since Parker had received complaints about Faiveley's wines from his readers, he decided to say something in his upcoming *Wine Buyer's Guide*. Pat didn't proofread it, as she did the newsletter. She said she would never have let those sentences pass, particularly not that "Ummm!" But no one at the publisher saw a problem. Parker had said many worse things. When the suit was filed Parker thought, well, okay, the lines do raise a couple of questions, but to him it was a classic overreaction.

So Parker consulted with his lawyer in Baltimore as well as a few people he knew well in Bordeaux. Following their recommendations, he picked the highly respected Eric Agostini, who was on the faculty of the University of Bordeaux, to represent him. Agostini's wife, Hanna, also a lawyer, spoke fluent English and translated the French legalese that Parker didn't understand.

At Simon & Schuster, then-chairman Dick Snyder was not a happy camper. Parker heard that Snyder was asking angrily, "Who is this guy Parker?" The lawsuit dragged on and on. To Parker it seemed endless. One of Faiveley's demands was that the publisher stop distributing the offending books, and since that would be costly Parker believed Simon & Schuster were dragging their feet, wanting to postpone a final settlement for as long as possible.

Both Faiveley and Parker gathered testimonials from importers, distributors, and others in the wine world, and Parker produced everything he had ever written about Faiveley.

From Burgundy, Parker received neither hate mail nor words of encouragement. He was surprised, disappointed, and a little hurt, too, that after *The Wine Spectator*'s report on the lawsuit appeared, not one wine journalist came to his defense. He knew he didn't have many friends among his wine-writing peers after years of bashing them in print, but still, he thought, why aren't they standing up for freedom of speech?

———

EVENTUALLY PARKER'S LAWYER, AS WELL AS THE LAWYERS FOR Simon & Schuster, agreed to settle the matter. If he hadn't had so much affection for Parker previously, Faiveley says, he would have taken the action much further.

"Robert Parker has agreed to what I asked," Faiveley told *The Wine Spectator* in October 1994, after the lawsuit had finally been settled out of court. As stipulated by the agreement, the *Spectator* reported, Parker and Simon & Schuster would eliminate the controversial lines from future editions of the book and would cease distribution of the current edition after the date the settlement had been signed. In the August issue of *The Wine Advocate*, Parker published a notice, in language hammered out by the lawyers, that stated he did not think Faiveley was cheating his customers. Despite widespread reports that Faiveley had demanded one million dollars in damages, he requested, he told *The Spectator*, only a symbolic payment from Parker: one franc. For Faiveley, the suit wasn't about money, but for old-fashioned concepts—honor and truth.

When Faiveley's close Burgundian friends gathered at one of their regular casual tastings and dinners, they were happy for him. When the subject of Parker came up, each one said, "Well, if Faiveley is not going to welcome Parker, neither will I."

In my interview with Faiveley, he said he had always refused to discuss Parker or the case with journalists before now because he didn't want to appear vengeful. To this day he holds on to his thick dossier of the case, which includes every note, letter, and document, as well as both copies of Parker's *Wine Buyer's Guide*—the one with the offending lines and a corrected copy—which he pulled from his shelves to show me. He continues to ask himself why Parker wrote those lines without talking to him first.

It was, at bottom, a culture clash.

———

AFTER THE SUIT WAS SETTLED, A HURT AND ANGRY PARKER HAULED cases of Burgundy up from his personal cellars, stacked them in his office, and told a friend that even though he loved drinking Burgundies, he planned to sell them off. As one of Parker's American friends told me, once Parker is your friend, he's a friend for life, but if you stab him in the back, he won't forget. Parker still believes lawyers blew what he wrote out of all proportion and that Faiveley overreacted, but he concedes, "I should have spelled it out. It should have been written in a different way." One of the ironies was that although the lawsuit ended up costing Parker more than $100,000 in legal fees, the publicity caused the book to sell like crazy in France. Essentially the extra royalties paid off most of the legal bills, and the exposure probably boosted the sales of all his subsequent books published in France. Unexpected support came from countries like Switzerland and Belgium, where the view seemed to be "Oh, you got sued; that must mean you're telling the truth."

Up to that point Parker had no liability insurance for lawsuits. Now he does.

AFTERWARD PARKER DID RETURN TO BURGUNDY TO TASTE A COUple of times, but although everything seemed fine on the surface, he could sense an underlying residual hostility. People tended to globalize the situation, Wasserman thinks. "No one could rest easy. If Parker can do this to Faiveley, people thought, if he can so misunderstand, then what could he do to me? It taught everyone the fragility of the system."

In turn, many in Burgundy felt Parker, too, remained angry at the region. When the lawsuit was finally resolved, says Devillard, "He intended to have Burgundians pay for that. We have been punished. Now, with years behind him he probably realizes that Burgundy didn't deserve the treatment." He feels that Parker should have been more "*grand seigneur*" afterward, that perhaps he was behaving "like a disappointed lover."

Parker simply added fuel to the fire in Burgundy when he wrote "A tongue-in-cheek guide to surfing your way through the Burgundy minefield" for his newsletter, which sounded more "sour grapes" than humorous. Ten "Côte d'Or lingo" sentences were listed, with Parker's interpretation of what they really meant. According to Parker, the subtext of "*C'est un vrai vin de garde*" (literally: This is a classic wine for long aging), as used by Burgundians and many American merchants of Burgundy, is "The wine is excessively tannic, and it will undoubtedly lose most of its fruit long before the tannin melts away." The eighth sentence, "*Monsieur Parker n'en sait rien.*" (Mr. Parker knows nothing), he claimed meant "We cannot influence him, nor can we bribe him."

In Parker's 1990 book, *Burgundy*, Faiveley had received a five-star rating; in the third edition of *Wine Buyer's Guide*, in 1993, it was demoted to a four-star domaine, though the average scores were the same; in the fourth edition, published in 1995, after the lawsuit had been settled, Parker demoted the domaine to three stars and wrote that he had now come to agree with two other writers, that Faiveley's wines were "tough" and "too earthy and coarse," after several years of bottle age. However, almost all the scores for the wines were as high (in some cases higher) for the 1991 vintage as for the wines in the previous edition.

Parker regrets that he and Faiveley, whom he liked personally, came to blows. He was always hoping, he told me wistfully, that on one of his many flights to or from Paris they would run into one another and somehow patch things up. Though he acknowledges that he should have worded what he wrote differently, he's still bitter about the fallout. "If the Burgundians thought they would break my power," he said, with bravado verging on arrogance, "they haven't succeeded at all. It increased people's awareness of what I was doing." He remains convinced "my work there was good. My interests for Burgundy were far more beneficial for their long-term interest as a region."

After he hired Pierre-Antoine Rovani to cover Burgundy for his newsletter a few years later, he tried to let bygones be bygones and arrange for him to taste at Faiveley, but Faiveley refused. That, says Parker, was "childish, stupid, and not very professional."

———

AUBERT DE VILLAINE, WHO MANAGES DOMAINE DE LA ROMANÉE-Conti, told me he once refused to be interviewed for an article on Parker because no one else in Burgundy had been willing to say what they really thought about him. But now he was willing to do so. In his whitewashed living room in Bouzeron, seated in front of the massive fireplace, he said, "The USA has so much power. The way Americans influence the wine world is through Parker. He has been extraordinary for wine, an incredible factor for expansion of wine in the world. But there is a bad side to his authority. His success is based on the fact that he says 'this is good, this is bad.' He offers a clear judgment, but when you give judgments that clear, you can make mistakes, like the family patriarch. Some members of the family may suffer from his authority. It is important for people to realize that this is the judgment of one man. People should be a little more critical about Parker's taste."

De Villaine acknowledged that since Parker stopped coming, "We feel more free, even if it makes things more difficult. Wine is not an exam. Now we don't have what seemed like an extraordinary machine, judging us. We can concentrate on *terroir*."

7

THE

EMPEROR'S

PROGRESS

*A*T 7:55 ON A MAY MORNING IN 2002, I SAT IN my car in the parking lot in front of the Napa Valley's Meadowood Country Club and Resort reception building, sipping a latte from the Oakville Grocery and waiting for Robert Parker. His scrawled note had specified 8 A.M., and I certainly didn't want to be late. I was tagging along on one of the emperor of wine's marathon days during his annual visit to northern California's wine country. It was cloudy and cool; already that week I'd heard the vines were two weeks behind where they should be in a normal growing season.

Meadowood, up a winding road off the Silverado Trail in the Napa Valley, was both convenient and highly private, perfect for a visiting celebrity like Parker, and also luxurious; rooms started at $400 a night. Owned by developer Bill Harlan, whose cult wine Harlan Estate was one of Parker's favorite California Cabernets (both the 1994 and 1997 rated 100), the resort had hosted the first Napa Valley Wine Auction in 1981. I remember it well: that June the vines had been well ahead of schedule due to the over-100-degree weather, and inside the tent it had been so hot that auctioneer Michael Broadbent kept his bare feet immersed in a tub of ice water hidden behind the draped podium as he exhorted bidders to buy wines many had never heard of. Few had heard of Robert Parker back then, either. The surprising fight that broke out between two bidders over a case of Opus One, which finally went for $24,000, was a portent of things to come. Two decades later Napa Valley wine culture had changed dramatically; the Napa Valley auction raised $55 million for charity, and absolutely everyone knew Parker's name.

Parker's note reminded me our day would be long, probably finishing around 8 P.M. with no stop for lunch. That's why I'd stuffed survival rations in a bag—an apple, a blueberry muffin, a hunk of Sonoma Jack cheese (well wrapped lest its odor contaminate the great man's sense of smell), energy bars, and two liter bottles of Crystal Geyser water, all of which I was prepared to share. Plus a good jacket in case Parker was dressed more formally than I thought he would be and a sweater for chilly cellars.

Parker pulled in a few minutes later in a modest white Taurus SES. I needn't have worried about what I was wearing. His hefty frame was packed into a pair of black REI shorts with zipper pockets and a black polo shirt with a red logo, on top of which he wore an open red-and-black windbreaker—color-coordinated, but definitely not a fashion statement. His old tattered brown oxfords and rolled-down socks looked like they could brave any muddy vineyard, and probably had. A black patella band hugged his leg just below the knee; the knee was giving him trouble—in fact, he was limping—but he just hadn't had time to schedule the arthroscopic surgery he needed.

His eyes looked a little red. He'd been crying in his room, he confided, because the wife of his best friend, Park Smith, was dying of a very aggressive brain cancer and he'd just been talking to him on the phone. The two had known one another since 1988; Smith, a great wine collector (his five cellars contained nearly 60,000 bottles) and part owner of Veritas Restaurant in New York, called Parker "Mongo," and Parker called him "The

Grand Parkster." At home they spoke on the phone every day. Parker worried that Carol Smith, known as "Ma," would die while he was on this trip and he'd have to cut it short and fly back for the funeral. But it wasn't anything he'd hesitate doing. "If I have to leave I'll just go," he said.

April hadn't been a good month for Parker. His mother, who'd been diagnosed with cancer in February, right after I had met her, had died on the eighth. His brother-in-law—not the one with whom he owned Beaux Frères—had had a quadruple bypass despite the fact that he didn't smoke, wasn't overweight, and exercised regularly. Now a good friend was dying. Other people might find traveling around to taste wine difficult or meaningless under those circumstances, but Parker welcomed this twelve-day trip as a distraction. His incredible focus when it came to wine kicked in on the tasting route. He said: "I just block all the sadness out."

Parker's treks to California always caused a buzz—though not as big a one as his trips to gossipy Bordeaux did—and dozens of wineries not on his schedule had sent sample bottles to his room, hoping to arouse his taste buds. Fifty-one lucky ones in Napa and Sonoma were on this trip's itinerary, a mix of "reference points"—people "making extraordinary efforts to make great wine"—and new producers and people he hadn't seen in a few years, like Hollywood movie mogul Francis Ford Coppola, who'd bought the old Inglenook vineyards twenty years ago and would meet with Parker personally. He'd made the list even though Parker had never been that crazy about his wines.

Parker often downplayed how much power he had, but when he'd sent out faxes with the date and time he wanted an appointment, not one winery asked him to reschedule. (No one today would dare refuse to see him, as Justin Meyer of Silver Oak did years ago, although Parker raved about his wines anyway.) Ed Sbragia, the chief winemaker at Beringer, one of the oldest and largest premium wineries in the state, would be interrupting his vacation to fly back to spend four and a half hours with him a few days hence. Most wineries were allotted only an hour. But Beringer's range of wines, like that of a few other giants who merited regular long stops, was huge. The only winemaker for a small property accorded a several-hour audience was Parker's favorite, "wine guru and goddess" (as he called her) Helen Turley.

I slid into his car. In the back, a case of Poland Spring bottled water rested alongside an enormous bunch of bananas. "I never eat while I'm on the tasting route," Parker explained, "but if I get hungry the flavor and texture of bananas don't seem to interfere with my taste buds." Dinner would be quiet, a salad and mineral water in his room. And then he'd call Pat; sometimes he talked to her three times a day. And he always took along three or four novels on a tasting trip so he could decompress after a day of focusing on his mouth. He leaned toward crime novels and thrillers, like those of Baltimorean author Stephen Hunter, who had been the film critic at the *Baltimore Sun*. Ordinarily, he'd shove in a Neil Young tape while driving and turn up the volume, but we were going to talk.

———

THE WINE SCENE IN THE NAPA VALLEY AND CALIFORNIA IN GENeral was far different in 2002 than it had been in the early years of *The Wine Advocate*, when Parker wrote blistering attacks on most of the state's wines and winemakers. In turn, they had complained bitterly that he had a strong bias toward French wines. From a couple of industry giants and a handful of cutting-edge boutique wineries who sold most of the wine they made locally, the roster of California wineries had grown to well over one thousand, many of whom exported their wines around the world.

By 1990 Parker had come to the conclusion that "only a fool could ignore that California is now producing many of the greatest wines in the world." Still, at the end of the decade he was convinced that fewer than one hundred wineries were making world-class wine. Many of them, he thought, were in the Napa Valley, which was why his annual treks were worth it.

Some things hadn't changed. Back in the 1970s, California wineries had sought out and come to depend upon the press in ways the French did not. It had been a struggle for tiny wineries dedicated to quality to get the attention of the world's wine drinkers—they had staged events and bold, sometimes outrageous comparative tastings, gone on the road, and pressed their wines on the media, hoping that a story and a favorable mention of their wine would result. They didn't have decades to build reputations, they needed to do it fast, and the media was the best marketing machine they could get in a country where 12 percent of the population drank 88 percent of the wine. The global media-ization of the wine world got its start in America, the media capital of the world.

In a field crowded with competitors, every winery needed the press more than ever to single them out from the pack; for a brand-new winery, media attention was essential to make a reputation. Many sold their wines in tasting rooms, but to persuade national distributors, retailers, and restaurants to carry their wines, either substantial production or good ratings from Parker and *The Wine Spectator* were required. A new crop of tiny wineries sprang up in the 1990s, many of whom sold all their small production via an exclusive mailing list and depended heavily on media buzz at the beginning to build a clientele for wines that ran upwards of $50 to $100 a bottle. They welcomed Parker with open arms, knowing that even well-heeled buyers still wanted assurance that they were getting a big bang out of those big-buck bottles.

Wine styles had shifted during the 1980s and 1990s, as winemakers gradually turned from copying French models to trying to find out what California did best, which for the most part turned out to be wines made from succulent, almost sweet fruit ripened under the warm California sun. Power, concentration, intense fruit, and plenty of tannin had been the hallmarks of many of the early California Cabernets that had won praise from Parker in the first years of *The Wine Advocate*—Silver Oak, Dunn, Caymus, Diamond Creek, Mayacamas—and the rich, fruity Chardonnays with a smack of spicy oak that made up in power what they lacked in subtlety and grace. With science and technology on their side, wineries believed they could leapfrog over the centuries of trial and error it had taken to match sites in Europe with the right grape variety. All-important were the wizard winemakers who were convinced the high-tech science of modern winemaking was the route to greatness; during the 1980s few of them had much interest in vineyard techniques—they started performing their magic only when the grapes were in. Early in the decade some had aimed for wines with more balance and restraint—dubbed "food wines"—and techniques once regarded as "stripping" were called "sculpting." When Warren Winiarski announced at one Four Seasons' Barrel Tasting Dinner that he'd filtered and fined his recent Cabernets, audible boos greeted this admission of what many regarded as wine abuse.

To Parker, despite a dozen or so stellar wineries, the state back then "led the world in making uninteresting, highly processed wines that are excessively acidified and have little aroma and stripped, washed-out flavors because the wines are eviscerated by multiple filtrations." Once again he was the stern lawyer prosecuting wine crimes, especially the most serious one: "destroying the joy of wine in the name of progress."

But by the 1990s, things had changed dramatically. Winemakers were heading for new regions like Santa Barbara, embracing new grape varieties like Syrah, and many had consciously gone back to making wines in a bold, up-front, full-throttle American style—ripe, rich wines with big flavors and awesome concentration. Fewer now had the ponderous, rustic, tannic character of the over-ambitious bottlings of the 1970s, but some were such intense, exotic, high-alcohol blockbusters that they still elicited criticism. Not from Parker. Despite a palate trained on French wines, he came out as a self-described "fruit fanatic" with a taste for the richness and intensity that came naturally to California. Parker called the eight years from 1990 to 1997 "history in the making" for the quality of California wine.

Terroir was once a word only a handful of California producers—such as Paul Draper at Ridge Vineyards—even uttered. But in the 1990s a new group of influential vineyard manager consultants, such as David Abreu and John Wetlaufer, emerged. The new hotshot winemaker-consultants, like Helen Turley and Heidi Peterson Barrett, began to pay more attention to what happened in the vineyard and and led the way toward making wine more naturally in the cellar. They took their philosophy from winery to winery and attained the status of celebrity chefs. Like Parker, many in the new generation of winemakers seemed to fit the mold of stubborn, ambitious mavericks, loners, often self-taught, who turned their backs on the establishment (in this case the oenology department at the University of California at Davis) and were convinced they knew the right way to pursue greatness. It was a story that Parker responded to.

France's golden decade for wine, according to Parker, had been the 1980s, and he had been there to praise the best. In the early 1990s, as Bordeaux suffered with poor vintages, American wines had suddenly shot ahead in quality. The top wineries, Parker believed, were now making great wines, but unfortunately some, whom he chastised regularly, were pricing their wines "in proportion to their egos." On his visits to the valley he was on the lookout for what was new and exciting—the first vintage of a potential cult Cab, Helen Turley's latest, the improvement at Clos Pegase since they'd hired another winemaker—and for whoever he thought was falling down on the job. In the past few years that list had included many wineries that Parker had once praised, like Clos du Val, run by French winemaker Bernard Portet; Al Brounstein's Diamond Creek; and even Robert Mondavi.

Parker had strong opinions on the right styles and winemaking techniques for California wines, just as he did for Burgundy and everywhere else. Woe to those who didn't agree—like Tim Mondavi, whom he'd be seeing in a few days.

Robert Mondavi Winery was one of the "reference points" Parker visited every year and had praised throughout much of the 1990s. But that hadn't kept him from scolding Robert's son Tim, who oversaw the winemaking, in the December 1999 *Wine Advocate,* for the direction Mondavi's wines had been taking recently. "It is obvious that the strength of California's finest wines lies not in delicacy and finesse, à la Bordeaux, but in power, exuberance, and gloriously ripe fruit," he wrote. Didn't Tim Mondavi's pursuit of "exquisitely nuanced wines of delicacy go against California's true strength"? Tim clearly didn't get the message, so in December 2000, Parker adopted a more strident, even emotional tone about what he called Tim's mistaken embrace of "Euro-elegance." By making "increasingly light, and to my way of thinking, indifferent, innocuous wines that err on the side of intellectual vapidness over the pursuit of wines of heart, soul, and pleasure . . . he is going against what Mother Nature has given California," Parker fulminated. Most of the scores he'd bestowed on the wines had slid down to the mid and low 80s, with a couple (gasp!) in the 70s. Parker had already decided that what California did best—and should continue to do—was produce "wines of ultra-ripeness, majestic richness, and provocative intensity." Everything else seemed "dead wrong."

After Parker's report came out, Mondavi's stock price fell 14 points.

As so often when others held a different view on wine, Parker didn't cast Tim's approach to wine style as simply different from what he himself liked; he fixated on an outside source that must be causing it. Six years earlier he had included Tim Mondavi on his "heroes" list, so someone had to be exerting a bad influence on him. Since the winery and Tim were involved in a half-dozen joint wine projects worldwide, Parker theorized that he was spending too much time with "some of the finest winemaking families in France, Italy, and elsewhere" who thought California wines were too alcoholic and over-the-top, "vinous examples of vulgar American culture." Six months later, Jim Laube, who covered California for *The Wine Spectator* and lived in the Napa Valley, echoed Parker's point of view in print.

The whole brouhaha escalated to the status of one of those gossip-driven industry feuds, kept going by reportage in two English wine publications and in-depth deconstruction in Sonoma's *Global Vintage Quarterly* and in *Wine Business Monthly*. Dan Berger and many other wine writers wrote Tim letters of outrage and support—"I just wanted to tell you he's offbase"—and retailer Gerald Weisl posted his own letter and Tim's reply on his store's website. For the most part Tim and his father, Robert, were blandly gracious about Parker when interviewed—"he's done much for the industry and he has his view and we have ours" was the drift—but both were hurt and offended, especially over Parker's personal comments about Tim and the implication their wines lacked "heart and soul," something guaranteed to enrage Robert in particular. The Mondavis requested a meeting with Parker, to discuss his comments.

But if the family smoothed things over in public comments, in private Robert Mondavi couldn't stop talking about Parker's slam of their wines. At the small birthday lunch his daughter Marcie gave for him at her apartment in New York, he told me, "We have a style of our own. We believe in it. If you put your heart and soul into a wine the people will prefer yours even over a better wine next door."

Tim, who, like his father, hardly ever stopped talking, his conversation wandering all over the landscape until you reined him in to the point, was convinced the problem started when he couldn't be at the winery for Parker's annual tasting in October 2000 because he was at the *Wine Spectator*'s New York Wine Experience. Genevieve Janssens, Mondavi's director of winemaking, told one interviewer that Parker arrived in a bad mood and was rude and abrupt, saying over and over as he tasted Mondavi's wines, "It's a problem, it's a problem, where's the pleasure?"

Parker, with a laugh, recalled it differently—he was disappointed, not in a bad mood, and wanted to discuss his view, but received no response. When he learned that Mondavi had hired the ubiquitous Bordeaux consulting oenologist Michel Rolland, whom Parker had known and respected for nearly twenty years, Parker surely must have assumed they had taken his words to heart and were trying to improve the wines he'd disparaged, as they had fifteen years before. From 1975 to 1985 Mondavi had filtered all their wines. Parker's withering comments had persuaded them to return to making wine as naturally as they could. Robert Mondavi had said so. Michel Rolland claimed that Tim had called him two weeks after the newsletter came out, but the official winery line was that Rolland was only called in to work on a specific Merlot project.

After Parker's slam of the Mondavi wines, Enologix, the lab in Sonoma that published *Global Vintage Quarterly*, set up a tasting panel of winemaking professionals to evaluate the Mondavi Cabernets blind against several others Parker had rated highly. No one else found the Mondavi wines inferior to those. Tim was quoted as saying, "I don't think he [Parker] would ever say he thinks there ought to be a single style, but I think effectively through the way he writes, that one could conclude that that is indeed what he believes."

But that had been last year. They had finally met and supposedly patched things up. This year's visit should be more satisfactory—at least to Parker.

———

HUNCHED OVER THE WHEEL, PARKER DROVE DOWN INTO ST. HElena, perking up as he began to psyche himself up into wine mode, then turned south on Highway 29, the north-south road that bisects this narrow valley. Workers were already in the vineyards, spraying, but traffic was light compared with harvest time, when the number of trucks transporting grapes could slow progress to a crawl. We headed to Rutherford and our first stop, Staglin Family Vineyard, but Parker, frustrated, was unsure of the turnoff road and had no map (a guy thing?). So I fished out the one I'd brought, with wineries marked on it. We turned down Bella Oaks Lane, and at its end, tucked into the hillside below Mount St. John in the Mayacamas mountains, was the iron gate across the winery road leading up through vineyards. It was locked. Parker pressed the buzzer in the stone pillar. "Hi, it's Bob Parker," he said. Shari Staglin's disembodied voice replied, "Great, could you bring the newspaper up with you?"

The emperor of wine rolled his eyes at me, as if to say, *See, I'm not such a big deal after all*, laughed, and picked up the paper. Parker liked the self-made, incredibly wealthy Staglins, who, he felt, had been "really beaten up" in James Conaway's latest chronicle of the valley, *The Far Side of Eden*, in which they were portrayed as the embodiment of the brash "new money" people who had come to the valley in the past decade to grow grapes and make wine to achieve status, satisfy their own egos, and "have a lot of lifestyle," as the Staglins put it. Well, they could be seen that way. They certainly had more money than most people would know what to do with. They'd built an 11,000-square-foot faux Tuscan country villa whose décor included a needlepoint rug that was a copy of the painted ceiling of an Italian Renaissance palazzo, started making Cabernet, and leased the property to Disney for the remake of *The Parent Trap*, to the dismay of their neighbors. But after all, they donated a lot of money, dinners, and wine to charity, especially to benefit mental health, and Parker admired that. He tried to do the same thing, though of course on a much smaller scale.

The valley was filling up with other entrepreneurs, dot-commers, and investment bankers who had made so much money so fast in the giddy boom years of the 1990s they could barely find enough projects to spend it on. They did just what the Staglins had done. The idea of starting a winery with your name on it seemed a hedonistic, easily attainable dream.

The Staglins' two Jack Russell terriers, Sami and Deuce (the canine hosts of a Doggie Dinner party that raised $28,000 at a Napa Valley Wine Auction), yipped wildly at Parker as we joined the Staglin family, including their two grown children, son Brandon and daughter Shannon, at the entrance to their brand-new 24,000-square-foot environmentally friendly winery, which was built directly into the hillside. The last time Parker had visited, two years before, they had still been making their wine at a custom crush facility nearby. It had taken them ten years and a design that hid the winery underground to get zoning approval to build it. People said their $1 million contribution to Robert Mondavi's baby, the Copia American Center for Wine, Food and the Arts in Napa, was what finally helped push it through. The Staglins could hardly wait to show it off to Parker.

Garen and Shari Staglin were both short, slim, and intense, he dark-haired and smiley, she blonde, pixieish, and talkative, and they used the same relentless in-your-face cheery energy in making and promoting their wine as they had in the businesses—a safety glass company and now eOneGlobal, an electronic payment systems company with 1,100 employees—that had made them enough money to afford all this in the first place.

You had to hand it to them. Whether through luck or smarts, they'd done everything right if they wanted to create a great wine. First, in 1985, they bought a historic 50-acre vineyard once owned by Napa Valley pioneer Georges de la Tour, the founder of Beaulieu Vineyard. It hadn't been easy prying it from his granddaughter, Dagmar Sullivan. Originally planted under the supervision of the famed Andre Tchelistcheff, the oenologist who transformed American winemaking in the 1940s and 1950s, the vineyard had contributed grapes to one of America's first great wines, Beaulieu Vineyard's Georges de la Tour Private Reserve Cabernet Sauvignon. This was the closest America could get to hallowed *terroir* with a track record and image.

As vineyard manager they hired the controversial David Abreu, whom Parker had listed as one of his "heroes" in 1993, to show they were serious. During the 1990s Abreu had worked as a vineyard consultant for a who's who list of the valley's cult wine producers. Parker had rated one of Abreu's own wines, 1997 Abreu Madrona Ranch, of which there were a scant three hundred cases, 100 points. Luckily he was absent that day. He'd been known to talk on and on for hours to Parker about every technical aspect of growing grapes—climate, vineyards, hang time, you name it.

The Staglins invested in the latest equipment, learned to talk the talk, and knew how to market themselves and their wines by capitalizing on all the elements that Robert Mondavi had used before them. Garen's Italian heritage (the family name had been Stagliano), the idea of family (their children were also involved; "it's the dynasty thing"), and their extensive art collection were all bound up with their wine. It was the right package for success. And two years ago, they'd hired winemaking consultant Michel Rolland to work with their winemaker. (That, Shari confided, had turned out to be far more expensive than they'd expected.)

Like many of the valley's wealthy newcomers they wanted to make only the best (no Americans set their sights on making a second growth, much less a *petit château* wine)—and, of course, to gain Parker's validation. Their 1990 Cabernet squeaked by at 90, but with their 1994 Cabernet they'd made the 90+ club. Parker called another of their wines, their Sangiovese, the best example in California. In keeping with Garen's Italian heritage, making a wine from the same grape used to make Chianti in Tuscany, they said, "added value to their story."

The new winemaker, Andy Erickson (who would leave in 2003), clad in a dark green V-necked sweater, took center stage to conduct the tour and tasting of twelve wines, deferentially trotting out his bona fides for Parker, who'd never met him: a stint at Harlan Estate, then Spottswoode, two Parker favorites, and for that matter his wife worked for Abreu. After a bit of small talk and a look at the five tunnels crossed by three, the room 30 feet high, 33 feet wide, and 100 feet long, and the seven fans that would go on automatically if the humidity jumped to over 80 percent, Parker happily dived into the winemaking minutiae of what else was new since his last visit. Erickson was happy to oblige: Abreu was lowering yields in the vineyard, trying new clones, changing the spacing of the vines where they replanted so the fruit ripened better, racking less and less when the wines were in barrel, manipulating the wine less and less (the "we think the wine is made in the vineyard" mantra)—in short, all the usual buzzwords of contemporary winemaking hummed in the air. Parker nodded, clearly approving, interrupting to deliver his own one-paragraph lectures on winemaking topics dear to his heart—the importance of low yields to ensure wines of character and of letting wines ripen fully in the vineyard, the problems with filtration. Shari Staglin stubbornly defended the idea of making a Sangiovese; Parker liked it, but from a global perspective wondered what was the point of growing it in California? It didn't do that well here compared to Tuscany. But the Staglins had sentimental reasons, and besides, it went with their story.

The tasting at a big round table set with crystal was punctuated by barks from Sami and Deuce, but Parker, with similar dog experience at home, was unfazed. He especially liked the 2001 barrel sample of Cabernet Sauvignon, Erickson's first vintage here. "We notched up the intensity," Erickson explained, "but we'll make the final blend when Michel [Rolland] comes." The 1999 Cabernet was unfined and unfiltered. "We followed Michel's advice: Have a reason for what you do. If there's no problem, why filter?" Parker couldn't have agreed more.

"Staglin is a work in progress," Parker told me in the car. "I keep coming back because I see the motivation to make world-class wines, and they're getting closer." He ended up giving the 2001 Cabernet barrel sample 94 points, their highest score yet. Andy Erickson had been a very good hire.

———

PARKER HAD BEEN VISITING FLORA SPRINGS, OUR NEXT STOP, FOR years. Julie Garvey, who owned it with her two brothers and whose husband was vineyard manager, greeted him warmly, but a little nervously. Parker usually tasted with the winemaker, Ken Deis, who couldn't be there. He and Deis shared a love of college basketball, and Garvey knew Parker was probably looking forward to an animated discussion about how the Terrapins, the University of Maryland team Parker rooted for, had won the NCAA men's title the month before. As a joke she handed him a small turtle and a miniature basketball she'd purchased in Napa and pointed out the Terrapins banner on the tasting table. "Hey, that's not an enticement for you to give us a good score, you know," she added, laughing. He seemed touched by this and her sympathetic comment, "I'm sorry about your mom." Parker had watched the game but without concentration because his mother was dying in the hospital. Just about everybody in the California wine world knew about his mother, and many had sent him personal notes.

Flora Springs was practically an old-timer, one of the wave of wineries started in the late 1970s before wine scores had taken hold of the American consciousness. It had a solid reputation but had never become fashionable or a superstar despite excellent wines. Garvey's parents had purchased an old homestead that came with two ghost wineries. Instead of retiring as planned, they planted vines and made wine, and were among the first to launch a Bordeaux-style Cabernet blend, which they called Trilogy. Now the family owned six hundred acres of vineyards and made more than a dozen different wines. Like many in California wine country they couldn't resist experimenting, taking a chance on some new grape variety. That love of experimentation had always been one of the big differences between America and France when it came to wine. It was what kept the business from becoming dull. "If we'd stuck to Cabernet and Chardonnay we'd have sold out long ago," said Garvey.

Flora Springs hadn't always had a great relationship with Parker. From the mid-1980s to the beginning of the 1990s he'd praised their whites but saw the Trilogy, at best, as "stylish and polite." They had to work hard to develop a following for it. Garvey hadn't forgotten. "It was a good wine, not as big in style as it is now. We had to put it into people's mouths. We didn't have a 99 from Parker." But Parker's view of the wines started to change with the 1991 vintage, and now he visited every year.

Garvey was warm and friendly, and Parker seemed especially comfortable with her; they traded information about their kids, complained about the new hassles of airline travel, and discussed the California trend toward bigger and bigger wines. Parker came down on the side of "wine as a beverage of pleasure," one of his favorite phrases, while Garvey said, "I guess the question to me is going to be . . . whether you get to a point of diminishing return in ageability. Are those bigger wines going to last? People are starting to do odd things in order to extract those kinds of characters that get rewarded by critics." In other words, was up-front, super-ripe fruit and high alcohol a guarantee that the wine would last, or that it would fall apart fast?

Parker listened, then interrupted, "You can make big wines, but they have to be balanced." It was only the first of several times he would say it that day.

Garvey had lined up her twenty-four wines varietal by varietal on a long table in the winery tasting room, but when Deis had checked in by phone just before we'd arrived, he'd said Parker liked to taste by vintage. Did he want her to rearrange them? No matter, Parker said. A tasting sheet contained all the details on each wine that he liked to have—grape variety, price, number of cases made, release date. Settled into the rolling chair she'd pulled from Deis's office, he worked his way down the line, pouring a third of a glass, favoring his right nostril as he inhaled, tasting, spitting into a red plastic cup, commenting—he liked the $12 Pinot Grigio—and asking questions. What was the oak program for the Chardonnay? (It was French.) Did Garvey see a trend away from malolactic fermentation to preserve more acidity in the wine? What about oak? "We're hearing people are tired of oak and richness," offered Garvey. And what was her take on the conditions during the past three vintages? While she talked, Parker took in what he wanted and blocked out everything else while he tasted, as if he were inside a silent glass bubble. In law school he'd studied with the stereo blaring; this was no different. He told me he could taste with "a hundred screaming kids in the room."

As a wine went into his mouth, the first impression that popped into Parker's head was textural, then a picture, a photograph of the wine, almost in three dimensions. He'd tried to describe the process many times. He knew it sounded like b.s. but he saw the wine in layers and textures, and in his mind he unpeeled them one by one. Sometimes he thought of it as pulling back curtains, at others as climbing up a building with ten stories. "I'm thinking about what's there and how much is there, where's the acidity vis à vis the tannin, the alcohol," he said. He usually asked questions about oak. If he knew a winery only used 50 percent new oak and the barrel sample tasted excessively woody, he knew they'd heard he liked lots of oak and had doctored the sample. "When that happens," he said, "I won't review the wine. Screw it."

Parker's left arm hooked protectively around his open fat black notebook as he wrote—he's a lefty. No peeking at the score. But I could see the Ninja Turtle stickers that a young Maia had stuck on her daddy's notebook.

After a couple of days of tasting, Parker had become convinced 2001 was going to be a big Cabernet Sauvignon year. "I like this vintage; you can see the sweetness. I think the great vintages everywhere are like this, great from the get-go," he said (the phrase "get-go" recurs frequently in Parker's speech). Flora Springs, like most California wineries in the past few years, was becoming fixated on *terroir*, just as they were in Burgundy. They'd initiated a new series of Cabernets from separate vineyards that "spoke of place." Parker's excitement—"Hey, this is new! Where's this from?"—was evident: tasting one that didn't even have a name yet, his face softened. The critic had a half-smile on his face; it was the way he looked, almost beatific, when something he liked hit the famous palate. Sometimes he smacked his lips, as he did after a sip of the Flora Springs Wild Boar Cabernet. "It's an old-style California wine, not completely politically correct," admitted Garvey. "God bless it," Parker said fervently, calling it "a wine for the Hell's Angels," though in his newsletter he would tone that down to "hard-core bikers."

Trilogy stood for Cabernet Sauvignon, Cabernet Franc, and Merlot, the three grapes used in the blend. In 1984, when Flora Springs first produced it, few were making a Bordeaux blend like that. "In fact, Bob didn't like it too much," Garvey said, turning to me. Her voice had a good-natured kidding tone, but a steely edge, too.

"Yeah, I was really tough on you," Parker agreed, seeming a little abashed.

She continued talking to me, while eyeing Parker. "And not only on that wine. He didn't like any wine we made after that. He thought they were too elegant."

"No, I thought they lacked texture and I thought they were too acidic."

"Um-hum. But the word you used was 'elegant.' "

"Well, too politically correct, too elegant, too polite. Now," shifting the subject a bit, "this [the 1999] is a quintessentially elegant style of wine. But I'm saying elegant in a positive sense. In those days—the food-wine period—the wines were so understated, so restrained. Elegant is not a bad thing. But so many people in my profession use the word 'elegant' to excuse diluted and thin, no character."

"Well, we never thought our wine was a wine of no character. But we weren't trying to make a big, saturated wine. And I think, honestly, the '90s were more generous. We've learned as we've worked and I think Bob's evolved as he's tasted," she laughed to lighten the jibe.

"But do you think . . . I always suspected there was a heavy hand with acidification."

"No, never. In fact that was the thing that was rather annoying because you talked about how we added acid and we overfiltered and we weren't doing either one of those."

"So you were picking too soon, then."

"I don't know. I think that we got to the point where we are picking at different rates of maturation than we did in the '80s. Everyone is. I think we know a lot more . . ."

"But certainly as you've seen my scores go up there's a lot more texture."

"No, there is. But . . . the biggest change is that we've cut back on the Cabernet Franc. Our Cabernet Franc has a grip to it, and it grips a part of your palate you're not used to. It kind of snags the whole package and makes it seem leaner . . . and we are picking at riper levels than we were in the 1980s. I agree that they are better than in the 1980s, but I don't think they were that bad in the 1980s."

"My rhetoric was more nasty back then."

"You said 'death to the winemaker.' That hurt Ken's feelings," she reminded Parker, who threw up his hands and turned to me.

"That is shameful conduct on my part. . . . You're sure I said that? I must have been really pissed off tasting those wines." Parker sounded upset.

Garvey pulled out bottles of Trilogy from 1985, 1987, 1990, 1992, and 1996 for us to taste to prove her point.

Later, still brooding about the exchange, Parker asked me, "Hey, does that sound like something I'd say? I have a very good memory for what I write. I can't believe I wrote 'death to the winemaker.' It's not me at all."

I tracked down the offending passage describing Trilogy in a 1991 issue of *The Wine Advocate*: "If Shakespeare lived today, he might well be saying, 'first let's kill all the lawyers, but then all the oenologists.' This emaciated, excessively acidified, tart, sterile-scented proprietary red table wine is an example of how California oenology can be abused."

No wonder they remembered it eleven years later.

There'd been more give and take, more difference of opinion at Flora Springs than at Staglin; a more relaxed, less worshipful view of Parker; and in fact, the closest thing to a challenge to his views and judgments that I heard all day.

Flora Springs had been around for a long time, survived poor scores, pulled forward, and established a reputation, though it was not cutting-edge like the cult wineries to whose wines Parker gave scores of 95 points and up. The competition was fiercer now; there were dozens and dozens of Bordeaux-style blends and great Cabernets.

"I'd rather just taste the wines privately and forget all the small talk," Parker admitted in the car. And maybe not have to put up with being challenged? "Their wines always show better in bottle. I don't agree with everything they do here, but there are many means to an end. Staglin is the French route. This is the old-style Joe Heitz way of making wine. And it works." He sounded almost surprised.

———

TO THE OWNERS OF PATZ & HALL, WAITING FOR US A COUPLE OF miles away at Honig Cellars, the winery where they made their wines,

Parker came close to being a hero or a saint—he'd saved them in the nick of time, more proof of his power to make or break. We arrived seven minutes late, unusual for Parker, who prides himself on punctuality. All four of the owners—winemaker James Hall and his wife, Anne Moses, also a winemaker, and Heather and Donald Patz, who handled administration and sales—crammed into the small, white kitchen-style tasting room with painted green cupboards. Sets of glasses, one more elegant than the others, obviously intended for Parker, were arranged on a square white table. Some were found for me, but not very graciously; I was an unexpected intruder into the coming performance. There were mugs for spittoons. We crowded in, sitting on uncomfortable rush-seated farm chairs. The tense atmosphere almost crackled with anticipation.

In the Napa Valley, few options existed for getting into the wine game if you didn't have the big bucks. The oft-repeated saying—to make a small fortune in the wine business, start with a large one—was all too accurate. During the 1980s land prices boomed, the cost of planting a vineyard ballooned, and since it took several years before you could make wine from the grapes, significant capital was required. The Patz & Hall solution, when they founded their company in 1988, was to become a *négociant*, purchasing grapes from the best vineyards, then making the wines at a "host" winery, in their case Honig Cellars, where Hall was then the winemaker. The new California concern with *terroir* dovetailed neatly with their aims, and now they made six Chardonnays and four Pinot Noirs with grapes purchased from some of the top growers in northern California.

In April 1991, though, they had been desperate and depressed. Their 1989 wines had flopped. They were running out of money, and after settling a huge tax bill they had $600 in their bank account and wine in barrels that they couldn't afford to bottle. Selling off some of it in bulk put enough cash in their pockets to bottle five hundred cases of 1990 Chardonnay. That was it; 1991 would be the make-or-break year. Then a miracle happened. A good friend, Bruce Gearhart, who also knew Parker, lobbied to get him to try the wine. He did. A rave appeared on the back page of *The Wine Advocate*. Patz & Hall's wine—Parker's new discovery—had "restored my faith in artisanal chardonnays in the Napa Valley." The score: 92.

Back east, their distributor, Lauber Imports, received a sudden rush of orders, unloading four hundred cases in just four days. A month later, a check arrived and suddenly, Hall said, "We were in the black, with enough money to establish a credit line, anticipate cash flow—go forward!" Their day jobs could wait; they almost wept with relief. After that, however much of their limited amount of wine Lauber wanted, Lauber got.

Understandably, the first time Parker visited, in October 1998, James Hall had felt nervous, as if he had to perform brilliantly on an oral exam for which he didn't know the questions. He'd met plenty of critics who were harrumphing curmudgeons. "I expected him to be demanding and difficult, that he'd ask trick questions," he recalled. "The first ten minutes were nerve-wracking, like sitting on eggshells. I didn't want to be too technical." Hall was relieved by Parker's warmth and the dawning realization that he liked their 1997s and they'd have an ongoing relationship.

On this day the owners appeared at ease, but jazzed up and totally concentrated on impressing Parker. There was no small talk. They enthusiastically plunged to the nitty-gritty of how much control they had over the growers they bought from; the complexity of acreage contracts for specific blocks of the vineyard that enabled them to call the shots on pruning, crop thinning, picking time; and the need for rapport with growers. Then on to the technical winemaking stuff, the virtues of indigenous yeasts, separating out press wine, bâtonnage, long malolactic fermentations, even their efforts to wiggle onto the French oak barrel A-list so they could obtain the very finest barrels. They even shared rumors they'd heard of special Pinot Noir clones from France being "suitcased" through Canada (i.e., vine shoots smuggled into the country).

By now the articulate Hall and Moses had the way to run a tasting for Parker down pat: they talked endlessly throughout, making Parker feel part of the team, patiently feeding him bit by bit their own view of the quality and character of the vintage, what the wines tasted like, how good they were and why, nodding happily when he agreed with them or echoed what they'd said. The 2000 Alder Springs Chardonnay reminded Parker of a French Bâtard-Montrachet, with hints of orange blossom and tangerine notes (later published score: 92). The same vineyard's Pinot Noir, he noted, "charged through the mouth." The Pisoni Vineyards' Pinot Noir, like all the other Patz & Hall wines, had remarkable "purity of fruit." That was what California had and should maximize, Hall insisted, a view that resonated with Parker, who frequently espoused the exact same thing.

Some cynical winemakers in California told me such talk was all-important in shaping how Parker wrote about and judged their wines—if they didn't conduct it, he just wouldn't get it—though they didn't think Parker realized they were trying to steer his impressions.

Did he? It was hard to say, because Parker played his part, too, sometimes acting more like a wine geek or groupie than a powerful critic, eliciting details and opinions with friendly chatter, journalistically throwing in wine observations gleaned from other wine people he'd visited to get a reaction, pontificating as a fellow winemaker by drawing on his experience with Beaux Frères, all as though he really wanted them to like him, to think of him as a regular guy. He clearly worked hard to make winemakers and proprietors feel comfortable with him. Lording it over wine people—in person, anyway—isn't his style. "I'm not an ogre," he told me.

He shared information about a type of oak barrel that had a rolling mechanism for convenience and offered to send a brochure; they told him the inside story of how they'd managed to score grapes from one of the state's trendiest vineyards, Pisoni (the Pinot Noir cost $7,000 a ton), and turned him on to a new hot sushi bar, Matsuri Sushi in Benicia, off Interstate Highway 780 on the way to Walnut Creek, where they'd "sucked down" the dishes accompanied by a magnum of Krug Champagne. On the spot, Parker made a plan to eat there on his way to the airport the following week and looked as if he were already salivating.

Parker's trips focused on tastings, not visiting vineyards, but Hall tried to tempt him: "If you wanted to visit Alder Springs Vineyard up in northern Mendocino, the owner, Stu Bewley, a sweetheart of a guy, would go completely unglued; he'd be hopping from foot to foot he'd be so excited. He has a program that if you give any of his wines above 90 points he gives everyone on that vineyard crew fifteen hundred bucks." The vineyard was 2,100 feet above sea level, seven miles from the Pacific. Maybe, Parker said, if he came out with Pat to stay on the coast for a vacation he could sneak away for a morning . . .

As we left, Hall handed me a camera and asked me to photograph the four of them with Parker, outside. They pressed in close, on each side of their savior, and flashed wide smiles under the sudden ray of sunshine.

THE REST OF THE AFTERNOON INCLUDED AN HOUR AT THE TASTING bar at Signorello, where slim Ray Signorello, his hair slicked back and sniffling with a cold, finally mustered up the courage to ask Parker, very shyly and tentatively, why he'd stopped recommending Signorello's wines; he'd virtually ignored them after a series of raves several years before. Parker looked semi-annoyed. That, apparently, wasn't a question you were supposed to ask. "Only 30 percent of the wines I taste make it into *The Wine Advocate*," Parker explained. When he headed for the bathroom, I noted that he'd carefully turned over the page in his notebook so that only a blank sheet was visible.

Shafer Vineyards, just down the road, offered a virtual love-fest tasting in a large white room with Mexican tile floors, a big open fireplace at one end, and vines visible through the windows. Light sparkled on the polished glasses lined up in careful rows on the white-clothed table. At one end tall, lanky pink-shirted Doug Shafer lounged back and rambled on disjointedly, as if he'd just smoked part of a favorite cover crop—"I love it when the French go apeshit over Turley Zins, etc.—" while his father padded around in jeans and sneakers and winemaker Elias Fernandez poured and stood at attention for questions. Parker had been effusing over Shafer wines since 1991, especially the Hillside Select Cabernet, which he considered one of the top dozen or so in the country. Here he seemed in his element; a smile wreathed his face after tasting four vintages of Hillside Select. The 2001 barrel sample won a sigh and a "wow" and later a score of 96 to 100, potentially perfect. This visit seemed the high point of the day to Parker. The wine tasted of sun and warmth, said Fernandez, precisely what Parker thought a great California Cabernet should taste like. In the Cabernet sweepstakes that took place in Parker's mind, Shafer was still in the rarefied top tier.

On the road with Parker, it was glaringly evident how powerful he'd become. For all, Parker's blessing was useful even if they didn't need to be saved from bankruptcy. The flip side to being discovered was suddenly being dropped. Three years after their Cabernet received the 100-point score Groth had been demoted, and a decade later Parker had dropped their wines from his *Wine Buyer's Guide* altogether. After praising Jim

Clendenen's Au Bon Climat wines for years, Parker lowered their scores. Clendenen was devastated. Distributors began asking him, "Why doesn't Parker like your wines?" and he had to grope for an explanation. Randall Grahm, the playful, eclectic winemaker at Bonny Doon, wrote plaintively in his Fall, 2001 newsletter, "It seems that I have most likely (and most un-fortunately) burned all my bridges with Mr. Parker, who seems to no longer review Bonny Doon wines. This puerile attitude—I can do it my-self, I don't need any help, thank you very much—is not one I can con-tinue to indulge if the business is to continue to prosper."

At 6:30 P.M., as we pulled into Meadowood, Parker was still going strong, ready for a tasting with Italian winemaker Angelo Gaja, who hap-pened to be in the Napa Valley with samples from his new winery venture, Ca'Marcanda, in Bolgheri, on the Tuscan coast, which he wanted to share with him. Sassicaia, the first of the Cabernet-based wines known as "Super Tuscans," was made in Bolgheri and had started a wine revolution in Italy.

Gaja, one of the most innovative, ambitious, and media-savvy Italian winemakers, had met Parker in 1982, when the American was still a complete nobody. He respected Parker for his generosity, which he re-garded as unusual in a wine writer. He had been entertained in Parker's home on many occasions. On a 1998 trip to Italy, Parker had even hosted a truffle lunch for winemakers, which, people said, must have cost at least $3,000. Gaja appreciated Parker's openness to and praise of both modern (Gaja) and traditional styles of wine in Piemonte. "When he suggests a wine has quality, there are immediate customers. Small producers who risk all their fortune for what they do receive encouragement and that multiplies our possibilities," he says. A growing new region could always use a supporter.

Still, Parker could be rather imperious in his enthusiasm. He'd visited Italy only three times in twenty years. Nonetheless, that didn't stop him from suggesting to David Shaw of the *Los Angeles Times* that he "under-stood the 1997 Barolo and Barbaresco wines better than the people who made them, some of whose families have been making wine for several generations . . . I don't think they really realize what they have."

―――

THERE WERE DOZENS OF STORIES ABOUT REGIONS PARKER HAD PUT on the map and wineries and winemakers he'd elevated to stars during the

1990s as the wine world heated up in California, as well as globally. As he often said, "I don't give a shit how long you've been making wine, what its pedigree is, where it's made, or whether it costs $4 or $400. If it's good, I'm gonna say so." The result was that where once it had taken decades, even a generation, to make a new region's or wine's reputation, he and the mediaization of the wine world had shortened that time to a few years, or even less.

In fact, Parker was proud of helping innovative producers making good wine succeed and considered rewarding the up-and-coming with attention and good scores one of the ways he contributed to the improvement of wine quality in general. The process was not much different from the way Parker had "made" the 1982 Bordeaux. Rule one: Parker liked to "discover" wines that others had neglected, never heard of, or failed to take seriously.

Take the Rhône Valley region in France, which Parker had brought to the attention of American wine drinkers not just in his newsletter, but through his book on the region published back in 1987 (updated a decade later), in which he hammered home his belief that these were great, underrated wines. He considers it his best and most original work and remains disappointed that it sold less well than any of his other books. The epigraph of his own words "When I drink a great Rhône, it's as if my heart and palate have traded places"—testified to a love affair with the wines that went back to Parker's law school years, when he first discovered them. With the exception of the great wines of Hermitage, they hadn't been considered on par with or as collectible as Bordeaux or Burgundy. In the past they had often been regarded as mere Burgundy boosters, brought in to give the region's pale Pinot Noirs the color and heft that Burgundy buyers in England demanded.

For Parker, the producer who first stood out for him was Guigal, which came into being in 1946, when Etienne Guigal left the *négociant* Vidal Fleury and he and his tall, birdlike winemaker son Marcel, who was rarely seen without his flat-topped country cap, started down the road to being the dominant *négociant* and vineyard owner in Côte Rôtie, an appellation in the northern part of the Northern Rhône. Marcel started the renaissance of the appellation and introduced significant changes to the way Côte Rôtie had traditionally been made—harvesting super-ripe grapes, aging the wine in new oak instead of large old *foudres* from which the taste of wood was long gone, and bottling wines from specific vineyards instead of blending them together, thus creating some of the first "luxury micro-cuvées." The concept spread to other wineries, and in the 1990s dozens of brand-new ones captured Parker's attention and high scores.

The Guigal Côte Rôtie wines from three small vineyards, La Mouline, La Landonne, and La Turque (added in 1985), absolutely stunned Parker, who called La Mouline "the single greatest red wine in the world." La Mouline was "one of the world's most intensely perfumed wines, offering in the great vintages nearly otherworldly aromas of bacon fat, pain grillé, cassis, white flowers, black raspberries, and occasionally Provençal olives." The "most supple and seductive" of the three, "offering voluptuously textured, hedonistic drinking for 15–20 years," it was "the Mozart of the Guigal portfolio," while the dense, massive, "almost intimidating" La Landonne was "the Brahms." Guigal reduced yields in his vineyards, used minimal intervention in winemaking, and above all didn't filter—at least that is what he told Parker.

Over the years Parker awarded Guigal more 100-point scores for these three wines than to any other producer in the world—La Mouline alone accounted for eight, more than any Bordeaux château—and their prices reflected that. By the end of the 1990s they'd more than quadrupled in price to $400 to $500 a bottle in the United States, achieved cult status, and were virtually impossible to find outside of the auction room. It was the old story of demand and supply: there were only three hundred to five hundred cases of each. Although sought after in France by two- and three-starred Michelin restaurants, the scramble for them and the prices here, Guigal's importer once said, were largely due to Parker. In the summer tourist season, sometimes one hundred people a day lined up at Guigal's doorstep in Ampuis to buy wines. They sent checks, begging for a bottle from an older vintage. Marcel Guigal didn't know what to do about them.

Guigal's financial success taught other producers that America would pay a premium for luxury cuvée wines from special prized vineyards, and ambitious producers like Chapoutier, and Domaine de Pegaü and others in Châteauneuf-du-Pape followed suit. Parker praised them, too.

Not everyone was a fan. Robin Yapp, a British merchant who had imported Rhône wines since the 1970s, thought they were too heavy, rich, and woody, lacking the finesse of true Northern Rhône wines. Kermit

Lynch and Robert Chadderdon, who brought in competing wines made in a very different, more elegant, traditional style, agreed with him. Even other Côte Rôtie producers said Guigal's wines were too oaky and alcoholic, but Parker dismissed their criticisms as coming from those who either made inferior wine or were envious. He couldn't say that about Pat, who would ask, "How can you like these?" when he pulled the cork on one to serve with dinner. She argued that they were too oaky. "I always get a very honest reaction from Pat about a wine," says Parker, "she's tough." Determined to change her mind, he poured a couple of ten-year-old examples of Guigal's top cuvées one night. Pat found she did like them—if they were aged. Parker, who told me the story, didn't mention whether he'd said, "I told you so."

With his immense profits, Guigal built a multimillion-dollar winery and enormous warehouse and bought a $50,000 Mercedes-Benz. The aura of the "La La" wines, as they were known, rubbed off on their simple but delicious Côtes du Rhône that sold for $10. Guigal rapidly expanded the amount of Côtes du Rhône they produced—a typical move to cash in on publicity—and soon, to get six bottles of the "La La" wines, importers had to take one hundred bottles of the Côtes du Rhône.

———

PARKER RATHER ENJOYED MAKING A WINE'S REPUTATION OVERNIGHT. Shortly after his marriage in 1990, Los Angeles restaurateur Manfred Krankl was sitting on his terrace in Ojai one evening thinking about how great it would be to be drinking a wine he'd made himself.

The tall, bearded Austrian-born Krankl, then managing partner of Campanile restaurant and La Brea Bakery, knew something about wine—he was in charge of the restaurant's wine cellar and had put together a stellar list—but neither he nor his wife, Elaine, had any winemaking training and they didn't own any vineyard land. And besides, Krankl was busy full time with a popular restaurant and bakery. But so what? He was tired of chasing after cult wines for the restaurant's list. If they purchased some grapes, they could make a few barrels of wine for the restaurant and their friends, and he could indulge his own concept of what wine should taste like: dense, rich, distinctive, complex, and completely unique. After all, Krankl reasoned, his beliefs about what was required to make a great wine—boundless passion, a heart-and-soul instinctual approach, and not too much interference with the process—weren't very different from those held by great bread bakers.

For a couple of weeks Krankl turned the idea of making wine over in his mind and liked it better and better. First they tried a barrel of Chardonnay, then a Barbera/Cabernet blend and a trial Syrah. But then Krankl made a few phone calls to vineyards in the Santa Maria Valley and eventually got the inside track to some old-vine Syrah from the Bien Nacido Vineyard. Like many beginning winemakers, Krankl was driven, passionate, and maniacal. Given to T-shirts and leather jackets, he could easily have passed for a California biker. Fortunately, his intensity was balanced by a wicked sense of humor.

So Sine Qua Non Winery was born. The name meant "without which nothing," in other words, "something indispensable." Its first wine was created with borrowed equipment and purchased Syrah grapes at somebody else's winery. Later it would be a rusty warehouse beside a junkyard off the Ventura Freeway. There was no business plan or office, because it wasn't really a business—at least not yet. Krankl named their 1994 dark, deep, ripe, fruit-driven wine the Queen of Spades, put it into absurdly heavy, oddly shaped bottles, and put his own artwork and gothic lettering on the funky labels. There were just one hundred cases stacked in their cellar. The Krankls sold some cases to Campanile and the restaurants of a couple of colleagues, gave away bottles to their friends, and started trying to figure out what they could do with the rest.

As a joke (he says), Krankl sent a bottle to Parker, along with a letter describing how he made the wine—completely natural, no filtering, new oak, etc. About a week later the phone rang. It was Parker. He loved the wine and was calling for details, like the number of cases made. When he learned that the contact number he'd reached was the Krankl's home phone, he advised Krankl to obtain a business number and call him back. "Trust me," Parker told him. "I've been through this before. You're going to get a ton of calls after the next issue of the newsletter goes out. You don't want me to print your home number."

Luckily Krankl listened to him, because just as Parker predicted, the phone rang and rang and didn't stop. Hundreds of calls flooded in from New York, Tokyo, London, and other cities, at all hours of the night and day to order the wine Parker called "a dazzling tour de force" and rated 95.

The rest of the wine sold out in two days. When it was gone, some people wouldn't take no for an answer. One CEO's assistant tracked Krankl down at Campanile to demand a case for her boss and refused to be put off, breaking into tears because she was afraid that she'd be fired if she couldn't obtain it. Finally, Elaine recorded a message in her distinctive high, almost squeaky voice: "Thanks for calling. We're sorry. The 1994 Queen of Spades Syrah is sold out, but leave your name and address if you want to be on the mailing list for our next vintage." Within weeks there were more than five hundred people on the list, Sine Qua Non became a viable business, and Parker's continuing high scores forced prices up and kept up demand for a slot on the mailing list. Each wine was a one-off, with a different offbeat name—"Impostor McCoy," "The Bride," "Backward and Forward"—and poked fun at the very exclusivity on which Krankl shrewdly capitalized. A few years later Parker would write that "even the least impressive cuvée electrified my olfactory and sensory circuits."

Parker warmed personally to the iconoclastic Krankl; they shared likes in wine—Rhône varietals, the bottlings of Helen Turley, Guigal, and, in Italy, Elio Altare—and both were small-town guys who'd made it in a wider world. Both had a well-developed sense of humor, a similar philosophy of wine, and a passion for food and wine that bordered on the obsessive. If he could allow himself to be friends with a winemaker, Parker told me, it would be Krankl. But that, he knew, wouldn't happen; he couldn't afford to let it. Parker felt he had to keep his distance from producers whose wines he reviewed; friendship might be construed as a conflict of interest and blown out of proportion—would people believe in the honesty of his assessments if he hung around socially with winemakers? Then, too, a friend might expect him to give his wine the benefit of the doubt. Parker had to be cautious; he was boxed in by his role. He didn't mention being worried about what I thought was obvious—that producers might pretend to be his friends for their own advantage. In fact, he did seem to have difficulty distinguishing friends from sycophants, as many powerful people do. It all seemed a little sad.

IN CALIFORNIA THE WINEMAKER/CONSULTANT WHOSE REPUTATION benefited most from Parker's praise was Helen Turley, whom he often referred to as "a goddess" and "a genius." Just as in Bordeaux, Italy, and elsewhere, the 1990s in California saw the rise of the wine consultant/oenologist. The six-foot-tall, blonde Turley, who wore her hair long and semi-unkempt, favored denim workshirts, and worked side by side with her workers, had been developing her philosophy of wine since coming to the Napa Valley in the 1970s. By the 1990s, she espoused a particular style of winemaking and wine—rich, exotic, super-ripe—and had become one of the most sought-after consultants in the state. Parker had raved about the wines during her tenure at Peter Michael, Pahlmeyer, Colgin, Bryant Family Vineyard, Martinelli, and Landmark, as well as those from her own winery, Marcassin, and hiring her meant Parker would certainly taste your wines and, it was assumed, probably rave about them, too. That was surely one reason the Texas millionaires who founded Blankiet, a new superstar-to-be—Parker said so—hired her. Famously reclusive, Turley rarely spoke to the rest of the press. She didn't need to. By 2002 Turley and her husband, viticulturalist John Wetlaufer, charged wineries fees in the neighborhood of $200,000 to $250,000 a year. And they were noted for insisting on a free hand and an open checkbook, too.

Parker acknowledged in one newsletter that "with the stratospheric numerical ratings, lavish professional praise, and immeasurable admiration this journal has expressed for Helen Turley and John Wetlaufer, two cutting-edge wine personalities, I can understand if readers suspect we share a Swiss bank account." Parker's high ratings for wines translated directly into job offers and higher fees for the consultants who made them. There were a handful of others in California that he respected—wizards-behind-the-curtain Tony Soter, Mia Klein, Heidi Peterson Barrett, Philippe Melka, Paul Hobbs, and more—and hiring them seemed a guaranteed route to getting Parker to taste your wines.

———————

AT THIS POINT IN HIS CAREER, PARKER'S PRIVATE LIFE AND WORK life continually overlapped; he worked every day of the week, popping into his office whenever he wanted to do some writing, check faxes, or just sit at his desk catching up on incoming information and e-mail. Much of his private life continued to revolve around food and drink. The Parkers entertained frequently, and every dinner with friends was an opportunity to pull out fabulous wines from one of his three cellars and set up a "mini-tasting." There was always more wine than anyone could possibly drink, but Parker set out dump buckets so no one felt obliged to finish each and every glass. By the end of the evening the table would be littered with bottles. The Parkers' friends were (and are) a broad and eclectic mix, from wealthy collectors to people in the wine business to successful professionals—doctors, lawyers, professors. What they all had in common was a love of wine and food. "Unless you like wine you won't be invited to a dinner here," Pat told me, laughing.

Some of those friends dated from law school days, others, from early tasting groups in Baltimore. Still others were major collectors, like Parker's friend Park Smith, the founder of an eponymous home furnishings company based in New York City, and his wife, Carol, the company's vice chairman. Parker and Smith had met in Washington, D.C., through a wine merchant who kept insisting they should meet; a few months later they had lunch, and they've been very close friends ever since. After Faiveley sued Parker in 1994, in a stand of solidarity, Smith gave away all the bottles of Faiveley Burgundies in his cellar at the time, some as Christmas presents, claiming he couldn't stand to look at them. In fact, Smith stopped collecting Burgundy altogether. He shared Parker's passion for Rhône wines, especially Châteauneuf-du-Pape, and certain California labels, like Harlan Estate and Sine Qua Non, but he also owned sixty cases of 1982 Mouton, a Parker 100-pointer. Parker and Pat occasionally spent the weekend at the Smiths' home in Connecticut; on one weekend Smith pulled one hundred wines from his cellar, but the two couples managed to get through only fifty-one.

For years Pat had been the exclusive chef *chez* Parker, but Parker was beginning to cook more and more. He seemed to find it relaxing. Both agree his recipe for crab cakes is superior; he is quick to acknowledge, however, that "Pat makes the best crab fritters." His specialties, as with many men, relied on grilling meat, especially aged strip steaks; stews and any meat dish requiring assemblage, like braised short ribs or shepherd's pie, were Pat's territory. He chops the onions because they bother Pat. Parker's repertoire also included a gratin of potatoes and spaghetti and meatballs; Pat says he has a heavy hand with butter and olive oil. In the evenings, for casual meals, they cooked together, talking over the day, but for dinner parties they found it too difficult and stressful to work together in the kitchen.

Parker's success had brought him economic security and the financial wherewithal to buy one hundred-dollar, six-ounce Kobe beef steaks from Lobel's as well as to indulge some of his other interests besides wine and food. One was off-road bikes. He used to run with Pat, but now he was too heavy, and besides, his knees had given out; instead he biked twenty-five or more miles a day when he could, usually at least 3,500 miles a year. He'd given up riding on the road and turned to a top-of-the-line off-road model. Parker liked the bumps; riding off-road was more thrilling. One of his favorite rides was the NCR Trail, an old ten-foot-wide railroad bed that extended for 60 miles, from north of Baltimore through Monkton and Parkton to York, Pennsylvania. The remote trail, which mostly followed a river, traveled through a scenic, pastoral landscape replete with wildlife and passed an old Civil War hospital. Here Parker could unwind and think. Pat didn't take to the trail with him. "I'm too aggressive, I ride at a macho pace," Parker admitted to me. "I'm the classic Type A."

Pat's involvement with Maia's school, immersion in her garden club work, and behind-the-scenes *Wine Advocate* help occupied her days. She was relieved not to be working with Parker in the wine regions, tasting alongside him. Her husband was too high profile now. Everyone knew "every damn move he made," and she hated the scrutiny and attention. But from time to time she met Parker in Paris or elsewhere at the end of his tasting trip, and they never tired of exploring new restaurants together.

Parker attended school events whenever he was home, and when Maia was in the sixth grade he visited her classroom to talk about France and French wine. When the world's greatest wine critic showed all the students how to open a wine bottle with a corkscrew it caused an outcry among their parents, who complained he was promoting the consumption of alcohol.

As Parker had become more prominent in the wine-and-food world, charities approached him regularly for donations to their events, and these, too, became a major part of his and Pat's social life. He rarely turned any requests down, especially when it came to funding heart or cancer research, and was "a sucker" when it came to research on cancer in children. He was happy to use his fame to raise money for those less fortunate. That was how he'd been raised, he told me. So when Ruth Bassin, the wife and now widow of Addy Bassin of MacArthur Liquors, who'd been so instrumental in helping him launch *The Wine Advocate*, entreated his support for a benefit for the American Heart Association in memory of her son, Parker didn't hesitate; he selected and recruited top Bordeaux properties for a tasting and dinner he'd host. Châteaux quickly responded to the request to attend and pour their wines. (Who would refuse, given Parker's power?) One of the most desirable donations was dinner with the great man at a restaurant, which generally drew large bids. Parker would choose the menu, bring the wines, and pay for it all. Pat sometimes came along. Most of the people were fun; Parker viewed the occasion as rather like a blind date. Oddly, when every year he offered to donate the same thing to the auction at Maia's school, they didn't take him up on it until she was a senior in high school.

But it wasn't always wine, wine, wine. Parker made time for family vacations, buying a beach condo and then a ski condo in Colorado, where he mountain biked during the summer and the family tried white-water rafting. Once a year he grew a full beard, which, he said, made him look like Rasputin. Pat didn't like it. For two to three weeks in August, Parker swore off wine altogether. When he started tasting again, he would always think, *How did I go so long without this stuff?*

IN THE 1990S, ADVENTUROUS AMERICANS PLANTED GRAPES IN new regions in California, in Oregon and Washington, and on the East Coast. Winery and château owners and brokers in Europe looked to little-known regions—Priorato, Ribeira del Duero, and Bierzo in Spain, Sicily and Bolgheri in Italy—and rushed into New World countries where the right conditions for planting vines existed and land was cheap. It was the grape equivalent of a new gold rush. International partnerships flourished, because it was easier and faster for the owner of a French château or American winery to partner with an existing local enterprise and create a special bottling from vines already planted.

The grapes France had made famous—Cabernet Sauvignon, Chardonnay, Pinot Noir, Syrah, and so on—had become the international grapes, the taste of modern wine, the ones people planted in the New World (and in some parts of the Old) so they could compete on the international market. In France, as in most European wine regions, grapes had always been tied to place. Centuries of trial and error had determined that Pinot Noir and Chardonnay thrived in Burgundy, for example, and the name of the grape remained subordinate to that of the region and vineyard.

By contrast, place had never had the same meaning in the United States and other New World regions, because no one yet knew what grew best where. It was the grape that ruled in America. The name of the grape was how American wineries identified most of their wines, and new ventures in other climes followed their lead. Wine lovers had learned the taste of contemporary Chardonnay and expected a similar taste from every bottle labeled with the grape name, whether the wine was from California, Australia, or Chile. The popular grapes in America were the familiar ones, which is one reason so many new wine regions planted them, and all this, along with the spread of scientific winemaking, contributed to wine styles all over the world moving closer to one another. The problem wasn't making fine wine in these places; it was selling it. In America the shrinking of the number of wholesalers hadn't helped. In 1963 there were 10,900; at the end of the 1990s, there were three hundred, and one-third of the wine went through only five.

———

THE WINE WORLD WAS GOING GLOBAL, BUT ALTHOUGH PARKER traveled annually to France and California, he planned no trips to new areas. The wines would simply have to come to him.

There were three main routes for getting a new wine to Parker. One, of course, was to send a sample to *The Wine Advocate* offices and hope that he included it in one of his regular tastings and was wowed. It happened, but this method seemed about as efficient as a writer sending the precious manuscript he'd worked on for years to a publisher in the hope that someone would pluck it out of the slush pile of unsolicited manuscripts and offer him a contract. A better bet was using the famous oenologist/consultant route, where the right name was worth its weight in platinum. The globalization of wine had been helped along by this new breed of experts-for-hire—there were one hundred or so, of whom about twenty had star status and only one, Michel Rolland, was a superstar—who jetted from region to region and sometimes continent to continent. These "flying winemakers," like world-famous architects or celebrity chefs, dispensed advice for stiff fees. They pronounced on what to plant, when to pick, and wine-making techniques, enabling wineries to make more sophisticated, up-to-date, internationally appealing, often better wines than they could on their own. Carrying their own vision and a portfolio of the latest in wine-making practices, they were changing the way wines were made interna-tionally. The most influential for fine wine was Rolland, the master of Merlot, who, after perfecting his theories of winemaking in Pomerol, eventually extended his personal reach to one hundred-plus wineries in twelve countries on four continents. By the end of the 1990s, his client list read like a who's who of the world's elite wine properties: sixty or so châteaux in Bordeaux, a dozen wineries in California, Chile's Casa Lapos-telle, Argentina's Bodegas Etchart, South Africa's Rupert & Rothschild, even a winery in India.

Since they first met, Parker had been a great supporter of Rolland's winemaking style, which aimed for round, fruity wines, especially reds, that were easy to enjoy when young. Rolland had been one of the pioneers of using riper grapes and more new oak, and was committed to avoiding filtration—one of Parker's mantras—and he brought the ideas every-where he went. Critics said that his emphasis on ripeness and oak was a formula that made all his wines taste exactly alike, which was not com-pletely true, though many shared a family-style resemblance. A new or struggling winery knew that Parker liked Rolland's wines and that hiring Rolland would attract his attention. One Bordeaux château compiled a list of the seventy-three wines from the 2001 vintage that had scores of 90 or above, and the names of their oenological consultants. Forty-seven had used Michel Rolland.

The third route was through a *négociant* or importer who had the ear—or rather, the taste buds—of Parker. Importer Leonardo LoCascio had in-troduced Rolland's Italian counterpart, Riccardo Cotarella, to Parker in 1995. Cotarella was a key figure in the Italian wine revolution that was transforming viticulture in the country's central and southern regions. He started with his family estate, Falesco, in Umbria, and with white wines, and rapidly expanded, eventually consulting to over fifty wineries. Cotarella credits an early discussion with Parker as essential in shaping his wine philosophy. "Parker showed me what consumers wanted," he told me, "and I understood what I had to do to make wine for the market." In the June 1996 issue of *The Wine Advocate*, Parker called him "the spiritual equivalent of Michel Rolland." Cotarella, too, loved the Merlot grape and made the same kind of plush, smooth reds that Rolland had made famous. Parker praised both Cotarella's own estate wines and those to which he consulted. As so often happened, the Parker effect was circular: Cotarella's scores brought him more consulting jobs, which brought him more high scores, and so on.

When an ambitious new winery chose an importer to represent them in the United States, one small importer told me, one of the factors they considered was whether the importer could get them good reviews from Parker. His scores were like batting averages in baseball. Certain importers were very successful in getting high scores, and thus selling their wine. This all helped with cash flow. An importer could get dumped if he didn't deliver. As one told me, "Cotarella added to LoCascio's reputation, which in turn attracted other producers and gave LoCascio a momentum that he hadn't had before." Plus Cotarella brought to LoCascio many of those he consulted for, like Morgante, a tiny Sicilian winery. When I visited, the winemaker, who spoke no English, had proudly displayed *The Wine Advocate*, with the 92 score for their best wine, 1999 Don Antonio, clearly marked. It was only their second vintage.

Of course, it worked the other way, too, with importers reluctant to take on a new winery unless they could be sure the wines would get high scores. I heard—off the record, of course—that a few showed samples to Parker before taking a chance on a new producer. If it received a high score soon after, they went ahead.

In the mid-1990s, several importers turned Parker on to the new-style wines coming from Spain. One was diminutive dynamo Jorge Ordoñez, who chain-smoked Marlboros, and whose star skyrocketed because of Parker's reviews in 1994 and 1995 of Remelluri in Rioja and Costers de Siurana from the emerging Priorato region. Standing before a band of retailers and distributors he hoped would buy, Ordoñez would wave in the air a bottle of a wine Parker had liked, shaking it vigorously, and call out in his heavily accented English, "94 Parker, 94 Parker!" Producers called him; he was seen as having the magic knack for getting high scores from Parker and he boasted that he was going to fly Parker around Spain.

"Thanks to Parker's review of Alejandro Fernandez and his Pesquera," importer Mannie Berk, who owned the Rare Wine Company, told me, "the region of Ribera del Duero became fashionable." Berk himself introduced Parker to another wine from the region, Pingus, created by Danish winemaker Peter Sisseck in 1995; Parker gave it a 98. At the beginning all of it was exported to the United States, but prices and scores finally triggered interest within Spain. By the year 2000, the 1995 Pingus sold in the United States for $500 a bottle and in Spain for $1,000.

IN 1993, PARKER HAD WRITTEN OFF AUSTRALIA AS "A SOURCE OF oceans of mediocre and poorly made wines," but by the end of the decade several small importers had convinced him otherwise. "Oh my god this is great," he told John Larchet after tasting the 1994 Clarendon Hills "Australis" for the first time. Dan Philips, whom Manfred Krankl had introduced to Parker, brought in a handful of super-rich, super-ripe, super-concentrated old-vine Shiraz from the Barossa Valley for his company The Grateful Palate. They were wines almost guaranteed to appeal to Parker, and when he awarded several of them 100 points, the news appeared on Australian television and radio and in newspapers. Parker had "discovered" Australia! One paper asked, "What took you so long, Bob?"

Parker's first-ever foray to Australia, two weeks in late June and early July 2001, was officially a vacation with Pat and Maia—they toured Sydney, ate at Tetsuya and in the city's other noted restaurants, and went diving and snorkeling at the Great Barrier Reef—but he carved out five days to visit a few tiny wineries in South Australia whose wines he had already canonized and to attend a massive tasting of Penfold's Grange, a red long considered the greatest wine made on the continent. All Larchet's producers were asking, "When is he coming? Can you bring him to see me?" The rest of the Australian wine industry felt neglected.

The first winery door Parker walked through was Burge Family Winemakers, in the Barossa Valley north of Adelaide, where Rick Burge had been up since dawn warming the office and preparing the wines. Since Parker had given his 1998 Draycott Shiraz Reserve 99 points, his export orders had jumped from 25 percent to 60 percent and the fax hummed at all hours. Understandably, he found Parker "a bloody decent bloke." Later, he joined Parker and three other winemakers for lunch. Prices for all their wines were astronomical in the United States, but it was the American importers who were reaping the profits, not the wineries. That made winemaker Chris Ringland, who'd also been at the lunch, so angry that he had upped the wholesale price of his 1996 Three Rivers Shiraz (100 points, "a wine of unreal concentration . . . akin to a dry vintage port") 923 percent, to $600, and it still sold out in two weeks.

As usual, Parker didn't hesitate to pronounce upon what Aussie winemakers should be doing and not doing—making Euro-style wines. "What I find offensive," said Tim White, the wine columnist for the *Australian Financial Review*, as we shared a couple of glasses of Stonier Pinot Noir over a late lunch in Sydney, "is the patronizing American attitude that Australian wine was nowhere until Parker and the United States discovered it and that 'you guys should stick to what you do well.' Parker's in the country for a week, he tasted with a handful of producers—Australia is not just the Barossa Valley." White, who had reviewed Three Rivers (favorably) and Clarendon Hills Astralis (unfavorably) several years before Parker had heard of them, pointed out that the continent did in fact have a 150-year history of grape growing. The Australian show system of wine competitions, he contended, had influenced the style for big, opulent wines that could stand out in a lineup—that had nothing to do with Parker, and predated him. Still, warmed up now, he went on, "My concern is about the 'Parker cola' wines like Cape d'Estaing—they're 'McWines' that taste sweet."

What he did appreciate about Parker was the idea that "if a wine was good in the glass, it was good." And besides, Parker made wines sound as though they "were something you'd want to stick in your gob."

Parker, as usual, had scant regard for the Australian wine press. He was sure, he told me, that they were all in the pocket of the four big wine companies.

PART OF WHAT MADE THE WINE MARKET GLOBAL WAS THAT PEOPLE with no history of wine drinking and no wine knowledge, especially in Hong Kong, Singapore, and Japan, were embracing the culture of the grape and French wines, and with them Parker and his point system. The catalyst for Parker's power and reputation among wine consumers in

Japan was a longtime American resident, Ernie Singer, who began importing Burgundy and Bordeaux in the 1980s. He'd talked to Marvin Shanken about publishing *The Wine Spectator* in Japanese, but ultimately saw Parker as more useful to building the market—his stance as a consumer advocate was key, and it helped that he'd made his reputation on blue-chip Bordeaux. "For wine to take off it would have to be a credible product," said Singer, who also set himself the task of getting Parker's book *Bordeaux* translated. Parker's scores, too, were an ideal shortcut for the status-conscious Japanese who had trouble reading Western labels, and he stood alone, the ultimate expert, an image Singer knew would play well in this culture. In 1985 and 1986, "Parkerpoints" gained currency, and highly rated trophy wines such as Domaine de la Romanée-Conti rose in value from $200 a bottle to $4,000 to $5,000. A fallacious report that red wine cured cancer boosted more sales, and after Shinya Tasaki, a sommelier in Tokyo's prestigious Hotel Seiyo Ginza, became the first Asian to win the world sommelier competition in 1995, interest in wine soared even more. Witness the hugely popular Joe Satake, the swashbuckling sommelier hero of a Japanese *manga* (comic) series: he solved crimes with the help of a stellar palate and his astonishing knowledge of *premiers crus* and obscure appellations, besting snooty European sommeliers who challenged him in blind tastings.

Singer and Bordeaux *négociant* Dominique Renard had long been urging Parker to visit the country and promote his book, arguing that for the most important person in the wine world to come, the only one that the Japanese read and trusted, would demonstrate that Japan had become an important part of the industry. It already was. Despite an Asian recession, French wine imports increased by 115 percent to 1.32 billion dollars' worth between 1997 and 1998 alone; the number of cases of California wine imported had jumped five times what it was in 1992 to nearly 30 million.

An opportunity came in 1998, when Parker spent nearly a month touring Asia. Almost three thousand people packed his tastings, and media coverage spread his name even more. The organizers of VinExpo Asia in Hong Kong gave Parker star billing at what one participant called "the French dog-and-pony show." He reminded attendees at one of his two lectures that "you have to be humble when you judge wines." Afterward people lined up patiently outside to ask for his autograph. One Asian newspaper article insisted Parker's scores had encouraged fat-cat would-be wine collectors to start drinking Bordeaux: if you took a high-scoring wine to a party everybody knew it was the best, because Parker said so.

Finally Parker flew to Singapore, and then Japan, where, said Singer, "He was like a rock star." Television cameras and the press followed him around, hoping for a quote from, as one called him, "the God of wine." Among the crowd that greeted him at the airport was a second-tier actress, Naomi Kawashima, noted in Tokyo circles for keeping her career alive by being photographed nude. Determined to attend the V.I.P. black-tie dinner after Parker's 1996 Bordeaux tasting and press conference, she pushed her way in under an assumed name. When Singer's wife, who was seated next to Parker on the dais, made her way to a microphone, Kawashima seized the opportunity and slipped into the vacant seat. "She was in a dress cut down to her belly button," recalled Singer. She leaned provocatively toward Parker, her cleavage guaranteeing that a photographer would take a picture of the two of them. "I don't think Parker realized what was happening, but when I did I called security and had her thrown out of the party," said Singer. The widely published photo made Kawashima's career; she became, Singer said, "the queen of the wine industry ads, a household name."

Parker had no previous experience with sake and knew little about it, but that didn't stop him from extending his methods and rating system to a comprehensive tasting of 225 of the best, in Tokyo, though he later described it as one of the most difficult he'd undertaken. The alcohol level in sake is 18 percent, higher than the average 13 or 14 percent for table wine, and about a third of the way through, sweat was pouring off his brow. Parker thought, *What have I got myself into?* He felt as though he would drop dead right there. From then on, if a sample didn't smell appealing, he passed. When he wrote up the tasting in his newsletter, fifty-two sakes merited scores between 87 and 92. He was on more familiar ground at a professional trade tasting at the Pan Pacific Hotel Yokohama, where ten of Parker's "perfect" wines were served.

───────

DURING THE 1990s, PARKER'S GROWING POWER AND FAME brought with it the usual mix of worship and criticism, parody and humor reserved for those with celebrity status. Nicknames for him abounded in the press: Le Grand Bob, Le Pape du Vin, His Bobness, Le Gourou, Mr. P.

Retailer David Raines of Gordon's Fine Wines & Spirits in Waltham, Massachusetts, who was known for his offbeat newsletters, had a dream in which a wholesaler concocted a fake issue of *The Wine Advocate*. Out of that came his amusing "The Wine Avocado," which showed Parker's corkscrew with an avocado and redefined his scores: "95–100, gobs of tropical fruit; 90–95, lots of tropical fruit; 85–90, some tropical fruit." He sent it to Parker, who wrote to him that he was "blown away" with amusement.

But for the most part Parker remained as hypersensitive as ever, and was not above using his immense power to smack flat those who stung him. There was, for example, the parody wine label Jory Winery in California put on some bottles of their 1989 Red Zeppelin, designed to spoof the Bonny Doon "Le Cigare Volant" label as an April Fool's Day joke—it was released on April 1, 1991. The bottles also sported a neck label with a caricature—a bust of a man with the nose covered—with the moniker Emperor Bob III (a reference to Parker's place in the line of critics before him, Bob Balzer and Bob Finigan). This was surrounded by the words "The Emperor Has No Nose." *The Wine Spectator* suggested the image was of Robert Parker. Parker's sense of humor deserted him when he heard about it: the winery received a letter from a New York law firm threatening to sue on his behalf. In his reply, the president and winemaker, Stillman Brown, insisted that the likeness on the label was not Parker's but modeled on Nero; in addition, he asked, "Is your client known as Emperor Bob III?" The threats dissolved. In the third 1993, edition of his *Wine Buyer's Guide*, Parker wrote that "I have never met a Jory wine that I have enjoyed," calling their "gimmicky-packaged" Rhône-style blends "anorexic, virtually charmless offerings devoid of pleasure." Jory, not to be outdone, put the same label on the next two Red Zeppelin bottlings, with the added line "The Emperor's Reserve."

Internet bulletin boards, in particular Prodigy, where Parker posted tasting notes and answered questions, allowed full-scale battles to flourish. One regular contributor, Stuart Yaniger, who had a scientific background, became frustrated over the way Parker put him off when he quizzed for specifics. He himself had tried two wines, one filtered and one not, blind, on two different occasions and didn't find the differences

Parker did. Could Parker tell him which wines he had used for his similar experiments? If Parker rated one wine 93 and another 92, did that mean he had tasted them against each other? And if he'd tasted one with a wine-maker at dinner and one later in a winery, how did he know he liked one better than the other? He soon learned such straightforward questioning of Parker's ideas or methods was unwelcome, interpreted as "argumenta-tive." If Parker bothered to answer, it would be a dodge. Then, said Yaniger, "that was followed up by thirty or forty or fifty e-mails from the Human Shield or the Parker sheep—'how dare you imply the great man is any-thing less than a paragon of honesty. Clearly you are a horrible person and how jealous you must be of the Great Bob.' " These, said Yaniger, "were more like religious reactions than anything else."

In 1996, Parker staged a challenge to show up, some said humiliate, one of his online critics, Robert Callahan, a wine clerk who had become a major "irritant to Parker"—someone who didn't play by the unspoken rules or respect the etiquette of the emperor's court. Parker was well into his 1990s pro-California phase, and Callahan, no fan of California wine, took him to task for his new positive reviews of the wines, charging that they weren't just wrong, but ridiculous. Then, in what Mark Squires, who'd managed the Prodigy bulletin board, called a "take-no-prisoners assault," he attacked the 1994 Beaux Frères. For Parker, that was appar-ently the last straw. "Mr. P," as he was called by his cyberfans, invited Calla-han (who invited Yaniger) and twenty of Parker's longtime faithful Internet followers to a blind tasting of thirty wines that he had rated 90 and above to see if they could pick out which were California and which European. He would pay for lunch for everyone and even offered to cover Callahan's air or train fare.

The event, titled by Parker "The Robert Parker Invitational: The Prodigy Tasting," took place in early December 1996 at his regular haunt, the Milton Inn in Sparks, Maryland. Sixteen "team guests" flew in from as far away as California. As usual Parker was a generous and hospitable host, laying on a six-course lunch "fit for Bacchus" and looting his cellar for stel-lar and very expensive wines, like 1983 Château Margaux. Parker had put together a printed menu listing the courses and wines. It also featured a quote from Shakespeare's *Othello* that was much remarked upon: "Beware of Jealousy: it is the green-eyed monster which doth mock the meal it feeds on."

Guests made their guesses as to which wines were French or California. Some did better than others; Callahan guessed 60 percent of the wines correctly. It turned out that one of his two top Pinot Noirs was the 1994 Beaux Frères he'd previously trashed. He felt utterly humiliated. Had Parker set him up? If the event was intended to silence Callahan, it didn't.

The following year, Callahan posted a critical note on alt.food.wine.com questioning Parker's review of a wine and the correctness of some information in his revised Rhône book, asking rhetorically whether Parker simply wrote it on autopilot without actually tasting it. Days later, a messenger delivered a letter from the senior legal counsel at Simon & Schuster to Callahan's boss at the wine shop where he worked. It charged that as "Mr. Parker's tremendous success as a wine critic derives from his ability to offer honest, meaningful, independent and uncensored evaluations of wines based on tasting them," Callahan's statements "to the contrary are false and defamatory, and are damaging to Mr. Parker in his profession . . ." If Callahan didn't delete what he had posted, Parker and Simon & Schuster would "consider further action." Fortunately for Callahan, his boss thought it was all very funny. Callahan posted the whole exchange, including the letter, on various Internet bulletin boards, where it has festered ever since.

It looked like Parker, now the most powerful critic in the wine world, was having a hard time accepting his celebrity status and the criticism, justified or not, that inevitably came with it.

═══ ❖ ═══

AS THE WINE WORLD EXPANDED, PARKER FOUND HIMSELF stretched thin. He knew that some wine regions received short shrift in *The Wine Advocate*. He wanted it to continue to be, as he thought it already was, "the most respected, reliable, comprehensive, pro-consumer wine buying guide in the world."

The Wine Spectator, his only real competition, now had specialists for various parts of the globe: Jim Suckling covering Bordeaux, Italy, and Port; Per-Henrik Mansson, Burgundy (though that would change); Jim Laube, California; Harvey Steiman, the Pacific Northwest, New Zealand, and Australia; Tom Matthews, Spain; and Bruce Sanderson, Germany, Austria, Alsace, and Champagne (and later, Burgundy). Hard-core wine-lovers often complained about the unevenness of the tasting notes and ratings in *The Wine Spectator*. They nicknamed it "*The Speculator*" and raised questions about whether big advertisers received better scores. But Marvin Shanken knew what he was doing. It was a business, and he ran it well. His plush offices on Park Avenue, with smooth, buttery-soft leather couches, a collection of antique humidors in the conference room, and a glass-walled wine room filled with stellar bottlings, testified to its success. *The Wine Spectator* was well on its way to becoming a "lifestyle" publication, filled with color photos and articles about wine-country living, winemaker profiles, hot restaurants, and ads for luxury goods, and by the end of the decade its circulation hit over 300,000. Still, when it came to Bordeaux it didn't move the trade quite the way *The Wine Advocate* did. And none of its individual writers had the same status and clout internationally as Parker. They were players on Shanken's team and Parker, the lone sage of Monkton, Maryland, was still the better media story.

In Parker's shadow, other wine critics looked like lesser lights. Steve Tanzer's newsletter, *The International Wine Cellar*, was respected, especially for its coverage of Burgundy, which many thought he understood better than Parker, but he had fewer subscribers and little influence on most of the trade. Frank Prial's stature derived from his position as a *New York Times* columnist, not from his wine opinions. West Coast writers like Dan Berger remained of regional importance, occasionally taking up the cudgels to call attention to some of Parker's limitations. But, as one writer (who asked not to be named) observed in a long phone conversation, there was simply no real way for other critics to criticize Parker. None of them was making as much money or had as much influence as Parker did, which inevitably led people—and Parker himself—to assume every bit of criticism was a result of simple envy and dismiss what were real questions and concerns about what was happening in the world of wine.

The United Kingdom had a gaggle of articulate wine writers. Some in the younger generation, like Andrew Jefford and Anthony Rose, actually thought Parker had been a positive force and respected his no-conflict-of-interest stance. But none of the British wine writers had much impact outside Britain. The judgments of Michel Bettane, called the "French Robert Parker," had only modest influence in France even though he then ran the country's most important wine magazine, *La Revue du Vin de France*.

Globally, Parker had no peers. But he still needed help, and he began seeking it.

In the December 1996 issue of *The Wine Advocate*, Parker announced to his readers that he had "taken on a tasting and writing assistant": Pierre-Antoine Rovani. The two had met in 1992 when Rovani, then twenty-eight, was working at MacArthur Liquors and Parker overheard him angrily chew out an importer on the phone in fairly explicit French. They both loved to joke, they both loved to eat, they shared a similar vision about good food and wine, and they soon became friends, tasting together frequently. Rovani, too, has a full face and expanding waistline, though not yet approaching Parker's girth. But their backgrounds are very different. Rovani grew up in Washington, D.C. His parents are French—his father worked at the World Bank, his mother at the French embassy—and he is bilingual, immersed in French culture from birth; his father had a wine cellar, though when we met I didn't detect an overlay of French savoir-faire in his manner. Rovani had learned to recite the names of the villages of Beaujolais at the age of five, he told me, and did a show-and-tell with the wines in first grade, giving everyone a splash in a Dixie Cup. His father was one of Parker's first subscribers.

After five years at MacArthur, Rovani was trying to decide among several offers from various importers and was about to leave for a wine-buying trip in the Rhône Valley. When he asked Parker for advice, Parker instead offered him a job.

A week or so later they hashed out the parameters of the position in a breakfast meeting at a Novotel north of Avignon, and Rovani joined the payroll. At last, Parker had someone to write about areas where he had little interest or little time to cover, like cool-climate regions that produced lighter wines, ones where winemakers who used to pray for his attention now realized their delicate bottlings were better off ignored and having escaped the crushing yoke of a low score. And handily, Rovani could cover Burgundy, where Parker was no longer welcome. He warned Rovani his acceptance there would take time. His first trip to Burgundy as a Parker employee was difficult. "So, you're Parker lite," one winemaker sneered. Although Rovani was helped along by his fluent French and reportedly a less confrontational style than Parker's, many Burgundians had scant respect for his tasting abilities. He was regarded as a Parker clone, with a similar predilection for big, rich wines and little understanding of delicacy and finesse. Many disagreed vehemently with his assessment of the 1998 vintage, assuming a directive to dismiss it as having come from Parker.

Rovani also covered Alsace, Germany, Austria, New Zealand, and the Pacific Northwest, with varying degrees of success. He both admired and liked Parker, and though he chafed at not having as much influence as he felt he deserved, didn't mind being number two. In fact, he could be tremendously protective when he felt his boss was being attacked and would rally the fan club to defend him.

Did he hope that he might one day take over the empire? That, he told me, would be something for Parker to decide.

Parker, however, reassured his readers that he was in no way "scaling back his efforts . . . My banners continue to be unfurled, and my mind set remains that of a horseman on the day of a cavalry charge." The Lone Ranger was still on the trail, only now Tonto was riding beside, or rather behind, him.

Parker also hired an assistant in France, Hanna Agostini. Her husband, Eric, had been Parker's lawyer in the Faiveley case, and after it was finally settled, Parker hired Hanna, who was trained as a lawyer, to do some part-time work for him, translating and sending out questionnaires for updates of his Rhône book. In 1996, she began translating his books for Solar, his French publisher. Gradually Parker came to depend on her more and more to set up his appointments in Bordeaux, gather samples for some tastings, and eventually to translate *The Wine Advocate* into French. He felt great affection for the couple and their daughter, and referred to Hanna as "La Grande Parkerette." He often dined with them at their home when he was in France, and once a year, for three years, spoke to her daughter's class at school about the American perspective of France.

———

PARKER WAS JUST AS PASSIONATE ABOUT FOOD AS HE WAS ABOUT wine, whether eating at his own table or in the world's best restaurants. He was opinionated about it, too, and as ready to condemn great restaurants that were falling down on the job as he did wines that didn't live up to their reputation. In 1997, he began writing extended reviews of his favorites in *The Wine Advocate*, in an end-of-year roundup of his best meals of the year. There were no scores. Over the past two decades, he and Pat had eaten in just about every two- and three-star Michelin restaurant in

France, but one of his favorites remained the great bistro L'Ami Louis in Paris. He'd eaten there some eighty times by 1997, and it still merited "three 'wows' on the Parker scale of decadence," especially for their huge portions of baby leg of lamb and garlic potato cakes and their perfect roasted chicken, which inspired his "holy grail" to find a flavorful chicken in America. But "the planet's greatest French chef" in his view was Daniel Boulud, who owned Restaurant Daniel in New York City, where Parker ate as often as he could.

High on his list of favorite food items were truffles, foie gras, Peking duck, and dim sum. Unfortunately there was a dearth of good places to eat in the immediate vicinity of Monkton, and even in Baltimore. The Milton Inn closed; Parker now met importers at the Oregon Grille in nearby Cockeysville. At least every three or four weeks he gathered a group of friends at his local Chinese restaurant, Mark's Duck House, in Falls Church, Virginia, for seafood dumplings, Hunan fried fish, and potstickers and bottles of old Barolos and Barbarescos from his cellar. To him, they were a perfect match.

In 1997, Parker turned fifty; it was an occasion for more eating. In Bordeaux he celebrated "with several others who had recently cruised past the half-century mark." They tasted twenty-eight bottles of 1947s; he gave three—Mouton, Cheval Blanc, and Pétrus—100-point scores. By now his fame as a wine guru reached well beyond the confines of the wine world. Parker had become a bona fide celebrity, even appearing in gossip columns that usually covered the antics of movie actors, rock stars, and famous felons. Partway into a ten-course birthday party lunch at Restaurant Daniel with a half-dozen friends, his head had started to spin. He broke out in a sweat, and a doctor at the table thought he was in the throes of a heart attack. Parker, sure he was not, asked for a Coke. When an ambulance finally arrived he was carried out by a rescue team, his sizable prostrate form on a stretcher alarming an incoming Governor George Pataki. "Don't eat the scallops," Parker cracked. The *New York Post*'s "Page Six" column headlined the incident "Parker Passes Too Much Port."

AT THE END OF THE DECADE, PARKER'S INFLUENCE WAS FELT everywhere around the globe. But despite his worldwide fame, his wine activities were regarded as dubious, if not illegal, in his home state of

Maryland. The repeal of Prohibition in 1934 left regulating liquor to individual states, and Maryland was one of many that prohibited the practice of shipping wine directly to consumers without passing through a wholesaler. Hundreds of unsolicited samples were delivered to the Parker household annually, plus the many more Parker ordered and paid for. Eventually, in 1998, Dr. Charles Ehart, then director of the Alcohol Tax and Tobacco unit of the State of Maryland's Comptroller office, took notice. He tried to shut Parker down, lecturing him that what he had been doing for nineteen years was illegal and insisting that they would have to qualify Parker as "a wine expert." This entailed a visit to their offices in Annapolis and filling out a questionnaire as well as obtaining the same type of permit that allowed hospitals to buy alcoholic products for curative reasons. "Basically they just wanted to make me jump through all their hoops," he told me. "I have a temper, and I lost it. Here it's easier to get a gun than have a bottle of Chardonnay shipped to your house." Parker was always ready to strike a blow against "the pleasure police," what he sees as the "vestiges of puritanism" in America that lurk behind crusades against alcohol. How different from France, where wine wasn't suspect; it was part of the national culture. A survey asking French citizens what constituted "being French" found "wine" among the top five items.

––––––––––

THE SAME YEAR, PARKER RECEIVED A CALL FROM THE FRENCH embassy letting him know that he had been proposed for the country's highest award, the Legion of Honor. He learned later that it had originated in Bordeaux, the French wine region with the most national clout, as had his previous award, and the complicated process consisted of a strict series of steps to determine whether there were any black marks against a nominee's name. Believing that he had many powerful enemies among those in the French wine world, Parker had doubted that the nomination would succeed. But he told his ailing father about it anyway before his death in mid-1998.

So when he received a letter from the French ambassador to the United States in April 1999, informing him that President Jacques Chirac himself would bestow the award on him in a ceremony at Elysée Palace in Paris on June twenty-second, he was, as he likes to say, "blown away." Though he had written jokingly in a very early issue of *The Wine Advocate* that he'd spent so much money purchasing bottles of Burgundy that the government of France *should* give him the Legion of Honor, this was, he said, "beyond his wildest expectations."

On the morning of the twenty-second Parker was in his hotel room in Paris, worried about being late for the ceremony. He, Pat, and Maia were staying at the Hôtel Scribe, across the street from the Opéra and only a ten-block walk from the Elysée Palace, but Parker had insisted on hiring a private car, something he'd never done before in Paris, to pick them up. It was late, and Parker was sweating. They were all supposed to be at the palace at 10:30. Finally, with only minutes to spare, the car arrived and off they went, invitation from the president of France in hand, but Parker was still sweating.

He'd been working in France for the last couple of weeks, and everyone had been so impressed that he would receive the award directly from Chirac that he finally came to understand this was something of a coup. French presidents gave the award personally to a mere 1 percent of recipients; most foreigners received it from the French ambassador in their own country. Parker would be right up there with Ronald Reagan, Colin Powell, Neil Armstrong, and Robert De Niro, among the few famous Americans who'd been so honored. "But," he assured me, "I would have taken it from anybody."

About one hundred guests gathered in the formal red and gold reception salon of the Elysée Palace and stood before large French windows partly covered by heavy dark red drapes, behind the line of honorees. Pat pulled out a small video camera, but she and Maia could see only Parker's back. Chirac was at the podium facing them all. Above, enormous glass chandeliers sparkled and light streamed in from half-moon windows above the rich gilt-covered wall molding.

Parker stood in the center of the line, his bulk in a nondescript dark suit, white shirt, and red and black patterned tie, his still-brown hair, with only a few streaks of grey, curled under in the back like a pageboy. Of the eight people receiving the Legion of Honor, Parker was the only American and the first wine critic ever. The *Légion d'Honneur* was created by Napoleon in 1802 to honor, as Chirac noted in his opening speech, "those who have served France by bringing prestige to the country with their particular gift." In Parker's case, Chirac said, this meant "being the man who taught America about French wine." He saluted the "infallible palate" of the man "who is the most listened to and most influential [wine] critic in the world" and recounted that former president Bill Clinton had mentioned Parker's judgment on a wine the two presidents had shared with a meal at the bistro L'Ami Louis. "This lover of France and its wines," he continued, "has enormously contributed to the promotion of French wines across the Atlantic and in the entire world."

Recollections of the experiences over the past twenty years that had brought him to this moment played through Parker's mind. A wine revolution had taken place, and he had played a major role in it. The years of eating great food and drinking two bottles of wine a day had also had their effect. Parker now weighed 265 pounds and had a distinct waddle as he stepped forward to receive the award.

As Chirac pinned on his lapel the five-rayed star around a gold center that hung from a scarlet ribbon, beads of perspiration formed on Parker's forehead. It was an intense moment, "an out-of-body experience," he recalled. His eyes welled up, and he blurted out, "*Merci, le Président.*" Apparently you weren't supposed to speak; no one else did, and Parker's words caught Chirac off guard. He air-kissed Parker on both cheeks, and the exuberant Parker again did the unexpected, planting two large kisses on Chirac's cheeks, to his evident amusement. Then they clapped one another's shoulders with Gallic cameraderie.

At the reception, the two discussed their favorite bistro, L'Ami Louis, and posed for several pictures. Parker keeps the one of Chirac shaking hands with Maia on his desk, next to the medal, which sits in a box.

Parker, Pat, and Maia ate a celebratory lunch at nearby Ledoyen, Paris's newest three-star Michelin restaurant, and then headed back to the hotel so Parker could shed his suit and tie and turn back "into a normal person." As they ambled across the Place de la Concorde they passed a whole battalion of gendarmes, who suddenly snapped to attention and saluted Parker. "That was as cool as it gets," he told me with pride. Surprised, he learned from them that "when you wear the medal, every policeman and every military person is required by French law to salute you."

When he stopped in the shop on Palais Royal to purchase the little red ribbon you can wear in your buttonhole to show you're a Legion of Honor, the woman behind the counter, all smiles, recognized him and brought out one of Parker's books for him to autograph. It was a far cry from his visit in 1993 to purchase the little blue ribbon for *La Croix du Chevalier de l'Ordre National du Mérite*. Refusing to believe a foreigner could have received it, the same woman had warned him that wearing it falsely was a felony.

Stories appeared in dozens of major periodicals, including the *New York Times* and *Le Figaro* (but not *The Wine Spectator*). One major benefit was that no one in France has sued or threatened to sue Parker since he received it. The French had evidently decided Parker was a fact of life that they would just have to live with.

Back in the United States, Parker and Pat threw a blowout celebration party for 150 people at Restaurant Daniel in New York City. Old friends and wine people who had believed in and supported Parker from the beginning were invited. *Négociants* Archie Johnston and Dominique Renard flew in from Bordeaux, winemaker Helen Turley and winery owner Bill Harlan flew in from California, and importers Bobby Kacher and Marc de Grazia came from Washington and Italy, respectively. Parker chartered a brand-new bus to bring the Baltimore contingent. All toasted him with 1990 Dom Pérignon and drank magnums of 1990 Château La Conseillante and bottles of 1982 Château Grand Puy Lacoste. Both Parker and Pat gave impassioned, sometimes weepy, speeches. Parker's mother gave Daniel Boulud a kiss on his cheek and told him, "I've never had food like this in my life." "It was," Parker's old friend Bob Schindler told me with emotion, "the greatest wine event of my life."

8

MAKING
WINES TO
PLEASE PARKER

*E*VERY TWENTY-FIVE YEARS THE BORDELAIS need shaking up, in everything, because life here is easy and pleasant," said Jean-Guillaume Prats, his thin face breaking into a grin as if the idea amused him. "Everyone grows complacent. Parker was the one to shake everyone up." In his early thirties, Prats is the estate manager of Château Cos d'Estournel, a second-growth château in St.-Estèphe owned by his family until 1998. He had taken on the job after training and working in the financial markets in London, which gave him the perspective of both an insider and an outsider on the wine world of Bordeaux. We were talking about Parker over dinner at La Tupina, a beloved bistro in the old part of the city that resembles a country inn, complete with fireplace, hanging copper pans, and hunks of meat rotating on spits, and noted for the heavy, rich, old-fashioned food of the region, like cassoulet. Driving there had been complicated; many of the city's streets had been torn up during a massive urban renewal, causing constant detours. In 2002 Bordeaux's wine world was in the throes of change, too. Whether this was good or bad depended on whom you talked to.

In the years since Parker had promoted the 1982 Bordeaux, Prats said, his comments and scores had pushed château owners to reach for higher quality and invest in new equipment, made the reputations of names and brands, and shaken up the whole region by "making it possible for more modest châteaux to think they could be part of the play." Parker was the first one to make it all simple—"one palate, one system, one world." He paused. "It was when the 1995 and 1996 vintages came on the futures market that Parker became so powerful. That was when he began to make a wine's price." It was the first time many châteaux decided to wait until after Parker's scores had been published in his annual Bordeaux issue at the end of April to set their prices.

"With the 1995 vintage the whole Parker commercial system started working," agreed *négociant* Bill Blatch, whom I saw two days after my dinner with Prats. If you had a good "mark" from Parker you could charge more. For a top estate, Blatch said, "the difference between a score of 85 and 95 was 6 to 7 million Euros." If a château received a score of 100, it could multiply its price by four. It was as simple as that.

All brokers and *négociants* track current sales and prices worldwide, but when I dropped by Erik Samazeuilh's spare, elegant, stylish office at Tastet & Lawton, in an old building on the Quai des Chartrons, I saw how intensely Parker's scores were scrutinized when it came to international trading. On Samazeuilh's bare glass desk stood a large, flat computer screen on which he could quickly pull up any figure from his immense database of eight hundred châteaux. For each it included the wine's opening price, the best price between the château and shipper, the variations from low to high, and the Parker score. In terms of economic impact, it was only Parker who made the market, Samazeuilh asserted: "*The Wine Spectator* doesn't reach the top of his socks." He told me about visiting a surgeon in the south of France whose cellar was organized not by region or type but by Parker scores and who proudly showed him all his 100-point wines.

The United States was Bordeaux's strongest market, and if Parker didn't prompt Americans to buy, the Bordeaux market was headed south. And Parker's influence now extended to Bordeaux buyers in England. His impact democratized parts of the wine business even in the United Kingdom, where merchants like Lay & Wheeler or Berry Brothers & Rudd had a long and illustrious history supplying the upper classes with fine wine. Driven by Parker's scores, the heady speculation in Bordeaux in the 1980s allowed a new group of wine brokers without old-boy connections to start up and succeed. That was the story of Farr Vintners, which, as partner Steve Browett explained, began in the front room of a partner's flat in 1978, and rose to importance in 1984 by offering 1982 futures. They subscribed to *The Wine Advocate*, bought what Parker recommended, started including his scores in their catalog, and their business grew rapidly. Parker, says Browett, "has the taste of the common man." Farr Vintners is now the largest seller of Bordeaux futures in Britain.

As I traveled around Bordeaux talking with *négociants* and château owners, everyone had a strong opinion about Parker. Even the waiter at the Restaurant Le Tertre in St.-Emilion weighed in, telling me the reason their cellar had so few half-bottles was because Monsieur Parker had caused the prices of wines to rise so high the French couldn't even purchase *demi-bouteilles*. Parker's power made even those who liked him personally and had benefited from his scores very uncomfortable. As importer Melissa Seré, who always knew the gossip, once told me, "The French can see Parker is magic for their wine, but they resent having to do things the American way."

In early March all the châteaux had been waiting for "Le Grand Bob." The official *en primeur* circus started the last week in March, but Parker and James Suckling of *The Wine Spectator* usually arrived a week or two earlier and sometimes were even gone before the hoi polloi descended— the hordes of journalists and buyers who now came from all over the world and packed the city's hotel rooms and jostled one another at the Union des Grands Crus tastings, where everyone's smiles showed purple-stained teeth. There was one tasting for each appellation to which châteaux sent barrel samples. Back in the 1980s, when these tastings had begun, Parker, too, had attended them, but since the mid-1990s he'd dropped out, concentrating all his time in private tastings at the same several *négociants* and in his hotel room and at appointments at a variety of châteaux, where he was treated like visiting royalty. During the week or two he was in Bordeaux, phones were busy; Parker was the hot topic of conversation. His notes, said one château owner, were more important for the wines than the weather at harvest time.

In 1995, the UGC had decreed that no one could taste barrel samples before April first. But when Parker mentioned to his friend Dr. Alain Raynaud, the organization's president, that he could only be there the last week in March, Raynaud had reportedly said, no problem, we will change the date, but just for you. Parker readily agreed not to release his notes until after April first, which he never did anyway. Then James Suckling insisted on tasting earlier, too. Now it was a free-for-all. Once, Parker had been the only wine critic to place so much emphasis on tasting barrel samples from the new vintage and delivering his detailed verdicts soon after, but now a half-dozen journalists, including Suckling and Britain's Jancis Robinson, posted their notes online well before *The Wine Advocate* even came out.

The *en primeur* tastings generated excitement and buzz (hype, if you were cynical)—always considered important in selling Bordeaux—but also plenty of grumbles and complaints, most of which now had something to do with what they called "*le problème Park-air*."

One was the authenticity of the all-important barrel samples. Gonzague Lurton, who is in charge of Château Durforts-Vivens in Margaux, one of the ten châteaux owned by his father, Lucien Lurton, had been one of the most outspoken on the issue. A few days into my Bordeaux stay we discussed it over lunch at the noisy Lion d'Or, a restaurant in Margaux that is always thick with wine people. "For years we heard about the special 'Parker barrels,' " Lurton told me. He was sure that many châteaux sent a "Cuvée Bob" to *négociants* for Parker's tastings, which didn't fairly represent the final wine that they would bottle. While many châteaux now did the *assemblage* by March, not all did, so those samples were only an approximation of what would be bottled more than a year later. Differences existed even among the barrels intended for the *grand vin*, with some always better than others. Lurton was not the only one convinced a high percentage of samples were a *crème de tête*, taken only from the very best barrel or doctored to look good so Parker would give the wine a high score. Many well-respected insiders with whom I talked, like Marcel Ducasse of Château Lagrange in St.-Julien, were convinced a fair amount of cheating was going on. "One joke you hear often in Bordeaux," he said, "is when you find a barrel that is better than the others, that is Parker's barrel."

Thin, intense Lurton seemed to have a schoolboy earnestness rare among the French; he sounded indignant, but also a little sad, when he told me how he had proposed setting standards for barrel samples but that the UGC ignored him. "In 1997 I made a little label on my computer guaranteeing on my honor that this sample represents 547 hectoliters of wine. I was the only one to do this." Everyone had laughed at his naiveté, but, he said, "People know I am like that." Many of them had told him he was right, but also, " 'if a team is winning, we don't want to change the players . . . Gonzague, don't speak too loudly.' " All the same, Lurton stopped sending samples. "To be part of Parker's big competition to find the biggest, the most color, the most fat—no, I don't want to do that." In the next edition of his book *Bordeaux* Parker struck back, attributing Lurton's "false" assumptions of adulterated samples to the fact that Durfort-Vivens didn't do well in blind tastings.

In fact, Parker did worry about being set up. He too had a lot at stake. He'd heard there were Parker cuvées, and maybe it was true. "But I taste samples many different times, in different contexts," he rationalized when I quizzed him on this, "and I follow up when the wine is in bottle. If I see significant differences I'll stop tasting the wine from that château." He admitted he had sometimes found differences, but usually the wine in bottle had been better. He was convinced the reports of widespread cheating were overblown. He fell back on his usual defense: "The myth of the 'Parker barrel' simply comes from competitors who are jealous of my success."

The problem was that even if the samples were "fair and accurate," the wine itself changed from day to day as it evolved in the barrel and according to the weather. "When the barometer is falling," Alexandre Thienpont, of Vieux Château Certan in Pomerol, told me a couple of days later, "a sample will not show well." It could, he said, mean a difference of as much as 10 points in Parker's score.

Then there were the complaints that Parker tasted upwards of a hundred samples at a go. The châteaux had everything to lose by these blind comparisons. A delicate wine sandwiched between two powerhouses would definitely appear to be a weakling. Also, no one knew how long the samples had been open. Prats had decided to follow the first-growths' lead, as several other second growths had, and only allow cask samples to be tasted in the château's cellars, even by Parker. "It's perceived as an arrogant demand," Prats said, but he felt it was the minimum effort a visiting journalist should make.

"It's a matter of what happens in a blind tasting," explained Thienpont, who had done the same. After Parker tasted at Vieux Château Certan, he'd raised the score of a wine previously sampled in a big tasting by 3 points.

Later in the week, as I sat in the small, elegant drawing room at Château Pichon Lalande, in Pauillac, May-Eliane de Lenquesaing, the proprietor, put a thick lavender looseleaf notebook on the glass coffee table. This was her Parker book, with each document stored neatly into a separate plastic pocket. She was elegantly turned out, as always: blonde hair pulled back, a red sweater, a perfectly tied red, navy, and white scarf, fashionable red shoes, and her well-behaved King Charles spaniel, named Picolo, curled on her lap. She had been running her family's château since 1978, tirelessly pushing for quality and promoting her wine, in Europe and America, where I had first met her. Her late husband had been a French army general stationed in Kansas. The Missouri River, she said, reminded her of Bordeaux's Gironde, which seemed to me an unexpected comparison.

Parker always praised Pichon Lalande, though not always as highly as de Lenquesaing thought he should have, and it was a particular annoyance to her that he often gave it a medium mark in March and the following fall would almost always say, "Oh, this is so much better than I thought." She knew him well. "I've seen in the past years all his mistakes, and I said, '*Attention, attention!*'" She objected to the scoring system, Parker's preference for "big wines"—"the modern wines have big body and no personality"—but most of all she was strongly against the press tasting barrel samples so early, in March. "They are like a little boy of four or five—how can you say he is going to have a brilliant career?" she asked. She didn't send samples to big tastings. Everyone had to come to the château, even Parker. "I'm going to be seventy-eight years old, I say what I think; if no one likes me, *tant pis!*" she exclaimed.

Professor Émile Peynaud had warned all the châteaux about this back in 1987: "They think they have something to gain. . . . And they don't dare refuse to let him taste, which they ought to do, for fear he will think their wine has some problem to hide."

To make sure barrel samples would look good for the press (read Parker) at the end of March, people were even shifting the way they vinified their wine. Everything depended on March. "The whole point is to get a month's start," *négociant* Bill Blatch had said. "Now people are putting wine into *barriques* as soon as fermentation is finished." They aerated wines with *microbullage* (micro-oxygenation), so they would taste softer and more obviously fruity.

"There's a lot of hypocrisy about the tastings being too early," insisted Parker. "The châteaux are the ones who want to sell as early as they can. I don't always want to be first. In fact, *The Wine Spectator* is already on the Internet for several weeks before my report, and so is Jancis. Unfortunately no one in Bordeaux cares." Which was all too true.

As more corporations had purchased châteaux and were looking for a good return on their investment, the pressure to obtain high Parker scores increased. One château that needed to make money right away was Château Lascombes. Once owned by Alexis Lichine and a group of investors, the now-underperforming property in Margaux had been purchased for $67 million by a consortium that included Colony Capital; Tony Ryan, the Irish owner of Ryan Air; and some investors from Goldman Sachs. The deal was organized by Dr. Alain Raynaud, who would manage the project, and he immediately brought in oenologist Michel Rolland. The combination was considered a surefire route to Parker. One broker told me over dinner that before the new Lascombes had offered their first vintage, he'd attended a lunch for brokers at which Raynaud had assured them that Parker was sure to give the wine a high score. If you had to sell a wine at a high price to cover your bank loans, you had to be sure to score 94 to 95. Even banks knew how important the score was.

Not everyone had to wait for the score to know whether to buy. In 2000, Guillaume Touton, an importer of French wines in New York City, received a telephone call in his office from a well-known Bordeaux *négociant*. "Parker just left with Hanna Agostini," he said. Parker had spent two and a half hours quietly tasting and spitting dozens of barrel samples from the 1999 vintage. The *négociant* had stood with his back against the wall, observing Parker. "Then," said the *négociant*, "Parker hit one wine and said, 'Wow!' I tried to control my excitement. That will probably be 93 points. You should buy it."

"Are you sure he said, 'wow'?" asked Touton. "Maybe it was 'whoa.' "

"No," said the *négociant*. "It was definitely a 'wow.' "

They argued for a few minutes about exactly how intense Parker's "wow" sounded. The price of the wine would be between $800 and $900 a case, and Touton had to make a decision in the next three minutes. "Just based on the 'wow' I made an $80,000 decision to reserve seventy-five cases," Touton told me, chuckling. "It was risky. But later Parker gave the wine 93+."

For the Bordelais, Parker's scores were not unlike the *Guide Michelin*'s restaurant stars. They waited in February to see who would be upgraded and downgraded after his January visit to retaste the wines he'd tried the previous March. In New York, at 11:30 P.M. on February 26, Emmanuel Cruse, whose family owns Château d'Issan in Margaux, was awakened by an arriving fax from his office—a three-and-a-half-page tabulation of all Parker scores of the 2000 vintage so far. First were listed his initial scores, which had appeared in his April twenty-third newsletter; then the update scores he'd given them when he retasted in January, from the just-mailed February twenty-eighth issue; and finally the difference between the two. Most had gone up or down by just a point or even half a point, but Château Phélan Segur had gone down 7.5 points. Luckily the château has sold their wine, Cruse told me, adding, "Now it will be impossible for the merchants to sell it for the next six months to a year."

Parker had changed the rules of the game. As his opinion of the quality of the vintage and individual wines increased his influence over prices, the influence of the *négociants* had decreased. In 1999 the Syndicat des Négociants de Bordeaux tried to fight back by issuing their own collective report on the *primeurs*, but it was too little too late.

Though he remained convinced he had been a catalyst for positive change, Parker was aware of the grumbles, and darkly assumed he had many enemies and few real friends. When he heard a rumor that disgruntled château owners were planning to alert the police to when he would be leaving a tasting in order to set him up for a drunk driving charge, he took it seriously. He purchased a digital alcohol measuring device from The Sharper Image and obsessively checked himself after every tasting.

———

WHAT CAUSED THE MOST CONTROVERSY ABOUT BORDEAUX IN THE late 1990s was the rise of the wines of the Right Bank, especially in formerly sleepy St.-Emilion, and the handful of revolutionary winemakers known as *garagistes*. The term referred to the fact that they made their wines, called *vins de garage*, in tiny quantities in sheds and garages rather than in traditional châteaux.

To many traditional châteaux owners these ambitious upstarts, like logger-turned-*négociant* Jean-Luc Thunevin of Château Valandraud or stonemason Michel Gracia of Château Gracia, seemed like barbarians at the gates. People complained that some of their wines were ultra-concentrated, flamboyant, oaky, and super-ripe, more like major-league New World Cabernets than traditional restrained Bordeaux, yet a few, like Valandraud and La Mondotte, actually sold for double the price of the first growths in the United States and Asia. What their vineyards lacked in pedigree, they made up for with super-low yields, hand harvesting, ultra-ripe fruit, controversial winemaking methods, lots of brand-new oak, and plenty of money thrown at tiny plots of land. These experiments and their overnight financial success most closely resembled the cult wines of California, and they inspired dozens of others. But they were highly vulnerable. They had no history or track record or reputation, only Parker's points to justify their high prices. Most of them, charged Erik Samazeuilh, "wouldn't even exist without Parker." Thienpont became more and more impassioned as he castigated them. "They seem like first-growths for three to four years," he said, and then "become rubbish."

The story of outsiders challenging the establishment with hard work and dedication to quality was one that always resonated with Parker, and the style of the wines did, too. He was convinced they were a wave of the future and that some of their winemaking methods would become commonplace in Bordeaux, a sentiment that struck fear and anger into the hearts of many château owners in the Médoc.

As Jean-Guillaume Prats and I compared the already fading fruit of the *vin de garage* 1996 Château Haut Condissas with the classically evolving, elegant 1995 Château Pape-Clement over dinner, he shared his own theory about why the *vins de garage* appealed to Parker: "America is a land of dreams and excitement," he said. "It is fun for him to discover something that is new." Prats himself yearned for a way to experiment, but it would not be in Bordeaux—he would look at Spain or South Africa. At Cos d'Estournel, it was important to "follow the rules."

The success of the *garagistes* that Parker had made possible had stabbed the very soul of Bordeaux and reignited long-simmering debates over tradition versus modern experimentation, *terroir* versus the winemaker's methods, connoisseurship versus sensation, wines with finesse and elegance versus wines with power and concentration.

It was as if the *garagistes* had brought a California virus to Bordeaux: the revolutionary idea that a *vigneron* didn't need a hallowed patch of ground, a slope of soil identified by centuries of toil, in order to produce superior grapes and wine. Given halfway decent fruit, a great wine could be brought into being through sweat, sheer ambition, vision, and daring, often aided, of course, by new technological advances and money.

Many château owners on both the Right and Left Bank found these wines far from the Bordeaux ideal of balance, complexity, and the ability to age. They, too, like Burgundians, believed in the specialness of *terroir*, that "hazy, intellectually appealing notion," as Parker once put it, "that a plot of soil plays the determining factor in a wine's character." It was the basis of their elaborate system of appellations and the various hierarchies of quality, and provided the "sense of somewhereness" of which their wines were supposed to taste.

Belief in *terroir* meant that it was the land that held the magic and shaped the wine, not the winemaker. To believe otherwise was to dethrone France, rob it of its pride of place among winemaking nations. To many Bordelais, like Pascal Delbeck, who manages Château Belair in Saint-Emilion, Parker was giving aid and comfort to the enemies of *terroir*, those who were ready to strip wine of context, culture, and the people who stood behind the label. Once that had been accepted, wine was reduced to a grab bag of effects in the glass and then, of course, it didn't matter how those were achieved—it was just a product of the cellar, the offspring of anonymous grapes and an oenologist's tricks, not a liquid that summed up . . . a *place*.

What it all boiled down to was that if you didn't make wines that Parker liked, you had a more difficult time selling your wine in America and had to sell it for a lower price. That was the story for Gonzague Lurton and also for Olivier Bernard of Domaine de Chevalier, one of the wines English critics believed Parker constantly underrated. The incentive to change your style and make wines in what was assumed to be the Parker style was great. The French called those wines *Parkerisé*—"Parkerized."

"There are a couple of dozen pure Parker wines and the number is growing," *négociant* Bill Blatch said, shaking his head. "And it is spreading to the Left Bank, too." In my travels people cited Château Haut Bergey, Clos L'Eglise, and Haut Condissas, and worried about Château Pape Clément, a Graves, because the new owner had hired as a consultant . . . Michel Rolland.

———

FOR YEARS, PARKER HAD BEEN HEARING THE ACCUSATION THAT HE only liked a certain style of wine. The topic seemed to depress him. Sitting in his messy office in late 2002, rolling his desk chair around as we talked, he shook his head wearily and then grew heated as he insisted, "People say I like these bombastic, oaky, fruit-driven wines, it's the uncivilized American taste, and I'm leading everyone to it—and it's such a myth. I'm a Francophile, and French wines by their very nature are elegant wines. . . . My European colleagues accuse me of being this globalist. Obviously when I look at wine quality you have sort of general ideas in your mind, a level of acceptability, but the most important thing is they have to reflect their place of origin and the grapes they come from. I want Nebbiolo to taste like Nebbiolo." He didn't want everything to taste alike. "My tastes," he said, "are too diverse." In fact, he was convinced that he appreciated and wrote about more different styles of wines than any other critic.

He claimed to be a wine-loving omnivore. Parker liked big rich wines, yes, but he liked the complexity of Château Haut-Brion, too. He liked the modern international-style wines of Angelo Gaja, but also relished the traditional-style Barolos of Bruno Giacosa, which he had bought regularly for his own cellar until, he said, they became too expensive. Indeed for lunch that day at his home we drank a lovely 1989 Trimbach "Clos St. Hune" Riesling and a supple, satiny 1970 Château Figeac, neither of them the blockbusters people assumed he favored, although perhaps the choices were calculated to make that very point.

Still, when I read through his books and newsletters, there seemed to be whole categories of wine that didn't much interest him: zingy New Zealand Sauvignon Blanc, herbaceous Loire reds, red wines with high acidity—in fact, lean, austere wines in general, even if they often went better with food. With few exceptions, the wines that received the 95+ scores and set the markets moving had certain things in common—rich texture, intensity and concentration, plush fruit, and, for reds, low acidity. In fact, they were mostly red. The 100-point wines were more often those Parker described as "massive and powerful" than "delicate and subtle." Some of his "perfect" and "near perfect" reds resembled a "dry vintage port," as he put it, although other palates might be more reminded of cough syrup. Many had high alcohol, even upwards of 16 percent. The sensation of ripe fruit was paramount. Parker was fond of asking, "How much enjoyment do you get eating a pear that's not ripe? If they're not ripe they're green, they're acidic, they're just not flavorful, and they're not fun." He himself had admitted in print that when it came to Bordeaux he had "a stylistic preference for more opulently textured wines with lower acidity." Reports that he was enamored of lavishly oaked wine seemed much exaggerated. After all, he was a fervent lover of Châteauneuf-du-Pape, most of which used no new wood. Parker fit neatly into British writer Steven Spurrier's body-type theory of wine preferences. Slim ectomorphs, he observed, liked delicate wines like *blanc de blancs* Champagne; big, burly mesomorphs (like Parker) gravitated toward big, brawny wines that matched their physique.

After many thousands of wines, Parker's concept of wine greatness in the late 1990s had come down firmly on the side of fruit, concentration, overall sensory impact, and sensuous texture. Was that so bad? In many ways, what the wine world thought of as "Parker's taste" reflected the zeitgeist. It was, after all, an age of sensation and immediacy. In David Brooks's *Bobos in Paradise*, his look at contemporary upscale culture in America, he pointed out that the educated elites in the 1990s "love texture . . . Everything the educated person drinks will leave sediment in the bottom of the glass: yeasty microbrews, unfiltered fruit juices, organic coffees." He could have added wine. The contemporary age, and not just in

America, seemed to want plush, easy, and somewhat obvious flavors, something you didn't have to work too hard to get. Pleasure meant loads of fruit with a hint of sweetness and soft tannins for drink-me-now consumption. In other words, wines for immediate gratification.

And, as Gerald Asher pointed out to me over a dinner in Paris, who could be surprised that subtlety and finesse seemed to be losing favor in wine when they weren't much in evidence elsewhere in the culture?

IN THE BEGINNING, MOST IMPORTERS, *NÉGOCIANTS*, AND WINE-makers simply hoped that Parker would give their wines a decent score. As his power grew, they began to pick up on his stylistic preferences, as Becky Wasserman did. Whether they tasted with him or not, it was simple: they could all keep track of which wines got the high scores. If they had a wine they thought fit his preference profile, they brought it to his attention, as Dan Philips did in 1997, when he started importing big, whacking old-vine Barossa Valley Shiraz wines from Australia (Parker's laudatory reviews inspired a story about Philips in *Time* magazine).

Some took a hands-on, proactive approach in the late 1980s and early 1990s. Importer Bobby Kacher, whose long, sleepy-eyed face belied his reputation as a wide-awake and persuasive salesman not only sought out big, rich, fruity wines in Burgundy, but convinced winemakers he worked with to use the techniques necessary to make them in a Parker-pleasing style. He urged Bertrand Ambroise to use much more oak, for example, even ordering the barrels for him, and pushed others to harvest grapes riper and change some of their techniques to get deeper color and more concentrated flavors. There were many in Burgundy who complained Kacher's wines were too big and woody, that they ignored *terroir* and didn't resemble traditional Burgundy at all. Estates that didn't want to co-operate, who had other ideas of what Burgundy should taste like, such as Etienne de Montille of Domaine de Montille, looked for another importer.

In Italy, Marc de Grazia, who had the look of a Renaissance poet—soft brown eyes, a delicate face, a prominent nose, rumpled black hair, and shirts with flowing sleeves—spent his creative energies on pushing the small producers he worked with in Piemonte and Tuscany to change the way they made wines, sometimes in the face of the older generation's strong disapproval. Winemaking families argued bitterly; one father stopped talking to his son for a month. But the result was more modern-style wines with more fruitiness and the taste of French oak that appealed to the international market. While Parker applauded the higher quality, others saw de Grazia as a meddler killing the taste and style of traditional Italian wines. They nicknamed him "Marco Disgrazia," using the Italian word for disgrace.

But were these wine-shifters true believers, who preferred the style they advocated, or just market-savvy importers and winemakers looking to make a buck off Parker's taste buds? Or did they just recognize the reality of what most modern wine-lovers liked to drink and what it took to compete internationally? The answer depended on which importer or wine-maker you talked to. The ubiquitous Riccardo Cotarella credited Parker with showing him the taste that consumers wanted. If you followed the formula, you could predict the results. Inevitably there was the temptation to crank up the wine to jump onto Parker's fast track and reap the financial benefits of a high score. As dozens of winemakers told me, "If I want to make a 95 Parker wine, I know what I have to do."

Each grape variety has a certain character and spectrum of flavors. But the clone of the grape, whether the vines are five years old or fifty, what kind of soil they are grown on, the slope and direction the vineyard faces, how the vines are pruned, whether they are sprayed with chemicals, what the weather conditions are, when the grapes are picked, the type of press used, whether fermentation takes place with wild yeasts or commercial ones and at what temperature, whether the wine is pumped or flowed by gravity, whether the wine is aged in oak and what kind and for how long, whether it is fined or filtered before bottling—all these factors contribute to the impression the wine will have in the glass.

The basic formula for Parkerization was well known, and any wine-maker could repeat it like a mantra: In the vineyard, thin the vine leaves (*effeuillage*) so that more sun hit the grapes to allow them to ripen more evenly and quickly. Reduce yields by "green harvesting," or cutting off many of the grape clusters while the grapes are still green so that yields will be low and all the flavor is concentrated in the remaining grapes. Harvest as late as possible, letting the grapes hang until they are super-ripe, ensuring substantial alcohol and body, and sweet tannins. Color comes from contact with the grape skins, so cold macerate them as long as possible. Smooth out the wine early in its life with malolactic fermentation in the barrel. Then do as little as possible: no racking, no fining or filtration before bottling. Keep it natural—that was the basic idea. (Of course, methods such as harvesting riper grapes could be pushed too far, and many winemakers in France and elsewhere insisted that super-ripe grapes made jammy-flavored wines that were too high in alcohol. An article in *Wine Business Monthly* warned that in pursuit of making "trophy" wines that succeeded in blind tastings, winemakers had created a new high-alcohol style of wine, "social wines" that were tiring to drink and didn't complement food. Average alcohol levels in Napa Valley wines had increased from 12.5 percent to over 14.5 percent since 1971, and "the spiralling of this social wine style seemed to be totally market driven.")

Beginning in the mid-1990s, new winemaking technology could be counted on not just for damage control in the cellar but for fine-tuning the final product. Now you could use dial-a-flavor (raspberry to tropical fruit) fermenting yeasts, add tannin, pigments, acid, and oak chips (cheaper than barrels), and employ treatments that were far more space-age science than little-old-winemaker. Reverse osmosis machines and other concentrators could remove excess water and create super-concentrated wines. Spinning cones could remove the excess alcohol that resulted from super-ripe grapes. Micro-oxygenation, in which tiny bubbles of oxygen are introduced into the tank of a barrel of wine to replace racking, could fix the color and speed up a wine's evolution so a barrel sample tasted better earlier, for critics. As one winemaker put it crudely, "Critics talk natural wines, but it is often the fucked-with wines that are the 93- to 95-pointers."

In fact, in a very ordinary one-story tan stucco office building on a street in Sonoma, California, wine consultant Leo McCloskey was taking the concept of making wines to please Parker to its logical conclusion. His office was square and plain; the view was lawn, not vines; and on the morning I spent with him his desk was clear but for a state-of-the-art laptop.

Perhaps it was only a matter of time before someone would come up with the idea of analyzing high-scoring wines in a lab and then devise a software program to predict whether a wine would please the big guy's palate and tell you what you had to change to make sure your wine's score would hit 90 or above. McCloskey was that man, to some the Dr. Frankenstein of the wine world. He liked to say, "We've solved the math of flavor for wine." His approach to winemaking was a kind of reverse engineering, starting with the end point—the score—and working backward to the cellar and vineyard.

McCloskey sounded half missionary and half bullshit artist, like so many in the wine business, and seemed younger than his age, then fifty-three. His yellowish-white hair was brushed straight up punk fashion and he had the habit of leaning back in his chair and liberally dropping in teenage slang (from his children) and the lingo of New Age economy books like *The Innovator's Dilemma* as he talked about how his firm, Enologix, used advanced chemistry and the power of computers to help his seventy-five-odd winery clients "cross the 90-point break." Most were small wineries following traditional ways to produce fine wines that were regularly reviewed by Parker, like St. Francis, Diamond Creek, and Hahn Estates. Help from Enologix, claimed McCloskey, had pushed Joseph Phelps's 1994 Bordeaux-style blend, Insignia, to a "breakthrough" score of 96. The company's ads promised "improved national critics' ratings," and often delivered.

In 1984, while working on his doctorate in chemical ecology at the University of California, Santa Cruz, and surfing in his spare time, McCloskey had researched techniques to measure the molecular structure of the chemicals in grapes and wine. Some four hundred to five hundred separate chemical compounds comprise a wine, but McCloskey had identified eighty-four—thirty-two in reds and fifty-two in whites—as "flags," the significant ones that in different combinations and proportions, he was sure, were related to its price and quality. He enlisted the head of the math department to help him create a mathematical model from his data.

A few tests convinced him he was on to the next hot thing. There was the tasting of a dozen Pinot Noirs with a group of winemakers in which McCloskey, based on the wines' chemical profiles, predicted how they would rank them. He was right. His software correctly predicted the relationship between chemical composition and price for a dozen vintages of Château Lafite. And in 1995, he realized he could use his Quality Management System, or QMS, software to predict the ratings of America's foremost wine critics. If you weren't getting the scores you wanted, QMS would help identify elements from vineyard to bottle to improve quality and, more important, the score.

The process started even before the grapes were harvested and took into account regionality. McCloskey crushed a bucket of grapes and created a "lab" wine, then used chromatographs and a spectrometer to determine the optimum time to pick based on a QMS projection of ultimate quality and aging potential. Later, when the wine was in barrel, he measured "flag" compounds in different barrels, linked these to vineyard and fermentation data in the database, and used QMS to help determine winemaking techniques and eventually to create a series of virtual mixtures to see how each would stack up in Parker's (and other critics') mouth.

Enologix's vast database of some 50,000 virtual wines, each broken down into the individual chemical compounds that are responsible for its flavor, texture, color, and aroma, was stored digitally on a secure server in a nearby warehouse. McCloskey, obsessed with trade secrets, guarded it closely and refused to divulge the science of it all, telling me, "We're like Los Alamos."

Of course, people had always made wines to please a market. The wines of Hermitage, in the Rhône Valley, had been added to Burgundy and Bordeaux to beef them up for the British back in the nineteenth century, at a time when Champagne-makers created special sweet cuvées that appealed to the Russians. In recent years giant California and Australian wineries test marketed their wines like the mass products they were, using the usual focus groups to determine the exact flavor profile that consumers wanted at a particular price point. Kendall-Jackson had scored big with their Vintner's Reserve Chardonnay, created in 1983 with a ripe, slightly sweet character designed to appeal directly to people just coming off Coca-Cola. These were commercial wines aimed at a mass audience, but the same techniques could be used to target niche markets.

Émile Peynaud believed that oenologists were responsible for changing wine styles by making wines that pleased themselves. "If tastes change, it is because the product changes," he wrote. "The public cannot imagine a taste which does not exist." Just as each period has its own style in music, clothes, and standards of behavior, it also has its own style of wine. "But,"

Peynaud added, "it is less a question of fashion than of conception, technology, organization of the production, and marketing."

The new wrinkle was that the production of fine wines, wine at its best and most diverse, was now being skewed to critics' tastes—especially Parker's. Industry giant Brown Forman gave sliding-scale bonuses to winemakers at its Fetzer Vineyards that took into account high marks from Parker and *The Wine Spectator*, and Dennis Martin, the director of winemaking, admitted that he factored in their preferences when he made Fetzer's $30 wines. How else would they move briskly in the marketplace?

Parker now fit perfectly into French sociologist Pierre Bourdieu's observation that certain people at the top of each specialty have the power to consecrate, to confer prestige and honor on favored individuals, subjects, and "styles of discourse," or ways of talking about them. Bourdieu was talking about intellectuals in the marketplace of ideas—Clement Greenberg's power as an art critic comes to mind—but just plug winemaker, wines, and winespeak into his model and it was clear that Parker = chief consecrator. Those that hold that power, observed Bourdieu, influence taste, favor certain methodologies, and define the boundaries of what will be talked about, all of which Parker certainly did.

But it wasn't just the taste of wine that Parker influenced, he had also changed the way people talked about it; he shifted the very vocabulary, which in turn changed the way people perceived wine, what they looked for, and how other people wrote about it.

Talking about smell and taste is difficult, mired in metaphor, and winespeak all too often seems a private language, confusing insider talk that requires some extended and elaborate initiation to penetrate. In fact, wine professionals had struggled for decades to come up with a meaningful vocabulary that had a modicum of objectivity and reflected true sensory analysis so that everyone would know exactly what the terms meant. In addition to its color, all wine has certain basic dimensions that are measurable—acidity, tannin, and sweetness, for example. And certain flaws, like the musty, cardboardy smell and taste known as corkiness that is the result of trichloroanisole (TCA) contamination, usually imparted by a tainted cork, can be identified by chemical analysis. It is not difficult to learn to identify these. But describing how wine tastes and smells is far more complicated. One prominent professor of oenology at the University of California at Davis, the late Maynard Amerine, made an attempt to come up with an analytic vocabulary in his 1976 book *The Sensory Evaluation of Wine*, sternly admonishing people to rely on his approved list of words, as well as providing a long list of those you should "avoid at all cost," abstract terms like "austere" or "lively" or "coarse" or "noble." He came down firmly on the side of objective science, against what he called the romantic or poetic emotional approach to talking about wine or anthropomorphizing it with terms like "manly" or "feminine." Another professor at Davis, Ann Noble, invented her own easy-to-use "Wine Aroma Wheel" with the goal of establishing a common vocabulary for tastes and smells to help tasters train their noses and brains. The Wheel is a flat laminated disk divided into colored pie slices; in an inner circle are the general descriptors, such as "floral," "fruity," and "spicy;" subsequent rings grow more and more specific—"fruity" leads to "berries" leads to "raspberry." Noble, too, emphasizes words that are "analytical" rather than those like "fragrant" or "elegant," which she calls "hedonic" or judgmental, terms that carry with them praise and blame. English professor Sean Shesgreen, in an article analyzing "winespeak for a new millennium," saw contemporary tasting notes as a new kind of pastoral that focused on "picturesque foods" and related the transformation of wine language to broad cultural changes. The old gender and class terms, he says, belong to the formal style of France, while the new food and pastoral images suggest the casual style of Italy that is more appealing to today's consumers. References to fruits and vegetables, he contends, remind "aging boomers" of "health foods."

However, Émile Peynaud had started the trend (in *Le Gout du Vin*) with his own list of categories of wine aromas, which Parker reproduced in some of his books. This formed the base of Parker's vocabulary, though he vastly expanded the range of terms.

Parker's tasting notes had evolved from awkward, clumsy descriptions to fluid, run-on riffs in the 1990s, recognizable for the pell-mell enthusiasm and breathless excitement that fired his readers into making frequent buying trips to their wine retailers ("run, don't walk" he advised periodically; with retailers it became a joke). Parker's language reflected his own taste preferences, his plain-English, tell-it-like-it-is approach, as well as the culture shift of late twentieth-century America. That language had taken a 180-degree turn from a generation before. The old class and gender terms didn't have much meaning to the new generation of middle-class Americans who were now educating themselves about wine; they suggested old-fashioned wine snobbery and male chauvinism. Gone was the social class terminology—"nobility," "breed," "refinement," "class"—once favored by the British. Gone were the most fanciful comparisons of wine to women ("Beaujolais is a young girl, Burgundy a woman of thirty," etc.) once favored by the French, who had, after all, come up with the term "legs" (*jambes*) to describe the drips of wine that run down the inside of a glass after the wine is swirled.

Instead, Parker's "imagistic database" included dozens of concrete terms, especially words for fruit and texture. He brought in references to just about every fruit ever eaten, including the most exotic—one writer called his tasting notes "unrelenting fruit mush." A typical note for a red detailed "overripe aromas of strawberry jam with a gorgeous blueberry, blackberry, creamy black raspberry flavor character." "Fruit bomb" and "jammy" were positives in his lexicon. He often spiced his descriptions by adding slang for large quantities, as in "gobs" or "oodles" of fruit. Food words also made a frequent appearance—chocolate, Asian spices, coffee, *pain grillé*, butter, vanilla, bacon fat, roasted meat, honey. No "whiffs," "nuances," "hints," or "echos" of flavors here; he was the master of adjective-fat wine prose.

The word "texture" (or more bluntly, "mouthfeel") was all-important, and Parker used a cornucopia of texture words: "dense," "thick," "concentrated," "silky," "fleshy," "velvety," "soft," "chewy," "viscous," "succulent," "lush," "unctuous." In an age when everyone strove to be lean, a "fat, fleshy, plump" wine was a prized commodity, though a "flabby" one was not. Parker employed a testosterone-charged vocabulary to describe the horse-power of wines: "big," "hefty," "massive," "huge," "powerful," "ferocious," "prodigious," "humungous." This muscular lingo clearly appealed to his readers, mostly American men who must have felt reassured that wine was a manly subject, just as TV chef Emeril Lagasse's emphatic "Bam!" accompanying his spicing techniques made it okay for regular guys to cook.

But Parker's vocabulary also reflected the tenor of the times as an age of sensation, luxury, and sex. In the "me" decades of the 1980s and 1990s, life was about *now*, about unalloyed pleasure. Parker liked to call himself a hedonist, a word that once had a quaint nineteenth-century flavor but now seemed to belong to the late twentieth century. He practically had a patent on the adjective "hedonistic," at least as a wine term. In a ten-year period he used it some 836 times to describe wines "designed to provide sheer delight, joy, and euphoria" that could only be faulted because they "provide so much ecstasy they can be called obvious . . . they are totally gratifying wines meant to fascinate and enthrall."

Terms evocative of luxury, like "opulent," "profoundly generous," and "decadent," made frequent appearances. Parker's definition of the last: "If you are an ice-cream and chocolate lover, you know the feeling of eating a huge sundae lavished with hot fudge, real whipped cream, and rich vanilla ice cream. If you are a wine enthusiast, a wine loaded with opulent, even unctuous layers of fruit, with a huge bouquet and a plump, luxurious texture can be said to be decadent." He favored action terms such as "explosive" and "exuberant" and references to sex: "sexy," "seductive," "sensual," "voluptuous," even "liquid Viagra." He once wrote that the 1999 Jamet Côte Rôtie was "pure sex," and he got into trouble with the owner of Château Tertre-Rôteboeuf, who was incensed over a description of his wine in the French edition; the incorrect translation said his wine resembled a whore, albeit a rich one.

Parker layered in his love of contemporary rock music, too. In the midst of the craziness of the wine world, music, he said, "keeps me sane." He drew especially on the songs of Neil Young, who had, Parker told me, "guitar strings attached to his heart." (Young's two greatest CDs, in his opinion, were the 1978 *Rust Never Sleeps*, recorded live at the Cow Palace, and the "tour de force" *Ragged Glory*.) An article in the music magazine *The Absolute Sound* had even compared Young's fight against industrial digital music with Parker's crusade against filtration in wine.

Parker's tasting notes were the opposite of pithy. He often went on for paragraphs about the scent and flavor and texture of a wine, chewing the words around, spitting out adjectives, all but smelling and tasting the wine for you. He calculated exactly how long the wine's taste lingered on his palate—a 70-plus-second finish for one looked like the record.

By reframing the language of wine, Parker also subtly shifted what people were looking for; he reframed the picture of how they expected a good wine to taste. Vocabulary is never neutral; it carries an implication of quality. In Parker's linguistic wine universe, "oodles of fruit" was singled out far more than subtlety. What some might call "refined," he described as "thin"; what others might call "over-the-top," he called a "blockbuster"; what one person called "nervy," he called "acidic."

But one of the most significant aspects of his way of describing wine was, in true postmodern fashion, to deconstruct an overall experience and impression by breaking it down into a combination of many individual elements—specific aromas and tastes. This approach fit perfectly with judging wine as simply a product, a liquid in a glass rather than a part of a wider culture. Often, though, the sum of his terms didn't seem to add up to a whole picture. The "surround" of experience was gone. And once critics began focusing on specific taste elements, winemakers could turn to technology to manipulate these elements to produce more of the desired flavors critics liked. A Chardonnay could be seen as just one part buttery, one part vanilla, and so on.

Still, this was a style of writing that other American publications with bottle-by-bottle wine reviews such as *The Wine Spectator*, had to adopt. If you were describing fifty Chardonnays, you had to make them all sound different. Not everyone approved. Hugh Johnson frequently complained about the "fruit salad" approach to describing wine, and Michael Broadbent found elaborate descriptions of dubious value. "There should be more drinking and less tasting," he'd once told me.

Although Parker's language presumably conveyed what he had tasted, did others understand it and find the same elements in the wines? When Becky Wasserman tasted with consumers, she found many confessed they couldn't detect all the flavors and smells Parker did. Which only served to confirm to many of them that he was, after all, the Super Palate. Retired linguistics professor Adrienne Lehrer conducted a series of experiments in which she asked wine lovers, both novices and experts from the University of California at Davis, to describe particular wines and then match others' tasting notes to a set of wines served blind. There was, she told me over the phone, little correlation, even among experts. Different people seemed to interpret smells and tastes differently.

Rhône Valley winemaker Michel Chapoutier liked to set up a similar kind of experiment, which he called "judge the wine critic." He would cut out six or seven wine writers' tasting notes on wines, leave off the wine's name, and put out fifteen to thirty wines. Then he'd ask friends and other wine people to match the note to the wine it described. Everyone found it difficult to do—few could match more than 25 percent correctly. But nearly everyone could recognize a note that had been written by Parker, even if they couldn't always identify the wine it was supposed to describe.

In 2003, wine researcher Patrick Rooney and his wife, Eva Simonsson, who'd been a sales manager at Chalone Wine Group, had the idea of analyzing and tabulating the wine word usage of several critics, including Parker, as a way of taste profiling. They called their company Wine Angels, and they were banking that vintners and wine marketers would pay $495 to find out the exact flavors and aromas critics mentioned most often in describing their 90+ wines. It turned out that Parker referred to smells and aromas more frequently than tastes, while two *Wine Spectator* reviewers, Jim Laube and Harvey Steiman, were far less likely to mention aromas at all. The words Laube used for his praiseworthy Merlots were "currant," "black cherry," "plum," "spicy," "cedary," "oaky," and "anise," a taste he seemed to have a thing for, while Parker's positive Merlot descriptors were "black cherry," "cherry," "jammy," "chocolate," "coffee," and "smoky."

Still, the most important part of the wine description remained the score. Parker's genius was to marry Americans' love of numbers with their equal love of hedonism. To write good wine criticism, he often said, you have to have both. The score and the description were reproduced on the "shelftalkers" wine shops now posted next to recommended wines. Importers and wineries lifted the score and tasting note from Parker's newsletter or *The Wine Spectator* and printed them up in a form that could be taped to the shelf as a mini-ad for consumers to read. In fact, *The Wine Spectator* sent out their own shelftalkers to major retailers a week before the issue appeared on the newsstand. In case a wine producer or importer didn't have time to comb the reviews, several companies had sprung into being that would do it for them and print desired shelftalkers for a fee.

But the score remained, as one New York writer put it, "the central icon" of the world of wine. And nowhere was that clearer than in the wine investment and auction market of the 1990s in America. In that financial climate, dot-coms boomed in Silicon Valley, the Nasdaq and Dow soared in New York, and IPOs everywhere made young entrepreneurs into multimillionaires overnight. The intense joy of competitive spending on luxury products like fine wine spread to new consumers, who relied on Parker's numbers to know what to buy. Henry Leung, the owner of Henry's Evergreen Café in New York, had caught the wine bug. He created a stellar wine list, and was noted for staging B.Y.O.B. dinners for which the price of admission was bringing a 90-, a 98-, or sometimes even a 100-Parker-point wine. It was the vinous version of the potlatch. Discussion often revolved around boasting about which 100-point wines you owned. They were trophies, and proved you were a player. It was the era of "1,000-point dinners"—meals featuring ten 100-point wines.

Wine auctions became another arena for conspicuous consumption emblazoned with Parker points. Selling wine at auction became legal in New York City in 1994, just in time to capitalize on the American financial boom, and the city quickly replaced London as the new epicenter of the wine auction market. Besides traditional houses such as Christie's and Sotheby's, two wine stores held regular auctions, a dozen online auction sites debuted in 1998, and hundreds of charity wine auctions raised millions. By the end of the decade, the value of wine being traded at auction in the United States had climbed to $100 million a year.

Parker's scores, which had helped boost sales of the 1982 Bordeaux, became the key indicator of any fine wine's investment value. In Christie's auction room, bidders—both silver-haired collectors in pinstripes and sprawling, long-haired software moguls in sneakers—consulted spreadsheets of scores on their laptops for the wines they were pursuing before raising their paddles. Bordeaux was still the blue chip, but show-off French *vins de garages* and California cult Cabernets had become the new collectibles, going for astronomical prices simply because everybody wanted to have something that so few could get their hands on. If measured by price escalation, the ultimate cult Cab was the Napa Valley's Screaming Eagle, whose first vintage was 1992. The 1997, a 100-point Parker wine, came on the market at $125 but rapidly rose to $2,500 a bottle at auction. A half million dollars was paid for a 6-liter bottle at the June 2000 Napa Valley wine auction. A spot on the mailing list of Bryant Family Vineyard, a cult winery whose wines Parker regularly rated 98, 99, and 100, went for $12,000 at a charity auction in New York to raise money for breast cancer research.

Parker's score was the only one that counted in the auction market. In recommending the best wine investment buys, John Stimpfig's "Collector's News" column in *Decanter* always featured the "Parker rating" as one of the most important factors to take into consideration if you wanted to flip for a profit. Once given, it stuck with a wine as though it were branded on the cork. Nowhere was his impact on auction prices more apparent than with Bordeaux from the 1989 and 1990 vintages. Parker favored the 1990, though many critics found the two vintages roughly equal. But at auction in 2002 the 1989 L'Evangile, a Pomerol with a Parker score of 89, sold for $125 a bottle, whereas the 1990, which Parker rated 96, sold for $240. When he saw that, wine consultant Mark Golodetz quickly sold off his cache of 1989s, fearing that without a high Parker score attached, its price would never go up. Parker rated the 1990 vintage of Châteaux Margaux 100 points; it remained at about $6,000 a case (by 2004 it went for $8,225); the 1989, rated 90, sold for less than half that despite a rating of 96 from *The Wine Spectator*.

Bordeaux Index, one of the London fine wine traders that created portfolios for individual and group investors, wouldn't include any wine that didn't have at least a 90 from Parker, and if it didn't have a famous name, the score had to be at least 95. "His scores," Bordeaux Index partner Dylan Paris told me in London, "are the price parameter for investors . . . the golden seal of approval."

Never mind that Parker was completely opposed to buying wine for investment, and said so frequently. He declined repeated invitations to appear on economic commentator Louis Rukeyser's popular *Wall Street Week* television show, he told me, because "All Rukeyser wants is to talk about investing in wine."

———

AS THE BELIEF IN PARKER'S INFALLIBLE PALATE AND ROLE AS taster supreme gripped the world, his critics raised questions about the limitations of his—or anyone's—nose and palate. True, friends and colleagues had seen him perform what seemed astonishing feats, like recognizing a wine blind that he hadn't tasted in a decade. His memory for tastes and smells seemed to be incredible and as a result his assessments appeared to be extremely consistent, meaning that whenever he tasted a particular wine blind he tended to give it a similar score and describe it in similar terms. But could he possibly, as he claimed, remember every one of the 200,000-odd wines he'd tasted in his career and the scores he'd given them? And what about palate fatigue when he tasted 150 wines at one go?

In the 1990s, new research on taste and smell was beginning to put the ability to taste in perspective. In fact, more was learned in a decade than in the previous two thousand years. Yale University professor and taste scientist Dr. Linda Bartochuk identified a category of "super-tasters," the 25 percent of the population (only 15 percent of Caucasian men) who had more taste buds (actually fungiform papillae—the structures that house the taste buds) on their tongues and experienced all tastes, especially bitterness, much more intensely. Parker, Bartochuk speculated, probably belonged to the super-taster group. He'd never tried her simple test, though: putting a small piece of paper impregnated with a thyroid medication called PROP on your tongue to see if you register unpleasant levels of bitterness. If you do, you're a super-taster.

Taste, smell, and touch overlap in distinguishing flavors in wine, and super-tasters also turn out to have a more intense sense of smell, so Parker probably does, too. We can perceive just five different basic tastes—sweet, salty, bitter, sour, and *umami* (Japanese for meaty or savory)—but ten thousand different scents. The sense of smell is one of the most primitive parts in our nervous system, but it is powerful, able to trigger intense memories. Until 1991, however, how it actually worked was a mystery. Then Dr. Richard Axel, now at Columbia University, and Dr. Linda Buck, at the Fred Hutchinson Cancer Research Center in Seattle, who won a joint Nobel Prize in 2004, discovered there is a family of about 350 individual odor receptors clustered in a one-inch square at the top of the nasal cavity. When a nose encounters a smell, the scent may activate a few or many of them—particular odors are detected by particular combinations of these receptors—and that perception is then transmitted to the brain. (People differ widely in their sensitivity to different smells; I know some wine people, for example, who immediately react with wrinkled noses to the tiniest whiff of sulphur in a wine's aroma, and others who don't seem to pick it up unless it is so blatantly obvious you'd have to have a bad head cold to miss it. These differences are thought to be partly genetic.) Very little is known about touch, what we think of as "texture" in wine terms, and which seems to be so important to Parker. The brain, however, turns out to be the most important factor of all. A May 2003 Italian study showed that while everyone possesses the sensory capacity to taste a wine, in trained experts, tasting triggers an intellectual response that is detectable by brain scans. Memory counts. Still, taste and smell preferences are largely a matter of culture, psychology, generation, and experience: One study showed that the French prefer flowery scents, the Japanese delicate ones, and the Americans bold ones. There is no universal palate.

Besides, as the late California winemaker Louis Martini liked to say, "The wet laboratory of your mouth is not always in the same state. You make your coffee every morning, using the same pot, the same water, the same coffee on the same stove. Most of the time it tastes good, but sometimes it tastes awful. The difference is in the chemistry of your mouth, not the coffee." Didn't Parker's palate ever fail him?

In fact, our noses and taste buds are quite temperamental, affected by humidity, hunger, and expectations. Constant exposure to a particular aroma or taste, researchers found, reduces your ability to detect it. In a lineup of oaky wines, for example, you might not even pick up a strong aroma of oak in the last few wines. At the Monell Chemical Senses Center at the University of Pennsylvania, Dr. Charles Wysocki studies "olfactory adaptation." He found that when a person is exposed to a strong smell, his nose gradually adapts, so that eventually he ceases to perceive it, and further, isn't able to smell it again unless he spends a number of minutes in an environment free of that odor.

Evaluating a long series of tannic red wines, like barrel samples, is particularly tricky because tannins accumulate in the mouth, causing carry-over effects from one wine to the next. The combination of acidity and tannin, Dr. Ann Noble was convinced, makes tasters less sensitive to aromas. For experts especially, memories and expectations can easily override the smell receptors being tickled by a particular aroma and the signals these send to the brain.

Then there is the irrefutable fact that sensory perception declines with age. A wine's bouquet increases with time; our ability to detect it, particularly after the age of seventy, does not, which may be the reason why it takes greater sensory intensity in wines to make an impact on a maturing palate. "The older I get," Parker cracked, "the younger I like them." He conceded to me that his tasting abilities might not be as razor sharp as they were when he was younger, but thought his amazingly well-stocked memory of smells and tastes more than compensated.

In California and Bordeaux no one claimed that Parker's tasting abilities were waning. But there were those who talked about how his palate had changed. As evidence, they pointed to the fact that the Parker of the twenty-first century now routinely praised over-the-top Cabernets that the sideburned publisher of the *Baltimore-Washington Wine Advocate* a quarter-century before would have slammed as overly alcoholic, corpulent Bordeaux wannabes.

9

SCORING PARKER

*I*N THE SPRING OF 2001, ROBERT M. PARKER, JR., was at the zenith of his career and power.

In the 1980s he had forged his reputation, opening up a new world to consumers and trumpeting a kind of criticism that woke up winemakers everywhere. In the 1990s he had staked out and extended his influence across the world's winelands, enlarging the market for fine wine and in the process financially benefiting many château owners, growers, and wineries. Now, in the first decade of the twenty-first century, he found himself an American icon, which meant, of course, that he had become a brand name.

The world of wine seemed to wax with his enthusiasms. On his annual visit to Bordeaux in late March, Parker was particularly exultant over the 2000 vintage, which had lived up to all his hopes and expectations as the millennium *millésime*. He had to pull out all the stops and ransack his bag of adjectives to describe the swashbuckling opulence of the wines, which rivaled the magnificence of the 1982s. The tone had already been set, but it was much more enthusiastic after the publication of his April issue, in which he called the vintage monumental, the "greatest year ever for Bordeaux," with more outstanding wines with "impressive strength" than he had seen in twenty-three years of tasting. Prices for these epochal reds soared into the stratosphere, and the French loved him for it.

As the one critic whose opinion mattered above all others, Parker was interviewed about the vintage on both *60 Minutes II* and *Charlie Rose*. Despite years of being lionized in the press and appearing on French television and in America on *Adam Smith's Money World* on PBS, Parker was nervous about how he'd performed on prime-time TV. So before he and Pat settled in to watch the telecasts, he pulled out a few fine bottles to help him get through the experience, including a 1990 Château Haut-Brion and a 1982 Château Gruaud Larose. Parker rated them 95 and 96, respectively, but not his own performance. He needn't have worried. He may have looked uncomfortable in a jacket and tie, but he kept his face coolly impassive as a genial, admiring Rose (a *Wine Advocate* subscriber) fractured the French names of châteaux. And he politely declined to reveal which wine from the 2000 vintage he had awarded 100 points. That would have to wait until his April newsletter came out. The two shows added to the stampede for the wines, and in the process pulled in two thousand more subscribers to *The Wine Advocate*. Joan Passman and his two other part-time secretaries were swamped.

―――――

ON SEPTEMBER 11, 2001, WHEN THE TERRORIST ATTACKS ON NEW York and Washington occurred, Parker was on his annual tasting tour in the Rhône Valley. He'd spent the day in one of his favorite haunts, Cornas, and had returned to his hotel in Tain L'Hermitage before his last appointment, a light dinner and tasting with Gérard Chave across the river in the tiny village of Mauves. Chave greeted him with tears streaming down his face, and they watched television for an hour in shock. Finally Parker said, "I can't watch anymore. Let's taste." True to form, Parker found distraction in focusing on wine, and though he says didn't enjoy the food and wine at dinner, felt his tasting notes were "reliable." He was deeply affected by what had happened. When he was finally able to return home to his anxious family several days later, he became uncharacteristically depressed and for the first time seriously questioned the relevance of what he was doing. In the vast scheme of things, how important was writing about wine? But he snapped out of his tailspin a few weeks later thanks to Julia Child, when both were honored by the American Institute of Wine and Food at an event in Baltimore. "Her spirit and appetite for life at the age of eighty-nine," Parker told me, pulled him back to his basically optimistic outlook. He had always, as he put it, "resided on the sunny side of life."

He channeled his concern into fundraising, rounding up wines and lending his name and presence to a $2,500-a-plate dinner in New York to raise money for the Windows on the World victims, where he and Marvin Shanken, who hadn't seen one another in ten years, greeted one another politely and shook hands. Parker welcomed guests, poured wine, and stayed late, socializing afterward with the crew of sommeliers, several of whom, like organizer Daniel Johnnes, were good friends.

His life was back to normal, focused on the pleasures of food and drink and his usual routine. He'd recovered enough to host the annual "Date with Decadence" dinner that he'd held every December for the past fifteen years. Parker pulled the best wines from his cellar—Champagne, the best California Chardonnays and the best white Burgundies, red Burgundy and American Pinot Noir (including, of course, Beaux Frères), Bordeaux, then the Rhône—for twenty close friends; they partied long and hard. Reenergized, he threw himself back into work on another edition—the sixth—of his *Wine Buyer's Guide* (2002, 1,635 pages); continued the rewrite of his massive tome on Bordeaux, whose publication was planned to coincide with the twenty-fifth anniversary of *The Wine Advocate* in 2003; and started planning a book on California wine, long awaited by his readers. Parker was in fine shape. Rumors of serious health problems (including throat cancer and heart disease) had swirled around him for years, but except for gout, which he had brought under control, and back problems, a result of lifelong scoliosis, his blood pressure was normal, his cholesterol a low 140, his weight stable. His doctor told him he was a walking ad for the Atkins diet and the health benefits of drinking red wine.

FOR A COUPLE OF YEARS PARKER HAD BEEN WORKING WITH A PART-
ner, an Atlanta wine lover and businessman who'd designed his "Cellar
Manager" software ("I fell under Parker's spell in 1987," he told me), to develop a website, which finally launched in late 2001. Parker had been slow
to get onto the web—he'd underestimated its potential, he said—but he
was "convinced it is the medium of the future for information and influence." He hoped that eventually the website would take over from his
newsletter, saving him some $400,000 in annual postage costs, almost
three times what he spent yearly on wine. For an online site, subscriptions
were expensive at $99 a year, especially when you considered that the notes
from *The Wine Advocate* appeared a month later than the newsletter edition, but it had a surprising number of useful search functions for the
faithful that made tracking down recommended wines and their scores
incredibly simple—you could retrieve all Parker's 100-point wines with a
click of the mouse. Though he often told critics of his scoring system that
the tasting notes were what counted, the site seemed to pander to the
point-chasers, listing wines in descending order of score.

One section that proved highly popular was "The Hedonist's Gazette,"
where Parker and Rovani posted accounts of their culinary and vinous experiences for subscribers' vicarious entertainment. And when his friend
Mark Squires's well-run bulletin board moved onto his site in March
2002, Parker was assured that www.eRobertParker.com would attract
some of the most knowledgeable and vocal wine geeks on the Internet.
The old Prodigy group reunited under Parker's umbrella. Threads offered
plenty of opportunity for passionate arguing over thought-provoking issues like old-style versus new-style wines, why *The Wine Spectator* wasn't a
source of serious tasting notes, the importance of *terroir*, and whether
Parker really understood Burgundy. One, "Parker, Burgundy, and
Knownothingitis," ended up with five hundred posts before Parker decided to cut it off. The Parker faithful, often led by Rovani, were sure to
spring into action to defend their leader's reputation, views, and presumed infallible palate, though newcomers often questioned what the
point of all the "pissing contests" was. Didn't they all love wine? At first
Parker rarely posted, but he soon found both following discussions and

responding almost addictive. The bulletin board unleashed something in him—at last he could easily and instantly answer his critics. From time to time he posted long rambles on many a topic, encouraged by those who quickly reacted to his rants with appreciation. On July 23, 2002, he went on to respond to birthday good wishes with thanks and wrote paragraphs arguing that his palate was more diverse than people said. A month later he was defending the methods of the *garagiste* winemakers in Bordeaux. Later, in October, he delivered his opinions about suggestions that high-alcohol wines wouldn't age well, and two months after that he was justifying his reviews of the 1993 red Burgundies. In turn, regular participants alerted him to retailers and importers misrepresenting his scores; Parker quickly took action, often calling on Rovani to investigate. More than ever he felt he had to respond to defend his reputation. He found the bulletin board useful in correcting the record and sharing his "random thoughts" on wine, and liked having the opportunity to show his readers that he was a down-to-earth, regular guy.

The truth was, however, that while he still had a low profile in his home-town (although a few neighbors did mention his appearance on *60 Minutes* with "Hey, I didn't know you were a wine expert"), in the world his profile had risen higher than ever. In France people approached him on the TGV, the fast train to Bordeaux, and told him they had his book. He was the subject of dozens of newspaper articles and cover stories of national magazines, a towering figure who cast a long shadow in books on Bordeaux and other wine regions, and was featured in a controversial documentary on globalization in the wine world, *Mondovino*. His beloved bulldog George and basset hound Hoover won the award for best dog performers in the same feature at the spoofy Palme Dog held at the 2004 Cannes Film Festival, beating out a French poodle. Parker was annoyed that *The Hollywood Reporter* cattily described them as ugly and flatulent.

His office now was a testament to his quarter-century in the wine world. The walls held framed copies of laudatory articles published in *Newsweek, USA Today, People*, and the *New York Times*, as well as a two-part front-page story in the *Los Angeles Times*, and the pictures of himself that illustrated a cover-story profile in the *Atlantic Monthly*. His awards are so numerous that some framed certificates are just stacked, though the *Diplôme de Citoyen d'Honneur de Châteauneuf-du-Pape*, an area so special to Parker, is on the wall, as are signed photos of two presidents, Reagan and Clinton. What takes pride of place is a souvenir from the dinner he hosted at Restaurant Daniel to celebrate receiving the Legion of Honor: a photo of his family with Jacques Chirac, mounted on white stock on which friends wrote good wishes. From his desk, images and symbols of his success in the world surround him.

Yet increasingly Parker is enveloped by the criticism that trails after him wherever he goes and that hums in the background of every wine industry event. It was there even at the beginning of his career, when his controversial scoring system had first come under attack. The 100-point rating system remains a hotly argued topic everywhere from the Seattle Vinifera Conference to a panel organized for the trade by Britain's *Harpers Wine & Spirit Weekly* to the pages of newspapers. Then there's the companion theme of continuing denunciations of Parker's taste and the debate about whether he is destroying diversity in the wine world—a paper at an international wine conference referred to it as "the Coca-colonization of wine." It had become clear that more than anything, it was his influence on winemaking and his power over the market that bothered so many in the wine world, even those who'd embraced him early on, liked him personally, and regarded him as both incorruptible and a superb taster.

It bothered him. Small minds, petty minds, jealous minds—they all wanted to rein him in, like Lilliputians tying down Gulliver. Parker had to watch his step everywhere. In Bordeaux, the gossip mill churned out stories about his alleged favoritism for a few Right Bank producers after his dinner at L'Aubergade, southwest of Bordeaux, in 2000. He was supposed to be meeting only Dr. Alain Raynaud, then president of the Union des Grands Crus, but Raynaud had also, unbeknownst to Parker, invited Gérard Perse, the owner of Château Pavie, and Michel Rolland to join them. The next morning Parker tasted a number of their wines. To the Bordelais it looked like cronyism on both his part and that of the producers, even though Parker paid his own way. Still, given that Parker gave high scores to Raynaud's new wine Quinault L'Enclos it didn't look that way to some of the UGC members, who reportedly accused Raynaud of taking advantage of his position and forced his resignation. Parker, as usual, put the best possible interpretation on Raynaud's intentions, but he acknowledged it was a lapse of judgment and announced that he would no longer have lunch or dinner with any producer while on any tasting route.

French newspapers and magazines seemed relentlessly critical, especially Bordeaux's own newspaper, *Sud-Ouest*, while the magazine *Cuisine et Terroirs* offered "*Pour ou contre Robert Parker*," giving six reasons for and six against. The editors rated Parker 85 out of 100. Châteaux also began to strike back. Château Bouscaut, whose 1999 received a devastating 79–82 rating from Parker, ran a much-talked-about newspaper ad in the form of a cartoon. In it a retailer told a customer, "A good wine with real *terroir*? An individualistic wine? It's a wine with a bad Parker score!!!!" Even châteaux whose wines Parker had praised objected to his power and took him to task. Château Soutard, a St.-Emilion whose traditional, tannic wines regularly received scores of 87 to 91, published a sarcastic poster titled "Pour Vos Fêtes de Fin d'Année *2001, Vis Parker*," which showed a "vertical tasting" of a series of screws of different lengths and widths, all with Parker-style scores, "guaranteed concentrated." And a self-styled "anti-Parker," Jean-Marc Quarin, started his own newsletter about Bordeaux.

A spate of articles decried Parker's championing of the Right Bank *garagistes*. Then, in April 2002, after returning from the *en primeur* tastings of the good but not flashy 2001 vintage, the *Financial Times*'s Jancis Robinson wrote that some Right Bank châteaux and *garagistes* at last seemed to be backing off from the big, bold oaky style Parker had praised, irking him no end. She included a direct slap at Gerard Perse, the former supermarket tycoon and multimillionaire outsider with a huge ego and arrogant manner who had purchased several estates in Saint-Emilion in the late 1990s, among them Château Pavie, which he floodlit at night, horrifying his neighbors. His new big, rich Pavie was a hit with Parker—100 points for the 2000, 94–96+ for the 2001 (and Perse had dared to complain!)—but roundly trashed by Robinson and many of the British—including Steve Browett of Farr Vintners—as a "caricature" wine, plush, portlike, and quite unlike the suave, supple Pavie of old.

Parker saw Robinson's criticism of these wines as a not-so-veiled criticism of his taste and influence. In May, as I drove around with him in California, Parker brought up her column. He was impatient and accusatory. "It's her British DNA taking over control of her logic," he said. "First of all, they couldn't make big rich wines in 2001 because Mother Nature didn't give them the raw materials. So they're spinning it like they're making more elegant wines." He thought Robinson had got Perse completely wrong: "She is just parroting the viewpoint of some of the old reactionaries in Bordeaux, and it is surprising that someone of that stature and intelligence would do that. And it's a way of cleverly taking a shot at me—saying Parker just likes these grotesque, overweight, overly woody wines, which is a total bullshit story." Perse was resented for his money and his arrogance, said Parker, but "that doesn't bother me. I think you have to look at what he's done." Everything told him they were going to turn into "prodigious" wines, and he was, he said, "100 percent sure of that. I'm more sure of that than I was about the '82 Bordeaux and I was pretty sure about those." But despite high scores, Perse priced the 2001 Pavie so high that little was sold.

WERE CRACKS APPEARING IN THE EDIFICE OF PARKER'S POWER? A sinister dimension to the ongoing criticism—the suggestion of criminal scandal—had been added in January 2002. During his regular visit to Bordeaux, Parker was stopped in the parking lot at Château Palmer and told of invoices for wine consulting on his letterhead. Though Parker thought this had to be a hoax, he met with the person who had the invoices, an M. Laporte, the manager of an obscure St.-Emilion château owned by the Geens Group, a Belgian company that owned a number of Bordeaux châteaux and extensive vineyards. Laporte showed him photocopies of several invoices for winemaking advice on *Wine Advocate* stationery, the sums payable to Hanna Agostini, Parker's French assistant and translator. That evening, when Parker met with Agostini and her husband, she admitted having done consulting work for wine clients in Bordeaux, to help them improve their wines, but she claimed to have billed them on her own stationery—for her company called Wine & Feathers—and showed Parker those identical invoices. She insisted that the photocopied invoices he had seen were counterfeits, and seemed to suggest how that might have happened when she also informed him that a few months earlier she had left an attaché case with some blank *Wine Advocate* stationery at an estate where she had been consulting.

Parker was upset with Agostini's poor judgment. If she wanted to continue working for him, he told her, she would have to stop doing consulting work that involved contracts for oenologists and giving vinification advice, and that the matter of the so-called fraudulent invoices had to be reported to the police for investigation.

Then, in the last week of April, just before Parker released his issue with ratings of the 2001 Bordeaux, a detailed article breaking the story of Agostini's consulting work, "L'Affaire qui ebranlé la maison Parker" ("The affair that brings the House of Parker down"), appeared in *Le Nouvel Observateur*. The information about Agostini and the invoices had surfaced as part of a much larger story about the Geens Group. The CEO had been fired by the company in mid-February and accused of embezzling 800,000 Euros; a full-scale investigation was underway. The Agostinis had had a long relationship doing legal work for Geens, and, the paper reported, Hanna Agostini had introduced Italian oenologist Riccardo Cotarella and Jean-Luc Thunevin of Château de Valandraud to Geens, which hired them to make a series of high-quality wines. Parker told *Le Nouvel Observateur* that he had first learned of Agostini's consulting in January, and insisted that she stop. Still, the story fueled gossip. After all, Agostini set up tastings for Parker in addition to her work as translator. He had tasted and rated those wines. Despite his belief in her, Parker prudently hired a French lawyer to monitor developments in the Geens case.

So the matter rested for a while, though occasional stories referring to it popped up in the French press and Parker was routinely quizzed on it when interviewed. Some château owners told me they thought Parker should have fired Agostini immediately and couldn't understand why he didn't.

I met Agostini in July 2002 in Bordeaux. Going into the interview I knew that Parker had sympathy for her because she was a woman in a man's world that often dismissed professional women and that he assumed at least some hostility was directed at her because she was dark-skinned and an outsider from Mauritius. I wasn't too surprised Parker had given Agostini the benefit of the doubt. He felt she had done a superb job for him in the past six years. By then, I had also heard comments from wine people who didn't like her and were convinced she took advantage of her position. Some didn't see how Parker could have been unaware of what she was doing.

We met in her office in Bordeaux, a large, square, old-fashioned, slightly dusty room with high ceilings and tall French windows, in a building Agostini and her husband shared. A plump, buxom woman with thick, short black hair and dark eyes, she seemed shy, suspicious, and uncomfortable. I knew she had agreed to see me only because Parker had asked her to. I was, she said, the first journalist she had spoken with, and no, I couldn't tape our conversation. I didn't have nearly as good an impression of her as Parker evidently did. Even after she warmed up and was more forthcoming about her work for him—he was easy to work for, "he never makes you feel you are his subordinate"—I couldn't shake the sense that there was more to the story than she was offering. A far as wine went, she explained she didn't even drink red Bordeaux, only Champagne, and to demonstrate, opened a bottle for us, and her husband, to share a glass before I left. Later, though, I wondered how much Parker had thought through using her as his assistant in deeply conservative Bordeaux. What did he expect?

———

THE BORDELAIS WERE NOT HAPPY WITH PARKER DURING THE SUM-mer of 2002. He had announced that there was really no reason to buy 2001 futures, and added his voice to a number of big trade buyers of Bordeaux who had been clamoring for prices to come down. Many châteaux didn't lower their prices—in fact, Château Lascombes came out with a 35-percent increase—and America followed Parker and didn't buy. May-Eliane de Lenquesaing, whose 2001 Pichon Lalande seemed to me just as good as, if not better than, her 2000, said, "He's hurt and destroyed the market. That is a very big responsibility." For many of the *garagistes* as well as Perse, there were few sales. The French wine writer Michel Bettane had little sympathy for the Bordelais, saying, "Parker is just doing his job as a wine writer. The wine trade helped create the problem. When he helped them with the 2000 vintage no one complained. Now he says the prices are too high, don't buy, so he's a bad guy. It's a human comedy."

Some American wine retailers, too, were now chafing at Parker's power. They had too many stories of customers who brushed off their advice and fixated on wine that Parker had recently rated 95+, usually a bottling made in such tiny quantities that it was virtually unobtainable. It took all the fun out of helping people find wines that they liked. The number of knowledgeable fine wine merchants in America had been small when Parker started *The Wine Advocate*, but the situation was vastly different now. While many retailers still counted on shelftalkers with Parker's or *The Wine Spectator*'s scores to sell their wines, others swore off scores altogether. A new breed of wine shop owners like Roberto Rogness of Wine Expo in Santa Monica had had their own European epiphanies, become passionate about wine, and developed their own taste and expertise, which they wanted to pass on to their customers. Rogness often used his newsletter to criticize Parker, likening the big man's favorite Zinfandels to "a drag queen in the middle of a great production of *Carmen*." "Thelonius Monk," he wrote, "has more to do with the way and the type of wines we select than Robert Parker does."

Importers, too, began to define themselves by their resistance to Parker, though many were reluctant to speak against him. One was Joe Dressner, a partner in Louis/Dressner Selections, who has a deadpan humorous shtick that seems quintessentially New York. "Am I going to regret this?" Dressner mused more than once. "I still have to put my kids through college." His attitude to Parker and his scores was best expressed at his annual portfolio tasting where, as a joke, an empty table held a sign: "Wines that received 96+ points in *The Wine Advocate*." Once he'd regularly shown his wines to the critic, but one of his specialties had become wines from the Loire Valley, especially the kind of savory reds with acidity that Parker didn't like. "He won't concede or delineate his limitations," says Dressner. "He still thinks of himself as a consumer advocate, but he went from pointing out bad wines to becoming a proponent of wine styles that wouldn't exist without him and undervaluing others. It's as if a theater critic only liked Shakespeare. He shouldn't mistake his predilections for objectivity." But he acknowledged that, "I couldn't grow big time without Parker's benediction. If I had a predominantly 20-dollar to 100-dollar range of wines and no reviews from R. P., I'd be out of business."

———

FOR PARKER, THE MONTHS FROM APRIL TO THE END OF THE YEAR were a sad time. His beloved mother had died just before I'd traveled around with him in California in May. Carol Smith, the wife of his best friend, Park Smith, died soon after he returned to Maryland from California. On his September trip to the Rhône Valley, accompanied by Pat and Smith, whom he was trying to cheer up, the three of them were caught on a flooded road in Châteauneuf-du-Pape, where their car flipped over at a 40-degree angle. When the water rose to the steering wheel, they had to abandon the car and then hike several kilometers through intense lightning. Parker lost his prize collection of Neil Young CDs that had been on the backseat. (After he bemoaned his loss in a post on his website, sympathetic readers flooded him with replacements, including rare recordings he had never known existed.)

But the worst was yet to come. When Parker was out biking at 6:00 one cold, crisp, windless morning near his home, a chipmunk ran out into his path. He hit the front brake to avoid it and flipped over the handlebars at 17 miles an hour, hitting his head and ending up with stitches, a black eye, a hairline fracture in his elbow, two badly sprained wrists, plus assorted bruises. The terrifying thought of how wine lecturer Harry Waugh had banged his head on the dashboard of his car in a fender-bender and completely lost his sense of smell put Parker into a panic. He limped home and splashed out some wine to sniff to convince himself that he hadn't endangered his most precious business asset, his ability to smell and taste. He was still suffering from bruises at the end of October when I saw him at VinExpo in New York; he and his pal Michel Rolland packed a hall with 650 people anxious to hear Parker pronounce on why he thought Bordeaux was better now than it had been twenty-five years ago. It took him a couple of months to completely recover.

In January 2003, soon after Parker had returned to Maryland from his regular tasting trip to Bordeaux, the Hanna Agostini story resurfaced yet again. On January 23, she was taken into custody by the Bordeaux police after a search of her home, and held and questioned for forty-eight hours on criminal charges of forgery and "profiting from a breach of trust," according to the police chief, Jean-Pierre Steiner. She was then released on $38,000 bail. Agostini and her lawyer insisted she was an innocent victim, and claimed to have proof that the invoices on official *Wine Advocate* stationery, on which she was alleged to have billed thousands of dollars of consulting work, had been fabricated to discredit both her and Parker.

The situation dragged on, with Parker making statements on his website and to me that he believed "everyone is innocent until proven guilty." Still, his patience was wearing thin, and by the end of January he began to consider whether he should "temporarily suspend" Agostini's activities for the *Wine Advocate*. More and more concerned, Parker consulted Pat, and, unusually, began asking people in Bordeaux, "What should I do?"

BACK HOME, THE LAUNCH OF HIS WEBSITE WAS A HUGE SUCCESS—
by 2003 some eight thousand subscribers had paid up. Parker had profited
handsomely from his writings over the years, but of course he had gained
nothing like the wealth of the wine producers and importers whose products he championed. Now he was looking for ways to cash in without selling out. A designer friend was working on "Parker in the Palm," a software
program for hand-held computers so enthusiasts could take his scores
wherever they went. "It would be especially useful for dining in restaurants," his partner told me. He had all kinds of ideas for how to spread
Parker's name and scores electronically and had added a half-dozen important links to the website. He'd arranged for Parker's scores to appear
next to wines for sale on an auction and wine-buying/selling website and
was discussing with potential partners how to put Parker's nose and
tongue in wine shops nationwide. The plan was to lease electronic kiosks
to merchants so their customers could access Parker's score and description for any wine they saw on the shelf.

Keeping up with the entire wine world for his subscribers, who were
clamoring for more coverage of both California and Italy, was becoming
more difficult, and at the beginning of 2003 Parker made a crucial decision. He spun off Italy to Daniel Thomases, to the dismay of many Italians
as well as importers like Leonardo LoCascio. He was looking to maximize
the coverage in *The Wine Advocate*, but the coattails of his power went only
so far. His fans were grateful he'd kept Bordeaux and the "big red" territo-
ries of the Rhône, California, and Australia for himself.

RUMOR HAD IT THAT PARKER WOULD BE A NO-SHOW IN BORDEAUX
that year, the first time in thirty years. Parker insisted he had his airline
ticket. He planned, however, to cover a smaller selection of wines than
usual. He was irritated by the Bordelais. The 2002 hadn't been a great vintage. Why did he have an obligation to cover the *en primeur*, anyway? But
if he didn't come, John Kolasa of Château Rauzan-Ségla said, "It would be
a disaster."

Then, after the Iraq war began, Parker changed his mind. Pat and Maia, remembering the scare when he was in France on September eleventh, entreated him not to go, and he wrote an open letter to all the châteaux giving that as his reason. But in France, some said it was because he feared he would be detained by the Bordeaux police at the airport in connection with the Hanna Agostini case. Later, Parker heard that Agostini herself had started that rumor. In any case, without him the 2002 wines languished, adrift in a market indifferent to a vintage lacking the all-important Parker scores. Would he come in June or September? Châteaux, especially those upstarts whose reputations were built entirely on Parker's attention, seemed desperate. "We hope he will come!" Jean-Luc Thunevin told a reporter.

MEANWHILE, PARKER FELT AGOSTINI AND HER HUSBAND WERE becoming increasingly evasive when he asked questions. During the summer, in order to gain access to the dossier compiled by the police during the investigation to see if any of the allegations against Agostini and her use of *Wine Advocate* stationery were true, Parker joined the lawsuit as a plaintiff. After his lawyer reviewed the evidence, he told Parker in August that Agostini had been lying to him. Parker offered to let her resign, and she did. A story damaging to Parker, quoting Agostini, appeared in the *New York Times*. Although error-ridden, it was deeply embarrassing to him. He used the convenient bulletin board to reassure his readers.

Parker finally headed for Bordeaux in September. Anxious to sort out the matter, he asked his lawyer to set up an appointment with Jean-Pierre Steiner, the Bordeaux police chief. They met on September fifth. The full story was complicated and sordid. Two months after being hired to translate *The Wine Advocate*, Agostini had signed a contract with the Geens Group to hire oenologists that Parker liked to, as Parker told me, "design wines to please him and to present them to him." For that she was being paid $50,000 a year. She had also approached Alain Raynaud, who turned her down. "It was appalling behavior," said Parker, who sounded sad and disgusted. "From the beginning, she obviously thought she could exploit the relationship for money . . . it was pure greed." She had set up tastings of the wines for Parker, but he had given most mediocre scores—only one had received 90 or better. She had even created her own *Wine Advocate* stationery by removing Parker's name from it, and she used these for billing some invoices. What she had told Parker about "fake invoices" was all lies.

"I tend to believe people are honest unless proven otherwise," Parker said. Though he felt his reputation had been damaged by all the press coverage—of course, *The Wine Spectator* had done a story—he had no regrets about the fact that he waited; he wouldn't have felt right firing Agostini without having evidence. He had heard reports that she was impolite, that she bullied people, that she was abrasive, but he had dismissed them as the response of reactionaries to a strong woman of color. Many believed that she had been abusing her position all along. Afterward, Jean-Bernard Delmas of Haut-Brion told Parker that Agostini had been very insulting to him. One Right Bank château owner had told me—off the record—that when he'd been disappointed with Parker's review of his wine, she organized a lunch with Parker for him, and he'd been grateful. But he knew she was pushing her friends in the tastings she set up for Parker. "We tried to tell him what was going on," Alain Raynaud told me, as did several of Parker's other friends on the Right Bank. "But he wouldn't fire her."

It was a perfect example of Parker's stubbornness and his blind side.

Unfortunately, Agostini still had a contract with Solar, Parker's French publisher, to translate his books, and according to French labor law couldn't be fired unless she was found guilty. As of February 2005 the lawsuit had not yet come to court, nor had a date been set. Parker, adamant that her name could not appear on any book, hired a second translator to go over her work. He closed down his French office and ceased publishing *The Wine Advocate* in French.

During that September 2003 visit, Parker was welcomed at châteaux and *négociants* effusively, as though he had been away for ten years. Many of the smaller *garagistes* were widely reputed to be near bankruptcy. Château Lascombes was rumored to be for sale. Parker found the 2002s lackluster, as he'd suspected they would be. Besides, the 2003 harvest had grabbed everyone's attention. The grapes, already ripe, were being picked early in freakishly hot weather. It was either a great year in the making or a global warming disaster certain to result in soupy, overblown wines.

Parker really didn't think it would have made any difference if he had come in March, he told Charlie Rose, when he had made a return appearance on his television show in August. Had Parker gone on television in part for damage control? This time, sitting across the table from Rose with no tie, he seemed more at ease, but also more arrogant. His attempt in an interview in Britain's *Guardian* to finally acknowledge his almost gravitational pull on the market after years of denying it ("Yes, I'm the most powerful person in the wine world") also came across as boastful rather than honest.

The French continued to have a powerful love-hate relationship with Parker. "They should be careful what they wish for," Parker commented wearily to me. The French government had been openly dubious of the American invasion of Iraq, and now some Americans were showing their displeasure at this lack of political support from an old ally by emptying bottles of Champagne into toilets. "Nobody," he said, "has promoted French wine the way I have in America." Which was true. For him these classic wines were still the points of reference, the wines that he drank, not just tasted. His three cellars held some 12,000 bottles of wine, and 90 percent of it was French, mostly from Bordeaux, Alsace, and the Rhône. He'd even bought bottles of the 2000 Burgundies.

But to the French, Parker had become a lightning rod for their greatest fears. The world wine market was changing. Wine producers in France were wringing their hands over falling wine consumption at home and the slippage of their share of the American market, which was increasingly dominated by New World wines—especially its own. For many years France had provided the models for what table wines should taste like, but led by critics, consumers had begun to look for newer, different styles of wine by the beginning of the twenty-first century. The country's wine industry was in crisis; six hundred to one thousand producers might have to close; many families were pulling up vineyards. The new wines from France's Languedoc region often carried the grape name on the label, just as American wines did. New World wines were in the ascendance; Old World wines, especially French, were fighting for market share.

Thirty years ago, in his book *La Mort du Vin* ("the Death of Wine"), French wine and food writer Raymond Dumay had predicted that France's sway as the greatest wine country in the world would eventually end, and that America would take over that role, influencing the way wine was made around the world just as France had.

It was all coming true. Since Parker had first published *The Wine Advo-*

cate, the American market had become the most important in the world and the American taste in wine now seemed destined to dominate. And Parker had presided over it all.

———

WHEN I CAUGHT UP WITH PARKER BY PHONE IN OCTOBER 2004, his voice sounded tired. It had been, he told me, a stressful year and it wasn't over yet. He was doing too much, a confluence of charity events he'd promised to attend and deadlines he had to meet coming together in a single year. He'd barely seen the friends with whom he and Pat liked to have dinner. He'd managed to wriggle out of his contract for a book on California—after sixty pages he'd decided he just didn't want to do it— and instead was penning a manuscript on the world's greatest wine estates, due in by the end of the year, and gearing up for another *Wine Buyer's Guide*. After that, he told me, no more book contracts. He planned to focus his attention on his website. That, he believed, was the wave of the future for wine criticism. Layered into his hectic schedule were the usual items of the father of a high school senior, including scouting colleges. Pat was more involved in gardening than ever, studying to be a garden design judge and working with the Garden Club of America. Parker's beloved bulldog, George, had passed away the same day Julia Child had; now he had a new puppy, Buddy, named after his father. His mother's sisters, eighty-three and eighty-one, were now handling the mail.

The Bordelais had been relieved that Parker arrived as usual in March 2004 for the *en primeur* tastings of the wines from the previous year's vintage, but even before he could publish his notes, the debate on Château Pavie heated up. On her website, Robinson called Pavie "porty sweet . . . a ridiculous wine," and Parker couldn't resist posting a four-hundred-plus-word comment on his website's bulletin board: "I had Pavie four separate times, and, recognizing everyone's taste is different, Pavie does not taste at all (for my palate) as described by Jancis. She has a lamentable and perplexing history of disliking not only all of Perse's wines, but virtually all of the *garagiste* wines of St.-Emilion. . . . Her comments are very much in keeping with her nasty swipes at all the Pavies made by Gérard Perse and mirror the comments of reactionaries . . . in Bordeaux." He went so far as to attack Robinson's credibility. Robinson hit back on her website: "Am I really not allowed to have my own opinion? Only so long as it agrees with Monsieur Parker's it would seem . . . What is the difference, I wonder, between a nasty swipe and a critical tasting note? Perhaps the former does not chime with the most powerful palate in the world while the latter does?"

It sounded like a catfight between two critics. English writers mostly sided with Robinson, Americans with Parker; articles on the disagreement appeared in print from Sydney to San Francisco. Farr Vintners saw a wider significance: "This debate. . . . is not just about this one wine but also about the power of wine writers and the way that certain producers appear to be making wines to score points from 'the most powerful critic of any kind anywhere.' "

On a trip to England, Christian Moueix of Château Pétrus, too, spoke out against the new superconcentrated wines in Bordeaux and Spain as well. He had welcomed Parker with enthusiasm at the beginning of his career; now he worried about where "Oncle Bob" was leading wine. Then *Decanter* published an article by the editor of the *Singapore Wine News*, Ch'in Poh Tiong, that was a clear attack on American taste, American critics, and Parker in particular. "Bordeaux," he wrote, "was never meant to be an American wrestling version of a smack-down wine."

Despite the wine world's grumbling and chafing under his imperial sway and people beginning to wonder how long they would have to endure his reign, Parker could take solace in the endless celebrations of his achievements—and global power. In mid-2004 Italy's prime minister, Silvio Berlusconi, and president, Carlo Ciampi, named him a Commendatore in Italy's Ordine al Merito della Repubblica Italiana, the country's highest honor, for his contributions to Italian wine. He became the first wine critic to receive it, joining American luminaries such as Steven Spielberg and Henry Kissinger.

In California, a weekend of lavish events in October to celebrate *The Wine Advocate*'s twenty-fifth anniversary and raise money to endow the Robert Parker Wine Advocate wine scholarships at the Culinary Institute of America in the Napa Valley seemed a virtual canonization. Winemakers from all over the world flew in to pay homage; the highlight was a tasting at the home of Ann and Gordon Getty that featured ten of Parker's 100-point wines.

In December, Parker flew off to Asia, where he had organized two "Hedonist Dinners" in Tokyo that would cost the lucky forty gastronomes who attended one million yen each. The menu, by Joel Robuchon, named France's chef of the century by Gault Millau, featured nineteen dishes intended to showcase nineteen vintages of some of the greatest wines in the world, including the last existing bottle of 1870 Château Lafite and an 1864 Lafite that had aged in the cellar of the Scottish castle that inspired Shakespeare's *Macbeth*.

Parker had come a long way from the days when he knocked on doors in Bordeaux and Burgundy; so had the wine world. His reputation seemed like a powerful 1982 Bordeaux, at its apogee, on a plateau that could last for decades.

WHAT STRIKES ME MOST ABOUT PARKER'S CAREER AT THE END OF 2004 is the polarization—the fierce *for* and *against* Parker, who has become more than a mere individual judging wines. He is now the symbol of a complex of ideas and the very context in which people today make and drink wine and define their own taste. Which, of course, is the reason so many people have devoted time to thinking about him, scrutinizing him, wondering about him, judging him, and developing theories about the reasons for his power and success.

I find myself taking one side and then the other over Robert Parker, the man, and what he has meant to the wine world. I had followed the trail of his remarkable rise, tracked down rumors, listened to the diatribes and accusations, complaints about his certainty and alleged favoritism for his friends, but I also heard gratitude and admiration, the respect he'd won through his work ethic and integrity. To me he seems quintessentially American, and I relate to that. His American qualities are ones perhaps not readily understood or appreciated by the British or the French: that exuberance and passion for what he is doing that seems almost excessive, even monomaniacal; that ruthless, action-oriented energy, the willingness to take on the establishment, charge in, change the status quo, and damn the consequences; that streak of earnestness about one's mission and drive to work harder than anyone else; that openness and impulse to democracy, ready to back some little guy or new region that dreams of making something great and becoming a star. America admires risk-takers willing to put their ideas on the line and bravely fight for them. Parker has surely helped to improve the quality of wines around the world by exhorting people to do better and recognizing when they do, and by castigating some winemaking practices that result in industrial wines with little flavor and no authenticity. His sheer enthusiasm and delight in wine have influenced a generation and convinced wine lovers that enjoying wine is not just for elitist snobs.

But I'm also on the other side. Those American traits have a downside, too, as do Parker and his influence. His success is surely due in part to another, not so attractive, American characteristic, the certainty that one is right. Power always gives the opportunity to push too far, to dominate, to crush the opposition, all of which Parker has done, whether he admits to it or not. I don't approve of the tyranny of one palate. I don't want to see traditions and wine styles worth keeping discarded simply because a single palate doesn't like them. I worry about wines with finesse and subtlety and savory, mouth-watering acidity disappearing, replaced by thick, rich, fruity wines that are better in a blind tasting than they are on the table, with dinner. I find scoring wine with numbers a joke in scientific terms and misleading in thinking about either the quality or pleasure of wine, something that turns wine into a contest instead of an experience.

At the end of twenty-five years, Robert Parker has the kind of power in the wine world that no other critic in any field has ever had, a power that the industry—and consumers—handed to him early on and continue to feed. He capitalized on it, yes, but it has grown largely beyond his control. Nevertheless, it's disingenuous for Parker to say that there is nothing he can do about the power and influence he wields. Rather than widen discussion and welcome in many other points of view, he defends his own in a way that brooks no challenge.

What will be the emperor's legacy to the world that he rules?

So much of what Parker says he stands for caused the opposite to happen. He argued for the democratization of wine, and yet became the very symbol of the elite expert pronouncing on unobtainable wines. He wrested the legitimatization of wines from the merchants, but put it in the hands of the media. Though he insisted he valued individual taste, the would-be consumer advocate became the supreme judge. He railed against high prices, but whenever he anointed a wine its price went up, and up, and up. He opposed the very idea of investing in wine, but those who collected his high-scoring bottles found the advice self-fulfilling and made handsome profits. He argued for diversity of styles, yet in many regions what he wrote ended up promoting wines that began to taste alike. His scoring system, intended as merely a quick assessment of quality, has encouraged the idea that only top-rated wines are worth drinking.

Still, his insistence that no reputation, no matter how long-standing, can substitute for what can be found in the glass, his belief that there are still undiscovered wines and regions that can challenge the wine establishment, and his fundamental conviction that wine can be judged and rated like anything else are lasting contributions.

———

AND WHAT OF PARKER THE MAN? HE IS WARM AND CARING, LOYAL to a fault, slow to recognize perfidy in others, but also quick to attribute dubious motives to those who don't agree with him and sometimes vindictive if you cross him. Are these sunspots? Mere motes that are outshone by the remarkable professionalism and hard work he's brought to his role—not to mention his generosity to friends, employees, and dedication to his readers? Or are they the result of stubbornness, self-righteousness, and a lack of self-doubt? Parker's all-too-human faults have been unduly magnified by his colossal success. Many other wine critics, had they been given a fraction of his awesome power, would have become monsters, their egos impossible to tether.

What seems like the ability to do whatever he wants carries the price that Parker can and will be scrutinized and criticized for whatever stand and actions he takes. It is something he clearly bristles at, though he has come to accept it. He claims never to be seeking extraordinary power. "It's a mixed blessing," he once told me. "And mostly negative." When it is given to him, he never feels totally comfortable with it—and in that he is also very American, finding it hard to accept that power has political and economic consequences, and carries with it extra responsibilities.

In some ways, Parker is stuck in his role. There are those at the very centers of wine opinion who say that his power is beginning to fade. But there are few signs of this yet, and his influence will surely ripple across the wine world for years to come. Will he retire? Ah, but this is a man who loves what he's doing; this is a man who dreams about wine.

It will be hard, if not impossible, for Parker to pass on his mantle of power, for his reputation is rooted in the idea of his specialness and the myth of his unique, semi-divine tasting ability. There will be others to follow him. But he has changed the wine world, and the unique circumstances that coronated him will never be duplicated.

There will never be another emperor of wine.

GLOSSARY

Acidity   A vital natural component in wine that preserves freshness and gives wines zing and liveliness; too little and wines taste soupy, too much and they taste sharp, even sour. Grapes grown in a hot climate or an unseasonably hot vintage year generally produce wines with lower acidity.

Aging   The process of keeping wines so that tannins become smoother and less harsh and the flavors and aromas develop complexity. Some wines are aged in barrel before bottling, and also aged after bottling. Only a small number of wines benefit from aging in the bottle.

Alcohol   A structural element of wine. During fermentation, yeast converts the sugar in grape juice into alcohol; the high degree of sugar in very ripe grapes will ensure a high level of alcohol in a wine. Alcohol in wine ranges from a low of 9 or 10 percent in some German wines to a high of 16 or 17 percent found in some California Zinfandels.

Appellation   A legally defined geographic zone; in some countries wines must be made in a specified way in order to use the appellation name on the label. The assumption—not always true—is that the more specific the appellation name, the higher the quality of the wine.

Assemblage   French term for the selection of wine from different lots to create the blend for a château's final wine.

Balanced   A balanced wine is one with all its elements—alcohol, acid, tannin, fruit, oak, and so forth—in harmony, so that no one component stands out.

Barrel   A wooden container used for storing, aging, and sometimes fermenting wine. It can be large or small; the standard small barrel used for fine wines in Bordeaux, Burgundy, and California holds 225 liters of wine. Most red wines and some whites are aged in oak barrels, though chestnut is also used. The barrel may be new or have been used before; new oak imparts a stronger taste to the wine.

Barrel tasting   Tasting unfinished wine that is still aging in a barrel or cask. To taste, a winemaker removes the bung that stops a hole at the top of the barrel and dips a long glass tube, or pipette, through the hole to extract a glassful of wine—e.g., a barrel sample. For a formal barrel tasting, when dozens of samples are needed, the wine will be put into small bottles.

Barrique   Refers to the standard Bordeaux barrel holding 225 liters of wine.

Bâtonnage   French for lees stirring, a winemaking practice that involves stirring up the deposits (lees) at the bottom of a barrel to add character.

Blind tasting   A tasting in which the labels of the bottles are hidden. If "single blind," the taster knows the wines, but not the order in which they are arranged. If "double blind," the taster knows neither which wines are included, nor the order.

Body   The perception of a wine's weight and texture in the mouth, which results from a combination of alcoholic strength and amount of extract. A full-bodied wine has higher viscosity and density; a light-bodied wine less.

Bottle age   The time wines spend in the bottle before drinking to soften and mellow certain elements in wine, especially tannin. Bottle aging also allows the aroma to develop. See also **Aging**.

Bouquet   Generally refers to the smell of a wine; the more specific meaning is the part of a wine's smell that develops as wine ages in the bottle.

Cask sample   Another word for barrel sample.

Cépage   French for grape variety.

Chai   French for a (usually) aboveground wine cellar in which wine ages in the barrel. An underground cellar is known as a *cave*.

Chaptalization   The addition of sugar to the grape must prior to or during fermentation, legal in France and Germany to boost the body and alcohol level in poor vintages when the grapes don't fully ripen.

Classification   Refers to one of several rankings of quality among vineyards and properties. The 1855 classification of some sixty Bordeaux châteaux is the most famous.

Clone   A group of vines propagated from a cutting, thus sharing the distinctive attributes of the "mother vine." Usually a "clonal selection" has been made for specific characteristics. A nursery may offer several different clones of a particular grape variety.

Concentrated Said of a wine that is dense and deep, with lots of fruit and rich flavors.

Commune French equivalent of a small district or village and surrounding vineyards. Particularly in France, wines are sometimes labeled by the commune in which they were made, such as Margaux or Puligny-Montrachet, rather than with a vineyard name.

Complex A wine with many layers and nuances of aroma and flavor, which makes it interesting to drink and analyze.

Corked, corky Used to describe a wet cardboard or moldy smell and taste in a wine that has been caused by a cork contaminated by the chemical compound 2, 4, 6-Tricloroanisole, also known as TCA.

Courtier French term for wine broker, used especially in Bordeaux. A *courtier* negotiates between the château owner and *négociants*.

Cru Literally, "growth," it refers to an individual property or vineyard; when preceded by the terms premier or grand, *cru* refers to the vineyard's rank. In Beaujolais, the term is used for the ten villages with their own appellations, such as Moulin à Vent and Brouilly.

Cru classé Literally, "classed growth"; a Bordeaux term for one of the châteaux listed in the 1855 classification or in another classification such as that of St.-Emilion and Graves.

Crusher The mechanical device that breaks the skin of the grapes and crushes them to release the juice. It also separates the leaves from the skins and grapes. In America, the term "crush" is slang for harvest.

Cuvée A French term used generally for a specific lot of wine; also used for the blend of wines from different vineyards or different grape varieties, or different barrels that make up the final blend in the bottle.

Effeuillage Removing leaves around the bunches of grapes in order to maximize exposure to sunlight for optimum ripeness; also increases exposure to wind, which reduces the incidence of rot.

En primeur A French term for wine sold as futures, before being bottled.

Extract The substances in wine—tannin, sugar, glycerol, and so forth—that contribute to the wine's color and flavor. Wines termed "extracted" have dense, concentrated flavors.

Fermentation   The natural process that turns grape juice into wine. In the alcoholic or primary fermentation, enzymes in yeast convert sugar into carbon dioxide and alcohol. For barrel fermented wines, this process takes place in a small oak barrel rather than a vat. Malolactic fermentation is a secondary fermentation that is sometimes encouraged to soften a wine's acidity and to contribute complexity to a wine.

Filtering   The process of clarifying wine before bottling by passing it through a coarse or fine filter to remove bacteria or other microorganisms that might spoil the wine as well as harmless sediment. Some experts believe that even very careful filtration can remove flavor from wine and strip out part of its character.

Fining   A winemaking process designed to clarify the wine by adding a fining agent such as egg whites, which draw suspended proteins and other particles to the bottom of a barrel or tank. The wine is then racked, or drawn off into another barrel, or filtered before bottling.

Finish   The final flavor and texture of a wine that remains in the mouth. A lingering finish is characteristic of fine wine.

First growths   The four *premier cru* wines in the 1855 classification—Châteaux Margaux, Lafite-Rothschild, Latour, and Haut-Brion—plus Château Mouton-Rothschild, which was elevated to *premier cru* in 1973. Usually the term is extended to include Châteaux Cheval Blanc and Ausone in St.-Emilion and Château Pétrus in Pomerol.

Flowering   The specific period when a vine produces small flowers, after which grapes begin to form. A successful flowering is critical to the development of grapes.

Futures   Wine sold while still undergoing the aging process in barrel, before bottling and shipping.

Garagiste   Refers to one of the group of small avant-garde producers in Bordeaux, mostly on the Right Bank, who began making minute quantities of wine in the 1990s. Their wines were termed *vins de garage* from the fact that they were made in sheds, as the producers didn't own châteaux and couldn't afford to build wineries.

Green harvesting   A form of crop thinning by cutting off bunches of grapes on each vine during the summer, well before the harvest. The purpose is to get the vine to concentrate all its energy on ripening fewer bunches, which will mean earlier ripening and more flavorful grapes.

Hang time   An American term used primarily for the length of time grapes remain on the vine before being picked. Many producers extend the time as long as possible to obtain grapes as ripe as possible, or physiologically ripe, without ending up overripe.

Lees   The sediment deposited at the bottom of a barrel or vat of wine as it ages; it is composed of small grape particles and dead yeast cells. Some wine-makers allow the wine to age on the lees instead of siphoning it off into another barrel periodically (known as racking). Their belief is that aging on the lees adds texture and character to the wine.

Legs   The viscous drips left on the inside of the glass after wine has been swirled.

Length   The persistence of a wine's flavor in the mouth after tasting.

Maceration   The period in which grape juice is in contact with grape skins and seeds, from which phenolics (color, aromas, tannins, and so forth) are extracted and become part of the wine. Some oenologists favor extra-long extended macerations for certain red wines, as this is thought to result in more mellow tannins. Another method thought to achieve better extraction is cold soak, or prefermentation maceration, a controversial method of keeping the must at very low temperatures for up to eight days before permitting fermentation to begin.

Maître de chai   French term, literally cellarmaster. He/she is in charge of all aspects of making and aging the wine.

Malolactic fermentation   Sometimes called secondary fermentation. A process in which bacteria soften a wine's acidity by converting tart malic acid to milder lactic acid; it also contributes complexity to wine. Encouraged in most fine red wines and some whites, especially top Chardonnays.

Médoc   The largest and most famous wine region in Bordeaux, located on the left bank of the Gironde River; encompasses the famous communes of Margaux, Pauillac, St.-Estèphe, and St.-Julien, as well as Listrac and Moulis.

Microbullage, or micro-oxygenation   A winemaking technique of introducing tiny amounts of oxygen into fermenting wine or wine aging in the barrel through a small tube. It has many purposes, but perhaps the most controversial is its use at some Bordeaux châteaux to soften the tannins of the wine so barrel samples will be more appealing to critics.

Mouthfeel   The weight and texture and tactile sensations of a wine in the mouth, mostly due to body and alcohol.

Must   Name given to the grape pulp and juice just after crushing, before it ferments into wine.

Négociant   A French term for wine merchant, a person or firm who sells and ships wine as a wholesaler.

Nose   The broad term for the smell of a wine.

Oak   The wood most often used to make barrels and large casks for aging fine red and some white wines, prized for the toasty, vanillan flavors and tannins it adds to the wine. New oak barrels impart the most oak flavor. Oakiness that overwhelms the other aspects of a wine is considered excessively "woody."

Oenologist   A winemaker or lab technician trained in the science of oenology. Many wineries and châteaux use a trained oenologist as a consultant to test samples and advise on technical details of winemaking.

Overripe   Tasting term for wines made from super-ripe grapes that have very high sugar levels and low acidity due to a long hang time on the vine, and which have begun to shrivel. In French, *surmaturité.*

Parkerized, also *Parkerisé*   Used primarily to describe wines that have seemingly been made to conform to a taste that would please Robert Parker.

Press wine   After fermentation, the wine that flows freely before pressing is known as free-run juice; press wine is the more tannic, darker wine that results from pressing the grape stems, seeds, and pulp in a wine press. Press wine may be used to add structure to wine that lacks tannin, but is not routinely added to the best wines. The French term is *vin de presse.*

Racking   Pumping or siphoning wine from one barrel to another in order to separate the wine from the sediment.

Reverse osmosis   A controversial practice in which water is removed from the grape must by a special machine to create a more concentrated wine. Widely used in Bordeaux when harvest rains have diluted the grapes. However, the process concentrates both flavor and any flaws.

Sediment   The solids in wine—fragments of grape skin, seeds, and so forth—that settle at the bottom of a tank or barrel or, if the wine hasn't been filtered, the bottle. Wines in bottle often throw a deposit as they age. A wine with sediment should be decanted before serving.

Shipper   Often used interchangeably with *négociant* in France. Generally a firm or individual who purchases wine while it is still in barrel, and, later, bottles, stores, and ships the wine. Since many individual estates now bottle their own wine, shippers store the wine once it is bottled and then commercialize it around the world.

Sommelier   French term for a waiter in charge of the wine in a restaurant.

Super Second   A Bordeaux château classified in 1855 as a second growth but which is considered to produce wine of much higher quality than others in that tier. For example, Châteaux Léoville-Las-Cases and Pichon-Lalande are both considered "Super Seconds."

Tannin   A group of complex molecules found in the skins, stems, and seeds of grapes; tannin helps red wines, which contain much more tannin than whites, to improve with age. A strongly tannic wine has a mouth-puckering astringency.

Terroir   A French term used to designate the complex of soil composition, microclimate, altitude, exposure, and so forth, or natural environment of the wine, which is reflected in the taste and quality of a wine made from grapes grown in that specific location. The highest quality wines are often thought to express an individual *terroir*.

Texture   The feel of a wine in the mouth. Dense, viscous, almost thick wines are said to have a rich, velvety texture.

Unfiltered   A wine that has not been filtered, though it has usually been racked and sometimes fined to remove small particles before bottling. Often labeled as such, unfiltered wines usually have a small amount of sediment. Some believe that unfiltered wines always have more flavor and body.

Unfined A wine that hasn't been fined, or purified, though other processes may have been used to remove small particles before bottling. Often labeled as such. Some believe that unfined wines, like unfiltered wines, have more flavor and body.

Varietal A varietal wine is named for the grape variety from which it is made, such as Chardonnay. Many European wines are labeled with the geographic names of regions, villages, districts, and vineyards.

Vertical tasting A comparative tasting of the same wine from different vintages.

Vigneron French term for wine grower.

Vine age The age of a vine is related to the quality of the grapes it produces; "old" vines, which generally means at least thirty years old, yield grapes and wines with greater intensity and concentration of flavor.

Vintage The year in which grapes were picked to make a wine. If indicated on the label, the wine must be from the stated vintage year.

Vineyard manager The person in charge of the vineyard.

Viticulture The science and study of grapes and grape growing.

Winemaker The person who is primarily responsible for making the wine; widely used in the New World. In France there is no word for "winemaker," generally the person in charge is a *maître de chai,* or cellarmaster.

Winemaking consultant Usually a very well-known and respected winemaker, who may also be an oenologist, and who acts as an outside consultant to a number of wineries rather than being employed by only one. These consultants may do everything from advising on vineyard practices to selecting the specific barrels to go into the final wine.

Wine trade Generally means the wine industry, especially the importing and retailing end of the business.

Yeast A single-celled organism essential in converting grapes into wine. During fermentation, yeast converts the sugar in the must to alcohol and carbon dioxide. Some winemakers inoculate the must with one of the commercially available yeasts, which have been scientifically selected for particular attributes that may add specific flavors to the wine. However, others are convinced that the yeasts that occur naturally on the grapes, known as native or wild yeasts, make more interesting, individual wines.

Yield   The amount of grapes harvested from a particular vineyard, resulting in a certain quantity of wine. In France it is measured as hectolitres per hectare. Generally the lower the yield, the higher the potential quality of the wine.

SOURCES

Much of this book is based upon interviews I conducted in the United States and abroad; the many individuals with whom I spoke are listed in the acknowledgments. I have also drawn heavily on Robert M. Parker, Jr.'s, writings, various interviews and speeches he has given, and his posted comments on his website. In addition, I have relied on a wide range of books; newspaper, magazine, and journal articles; wine newsletters, and other wine publications, primarily in English and French. Those I found particularly useful for historical background on wine and the wine industry, and information about, perspectives on, and assessments of Parker and his influence, are listed below.

Works by Robert M. Parker, Jr.

The Wine Advocate, vol. 1 (1978) to vol. 156 (2004).

www.eRobertParker.com

Columns for *The Washington Post, Food & Wine, Connoisseur, Money, The Wine Enthusiast*.

Speech at Boston University's Sherman Union, January 25, 1999, published in *Massachusetts Beverage Business*, May 1999.

Bordeaux: A Definitive Guide for the Wines Produced Since 1961. New York: Simon & Schuster, 1985.

Parker's Wine Buyer's Guide 1987–1988. New York: Simon & Schuster/Fireside, 1987.

The Wines of the Rhone Valley and Provence. New York: Simon & Schuster, 1987.

Parker's Wine Buyer's Guide 1989–1990. New York: Simon & Schuster/Fireside, 1989.

Burgundy: A Comprehensive Guide to the Producers, Appellations, and Wines. New York: Simon & Schuster, 1990.

Bordeaux: A Comprehensive Guide to the Wines Produced from 1961–1990. New York: Simon & Schuster, 1991.

Parker's Wine Buyer's Guide. 3rd ed. New York: Simon & Schuster/Fireside, 1993.

Parker's Wine Buyer's Guide. 4th ed. New York: Simon & Schuster/Fireside, 1995.

The Wines of the Rhone Valley. Revised expanded 2d ed. New York: Simon & Schuster, 1997.

Bordeaux: A Comprehensive Guide to the Wines Produced from 1961 to 1997. New York: Simon & Schuster, 1998.

Parker's Wine Buyer's Guide. 5th ed. New York: Simon & Schuster/Fireside, 1999.

Parker's Wine Buyer's Guide. 6th ed. With Pierre Rovani. New York: Simon & Schuster, 2002.

Bordeaux: A Consumer's Guide to the World's Finest Wines, 4th ed. New York: Simon & Schuster, 2003.

Foreword to *The Heart of Burgundy: A Portrait of the French Countryside,* by Andy Katz. New York: Simon & Schuster, 1999.

Foreword to *Hachette Atlas of French Wines & Vineyards.* 2d ed. Edited by Pascal Ribereau-Gayon. Paris: Hachette Livre Direction, 2001.

Foreword to *Daniel Johnnes's Top 200 Wines,* by Daniel Johnnes with Michael Stephenson. New York: Penguin, 2003.

Books

Adams, Leon D. *The Wines of America.* 3d ed. New York: McGraw-Hill, 1985.

Amerine, Maynard A., and Edward B. Roessler. *Wines: Their Sensory Evaluation.* New York: W. H. Freeman, 1976, 1983.

Barr, Andrew. *Drink: A Social History of America.* New York: Carroll & Graf, 1999.

———. *Wine Snobbery.* New York: Simon & Schuster, 1988, 1992.

Bourdieu, Pierre. Translated by Richard Nice. *Distinction: A Social Critique of the Judgement of Taste.* Cambridge, Mass.: Harvard University Press, 1984.

Brillat-Savarin, J. A. *The Physiology of Taste, or Meditations on Transcendental Gastronomy.* New York: Dover Publications, 1960.

Broadbent, Michael. *Michael Broadbent's Wine Vintages.* London: Mitchell Beazley, 2003.

———. *Vintage Wine: Fifty Years of Tasting Three Centuries of Wines.* New York: Harcourt, 2002. Previous editions published as *The Great Vintage Wine Book* (1980) and *The New Great Vintage Wine Book* (1991).

————. *Winetasting*. London: Wine & Spirit Publications, 1968. Three revised editions were also useful: *Winetasting enjoying understanding* (London: Christie's Wine Publications, 1977); *Michael Broadbent's Pocket Guide to Wine Tasting* (New York: Simon & Schuster, 1982); and *Michael Broadbent's Wine Tasting. New and Revised Edition* (London: Mitchell Beazley, 2003).

Brook, Stephen. *Bordeaux: People, Power and Politics*. London: Mitchell Beazley, 2001.

————. *Wine People*. New York: The Vendome Press, 2001.

Brooks, David. *Bobos in Paradise*. New York: Simon & Schuster, 2000.

Clemens, S. B., and C. E. Clemens. *From Marble Hill to Maryland Line: An Informal History of Northern Baltimore County*. Privately published, 1976, rev. ed. 1999.

Coates, Clive. *Côte d'Or: A Celebration of the Great Wines of Burgundy*. Berkeley and Los Angeles: University of California Press, 1997.

————. *Grands Vins: The Finest Châteaux of Bordeaux and Their Wines*. Berkeley and Los Angeles: University of California Press, 1995.

————. *The Wines of Bordeaux: Vintages and Tasting Notes 1952–2003*. Berkeley and Los Angeles: University of California Press, 2004.

Conaway, James. *Napa: The Story of an American Eden*. Boston: Houghton-Mifflin, 1990.

————. *The Far Side of Eden: New Money, Old Land and The Battle for Napa Valley*. Boston: Houghton-Mifflin, 2002.

Dickinson, William B., Jr., and Janice L. Goldstein, eds. *Watergate: Chronology of a Crisis*. Washington, D.C.: Congressional Quarterly, 1973.

Dumay, Raymond. *La mort du vin*. Paris: Editions Stock, 1976.

Echikson, William. *Noble Rot: A Bordeaux Wine Revolution*. New York: W. W. Norton, 2004.

Faith, Nicholas. *The Winemasters of Bordeaux: The Inside Story of the World's Greatest Wines*. rev. ed. London: Prion Books, 1978, 1999.

Feret, Claude. *Bordeaux et ses vins*. 13th ed. Bordeaux: Editions Feret, 1982; English ed., 1986.

Fielding, Temple. *1963–64 Fielding's Travel Guide to Europe*. New York: William Sloane, 1963.

Finigan, Robert. *Essentials of Wine*. New York: Alfred A. Knopf, 1987.

Frommer, Arthur. *Europe on $5 a Day. 1965–66 ed.* New York: Arthur Frommer, 1965.

Ginestet, Bernard. *La Mémoire des oenarques*. Bordeaux: Mollat, 1998.

Guy, Kolleen M. *When Champagne Became French: Wine and the Making of a National Identity*. Baltimore: Johns Hopkins University Press, 2003.

Hanson, Anthony. *Burgundy*. London: Faber & Faber, 1982, 1995, 1999.

Jefford, Andrew. *The New France: A Complete Guide to Contemporary French Wine*. London: Mitchell Beazley, 2002.

Johnson, Hugh. *Wine*. New York: Simon & Schuster, 1966.

————, and Jancis Robinson. *The World Atlas of Wine*. 5th ed. London: Mitchell Beazley, 2001. Previous editions in 1971, 1977, 1985, 1994.

Kammen, Michael. *American Culture American Tastes: Social Change and the Twentieth Century*. New York: Alfred A. Knopf, 1999.

Korsmeyer, Carolyn. *Making Sense of Taste: Food and Philosophy*. Ithaca, N.Y., and London: Cornell University Press, 1999.

Kramer, Matt. *Making Sense of Burgundy*. New York: Quill/William Morrow, 1990.

Lehrer, Adrienne. *Wine & Conversation*. Bloomington: Indiana University Press, 1983.

Lichine, Alexis, with Sam Perkins. *Alexis Lichine's Guide to the Wines and Vineyards of France*. New York: Alfred A. Knopf, 1979, 1982, 1984.

————. *Alexis Lichine's New Encyclopedia of Wines & Spirits*. New York: Alfred A. Knopf, 1969, 1974, 1981, 1985, 1987. First published as *Encyclopedia of Wines and Spirits* in 1967.

————. *Wines of France*. New York: Alfred A. Knopf, 1951. Many subsequent editions.

Loftus, Simon. *Anatomy of the Wine Trade: Abe's Sardines and Other Stories*. New York: Harper & Row, 1985.

Loubère, Leo A. *The Wine Revolution in France: The Twentieth Century*. Princeton: N.J.: Princeton University Press, 1990.

Lukacs, Paul. *American Vintage*. Boston: Houghton-Mifflin, 2000.

Lynch, Kermit. *Adventures on the Wine Route: A Wine Buyer's Tour of France*. New York: Farrar, Straus & Giroux, 1988.

Markham, Dewey. *1855: A History of the Bordeaux Classification*. New York: John Wiley and Sons, 1998.

Matthews, Patrick. *Real Wine: The Rediscovery of Natural Winemaking*. London: Mitchell Beazley, 2000.

McCoy, Elin, and John Frederick Walker. *Thinking about Wine*. New York: Simon & Schuster, 1989.

Mondavi, Robert, with Paul Chutkow. *Harvests of Joy: My Passion for Excellence*. New York: Harcourt Brace, 1998.

Osborne, Lawrence. *The Accidental Connoisseur: An Irreverent Journey through the Wine World*. New York: North Point Press, 2004.

Penning-Rowsell, Edmund. *The Wines of Bordeaux*. 4th rev. ed. London: Allen Lane, 1979.

Peppercorn, David. *Bordeaux*. London: Faber & Faber, 1982, 1991.

———. *Wines of Bordeaux: Mitchell Beazley Wine Guides*. London: Octopus and Mitchell Beazley, 2002. First published in 1986 as *David Peppercorn's Pocket Guide to the Wines of Bordeaux*.

Peters, Gary L. *American Winescapes*. New York: Westview Press, 1997.

Peynaud, Émile. *Knowing and Making Wine*. New York: John Wiley & Sons, 1984.

Phillips, Rod. *A Short History of Wine*. New York: Ecco Press, 2000.

Prial, Frank. *Wine Talk*. New York: Times Books, 1978.

———. *Decantations: Reflections on Wine by the New York Times Wine Critic*. New York: St. Martin's, 2001.

Robinson, Jancis. *How to Taste: A Guide to Enjoying Wine*. New York: Simon & Schuster, 1983, 2000.

———, ed. *The Oxford Companion to Wine*. Oxford: Oxford University Press, 1994, 1999.

———. *Tasting Pleasure: Confessions of a Wine Lover*. New York: Viking Penguin, 1997, 1999.

Root, Waverley. *Paris Dining Guide*. New York: Atheneum, 1969.

Rosenblatt, Roger., ed. *Consuming Desires: Consumption, Culture, and the Pursuit of Happiness*. Washington, D.C.: Island Press/Shearwater Books, 1999.

Rouby, Catherine, et. al., ed. *Olfaction, Taste, and Cognition*. Cambridge: Cambridge University Press, 2002.

Rubenfeld, Florence. *Clement Greenberg: A Life*. New York: Scribner, 1997.

Russell, Cheryl. *100 Predictions for the Baby Boom*. New York: Plenum Press, 1987.

Schoonmaker, Frank. *Encyclopedia of Wine*. New York: Hastings House, 1964.

Sokolin, William. *Liquid Assets: How to Develop an Enjoyable and Profitable Wine Portfolio*. New York: Macmillan, 1987.

Swinchatt, Jonathan, and David G. Howell. *The Winemaker's Dance: Exploring Terroir in the Napa Valley*. Berkeley: University of California Press, 2004.

Tower, Jeremiah. *California Dish*. New York: Free Press, 2003.

Waugh, Harry. *Diary of a Winetaster: Recent Tastings of French and California Wines*. New York: Quadrangle Books, 1972.

————. *Harry Waugh's Wine Diary*, vols. 6–9. London: Christie's, 1975, 1976, 1978, 1981.

Waters, Alice. *Chez Panisse Menu Cookbook*. New York: Random House, 1982.

Wilson, James. *Terroir: The Role of Geology, Climate, and Culture in the Making of French Wine*. California: Wine Appreciation Guild, 1998.

Wine Angels. *Wine Marketer's Companion*. Wine Angels at wineangels.com, 2003.

WINE PERIODICALS AND WEBSITES

Connoisseur's Guide to California Wine (CGCW)
Decanter
Food & Wine
Friends of Wine
Gambero Rosso
Global Vintage Quarterly (GVQ)
Harpers Wine and Spirit Weekly (Harpers)
Impact
Rich Cartiere's Wine Market Report
Robert Finigan's Private Guide to Wines
The Vine
Vintage
Wine and Spirits
Wine Business Monthly (WBM)
The Wine Enthusiast (WE)
The Wine Spectator (WS)
Wines & Vines
Financial Times (FT)
The London Sunday Times (LST)
The Los Angeles Times (LAT)
The New York Times (NYT)
Washington Post (WP)
www.bipin.com
www.datamantic.com/joedressner
www.farrvintners.com
www.jancisrobinson.com
www.weimax.com
www.westcoastwineweb.com

www.wineanorak.com
www.winecrimes.com
www.wineint.com
www.wine-pages.com
www.wine-people.com
www.wine-lovers-page.com

ARTICLES

"A Life Less Ordinaire," Andrew Jefford, *Waitrose Food Illustrated*, 2003.

"A Nobel for Explaining How the Nose Knows," Thomas H. Maugh II, *LAT*, Oct. 5, 2004.

"A Nose for Wine," Harvey Steiman, *WS*, March 31, 1994.

"A Nose for a Winner," Jancis Robinson, *London Sunday Times*, 1986.

"A Sip Too Far," Tim Atkin, *Decanter*, April 1998.

"Accusations Rise from Wine Cellars of Bordeaux," Craig S. Smith, *NYT*, Aug. 10, 2003.

"Aftershock: Revisiting the 1976 Paris Tasting," Brenda Maitland, *Culinary Concierge*, 2003.

"An Ombudsman for the Wine Consumer," William Rice, *WP*, September 28, 1978.

"Back Talk," Lettie Teague, *Food & Wine*, January 2001.

"Baker v. Parker: A View from the Sidelines," Clive Coates, *The Vine*, July 1987.

"Beyond Sweet and Sour," Danny Kingsley, *ABC News*, Feb. 27, 2002.

Bonny Doon Vineyard newsletters, 1999–2002.

"Book Reviews," David Peppercorn, *Decanter*, December 1986.

"Bordeaux Eyes the Sky as Parker Arrives," Adam Lechmere, *Decanter*, Sept. 6, 2003.

"Bordeaux: l'affaire qui ébranlé la maison Parker," Jean-Jacques Chiquelin, *Le Nouvel Observateur*, April 25, 2002.

"Bordeaux Remembers: 'The American Vintage,'" Julia Mann, *WS*, Nov. 30, 1998.

"Burgundian Vintner Sues Robert Parker for Libel," Per-Henrik Mansson, *WS*, May 31, 1994.

"California Crush," *Decanter* supplement, August 1981.

"California's New Stars," Bruce Schoenfeld, *Cigar Aficionado*, July/Aug., 1999.

"Case for the Defense," John Stimpfig, *Decanter*, 1999.

"Château La Pique," Malcolm Gluck, *The Guardian*, June 10, 1994.

"Clawing their way to the top," William Rice, *The Chicago Tribune*, Oct. 16, 1986.

"Consumers Hold the Cards: This 'Fad' Won't Fade," Robert B. Morrisey, *Advertising Age*, March 29, 1982, vol. 53, issue 14.

"Critic's Judgment Puts Bordeaux Noses Out of Joint," Adam Sage, *London Times*, May 25, 2002.

"Cyberspace Comedy in 3 Parts: R. Parker, Producer," Mark Squires, *Mark Squires E-Zine on Wine*, marksquires.com. Nov./Dec. 1996.

"Distaste for Tastings," Hugh Johnson, *Decanter*, June 1992.

"Does UC-Davis Have a Theory of Deliciousness?" Clark Smith, *Vineyard & Winery Management*, July/Aug. 1995.

"Emile Peynaud," Colin Parnell, *Decanter*, March 1990.

"For Guigal, It's Anything but Lonely at the Top," James Suckling, *WS*, Mar. 15, 1990.

"Frank Schoonmaker—Visionary Wine Man," Frank Johnson, *winemouse.com*, n.d.

"French Have a Beaune to Pick," John Lichfield, *Canberra Times*, July 25, 1998.

"French Wine All at Sea," Jo Johnson, *FT*, Nov. 10, 2001.

"Future Bright for 2000 Bordeaux," Adrienne Roberts, *FT*, May 30, 2001.

"German Wine-Tasting Melts His Fillings But Wins His Heart," Michael Dresser, *Baltimore Sun*, Nov. 8, 1992.

"Great Bordeaux Debate," Terry Robards, *WS*, August 1983.

"He Sips and Spits, and the World Listens," David Shaw, *LAT*, Feb. 23, 1999, and Part 2, "For Wine Critic, the Truth Is in the Glass," Feb. 24, 1999.

"His Palate Is Insured," Mary Corey, *Baltimore Sun*, Mar. 28, 1993.

"How Château Pétrus Became Bordeaux's Most Coveted Wine," James Suckling, *WS*, Feb. 15, 1991.

"How Rating Systems Stack Up," Jerry Shriver, *USA Today*, Mar. 12, 1993.

"Hugh Johnson, The Great Communicator," *Decanter*, March 1995.

"I Am the Most Powerful Person in the Wine World," Suzanne Goldenberg, *The Guardian*, July 23, 2003.

"Il repointe son nez en France," Pauline Damour, *Challenges*, August 2003.

"In Defense of English Wine Writers," Robin Young, *Decanter*, April 1988.

"Inside the Mind of Michel Rolland," John Stimpfig, *Decanter*, January 2004.

"International Wine Expo Opens in Hong Kong," *Business World*, June 6, 1998.

"In Vino Veritas," Lettie Teague, *Food & Wine*, September 1999.

"Is Terroir Dead?" Karen MacNeil, *San Francisco Chronicle*, Sept. 18, 2002.

"Judgment of Paris," *Time*, June 7, 1976.

"Kacher takes a hands-on approach," Thomas Matthews, *WS*, Mar. 31, 1991.

"La Représentante en France du critique de vins Parker a été mise en examen pour 'faux et usage,' " Pierre Cherruau and Jacques Follorou, *Le Monde*, Feb. 1, 2003.

"L'Affaire Geens Rebondit," Fabien Pont, *Sud Quest*, Jan. 24, 2003.

"Le Grand Bob," Jean Durand, online questionnaire in sur.fr.rec.boissons.vins, March 1998.

"Le Grand marathon des vins du monde," *Gault-Millau*, October 1979.

"Les Decorés de Chirac," Anne Fulda, *Le Figaro*, June 23, 1999.

"Les Policiers interrogent," *Le Monde*, Jan. 31, 2003.

"Letters to the Editor," *Decanter*, many issues.

"Lichine Shuns Critical Press; Group Praises '82 Bordeaux." James Suckling, *WS*, August 1983.

"Magnum Force," *Los Angeles Magazine*, December 1998.

"Mais Non, Monsieur Parker," Elizabeth Bryant, *San Francisco Chronicle*, Jan. 16, 2003.

"Marvin Shanken Looks Back at 20 Years of Wine Spectator," *WS*, April 1996.

"Mechanisms of Terroir," Jamie Goode, *Harpers*, Sept. 12, 2003.

"Merchants Flooded with 1982 Futures Orders," *WS*, August 1983.

"Michel Guillard interroge le professeur Émile Peynaud," *Amateur de Bordeaux*, 1987.

"Michel Rolland: Français, reveillez-vous," interview in *La Revue du vin de France*, December 2001/January 2002.

"More Power to the Elbow," Patrick Matthews, *Harpers*, Archives, 2003.

"Moueix Hits Out at Over-Concentration," Josie Butchart, *Harpers*, Archives, 2003.

"Much of Bordeaux Goes Begging," Frank Prial, *NYT*, Aug. 11, 2004.

"New Home for Guigal's Prestige Rhône Wines," Per-Henrik Mansson, *WS*, January 1994.

"Not Exactly for the Faint of Palate," Alice Feiring, *NYT*, June 9, 2002.

"Noted Wine Critic Explodes," *The National Vinquirer, Bonny Doon Vineyard* newsletter, April 1, 2003.

"On the Blind Tasting of Wines: A New Method of Analysis and Beyond," Domenic V. Cicchetti, Ph.D., www.hicstatistics.org/2003StatsProceedings, 2003.

"Parker décoré par Jacques Chirac," *Agence France Presse*, June 22, 1999.

"Parker on Food," Fiona Beckett, *Decanter*, July 2003.

"Parker Parade Comes to Town," Stephen Brook, *Decanter*, June 1996.

"Parker Slams Bordeaux over Fraud Case," Adam Lechmere, *Decanter*, July 24, 2003.

"Parker Too Influential Says Lencquesaing," Adam Lechmere, *Decanter*, Nov. 3, 2001.

"Parker Will Not Go to Bordeaux," Adam Lechmere, *Decanter*, Mar. 21, 2003.

"Parker's Retort," Panos Kakaviatos, interview in *Wine International*, July 2003.

"Parker's 25-year Love Affair with Bordeaux Could Be Over," Kathleen Buckley, *San Francisco Chronicle*, Mar. 6, 2003.

"People to Watch: Dan Philips," Joshua Cooper Ramo, *Time*, May 22, 2000.

"Perfect Grapes of Groth," Jancis Robinson, *FT*, March 31, 1990.

"Pierre-Antoine Rovani," Mark Golodetz, *WE*, July 1998.

"Playing the Rating Game," *The Economist*, Sept. 18, 1999, U.S. edition.

"Pour ou contre Robert Parker," J-F. W., *Cuisine et Terroirs*, Fall 2000, no. 3.

"Preliminary 2000 Bordeaux Vintage Report," Bill Blatch, in *Wines & Vines*, March 2001.

"Present at the Revolution," Matt Kramer, *WS*, Nov. 15, 2001.

"Pulling Out All the Stops," Jamie Goode, *Harpers*, Archives, 2003.

"Quality Scores, Pt. 1: Biology & The Wine Critics," Jordan Ross, *GVQ*, Nov. 1, 2002.

"Quick Tasting Opens Door to a New Market," Nigel Austin, *The Advertiser*, July 7, 2001.

"Rating Robert Parker," Natalie MacLean, nataliemaclean.com, Dec. 12, 2002.

"Red China," *South China Morning Post Sunday Magazine*, Nov. 23, 1997.

"Relishing the Flavor," *The Economist*, June 20, 2002.

"Rencontre avec le guide supreme: Robert Parker," Jean-François Chaigneau, *Paris Match*, Sept. 11, 2001.

"Robert Parker," Bill Baker, *The Three Course Newsletter*, 1987.

"Robert Parker: A Taste for the Good Life," Nora Frenkel, *The Sun*, Oct. 18, 1987.

"Robert Parker and the Palate of Gold," Edward Guiliano, *Wine Times* (precursor to *Wine Enthusiast*), Part 1, May/June 1989, and Part 2, Sept./Oct. 1989.

"Robert Parker Attacks Free Speech," Robert Callahan, News.admin.censorship newsgroup, July 10, 1997.

"Robert Parker: 'Habría que empezar por matar a los enólogos,' " Victor Rodriguez, *Sobremesa*, no. 132, 1996.

"Robert Parker: Interview on BBC transcript," Andrew Jefford, March 3, 1995.

"Robert Parker, Jr.," Dan Sperling, *USA Today*, March 1, 1988.

"Robert Parker, le Bordeaux dans le nez," Jose-Alain Fralon, *Le Monde*, Sept. 20, 2002.

"Robert Parker Looks Ahead," Edward Guiliano, *WE*, Part 1, June 1993, and Part 2, Aug. 1993.

"Robert Parker Pops the Corks . . . ," Linda Murphy, *San Francisco Chronicle*, Oct. 21, 2004.

"Robert Parker, Winosaur," Tony Hendra, *Forbes FYI*, Fall 1997.

"Robert Parker's Oregon Pinot Noir Released," *WS*, Feb. 28, 1994.

"Saint Emilion Winemakers recant on extremism," Jancis Robinson, *FT*, Apr. 15, 2002.

"Secret Rites of Linen Bibs and Dinner Bells," Phyllis C. Richman, *WP*, May 3, 1981.

"Settling Scores: Hugo Rose v. Steve Browett," *Harpers*, Jan. 11, 2002.

"Settling the Score," Tina Caputo, *Wines & Vines*, October 2002.

60th Anniversary issue, *Gourmet*, September 2001.

"So What's the Score?" *Harpers* conference, 2002.

"Taste Tyrants," Frank Prial, *NYT Magazine*, February 17, 1991.

"Tasting St-Emilion's Thoroughbred Wine," James Suckling, *WS*, Feb. 15, 1991.

"Tempest in a Wine Barrel," Dan Berger, *LAT*, June 23, 1994.

"The American Century" issue, *Bon Appetit*, September 1999.

"The Boys from Garlic Mountain: A Talk with Renzo and Riccardo Cotarella," Tom Maresca, *Wine & Spirits*, October 2001.

"The Class of Claret," Peter Melzer, *WS*, July 31, 2002.

"The Day California Wines Came of Age," Thane Peterson, *businessweek.com*, May 8, 2001.

"The Democracy of Wine," Lynn Stegner, *GVQ*, Oct. 1, 2000.

"The Four Seasons California Barrel Tasting," Elin McCoy and John Frederick Walker, *Food & Wine*, July 1981.

"The French Way," Norm Roby, *Decanter*, April 2002.

"The Future of American Wine Ratings," Jeff Morgan, *GVQ*, Apr. 1, 2000.

"The Gallo Legacy Updated," Robert Lawrence Balzer, *WE*, October 1993.

"The Grapes of Math," William Neuman, *Wired*, November 2001.

"The Great and Powerful Shnoz: Does the emperor of wine have any clothes?" Michael Steinberger, slate.msn.com, June 17, 2002.

"The Hit Man," Huon Hooke, *Australian Gourmet Traveller*, October/November 2001.

"The Imperial Palates," Jeff Cox, *GVQ*, Sept. 1, 2001.

"The Man Who Turns Wine Into Bread," John Stimpfig, *How To Spend It (Financial Times)*, November 2001.

"The Man with the Paragon Palate," John Elson, *Time*, Dec. 14, 1987.

"The Merciless Man of Wine," *Newsweek*, Dec. 14, 1987.

"The Million-Dollar Nose," William Langenwiesche, *The Atlantic Monthly*, December 2000.

"The Million-Dollar Nosey Parker Interview," Helena de Bertodano, *Sunday Telegraph*, March 10, 1996.

"The Monk Who Loved Gloria Swanson," Tim Fish, *WS*, July 31, 2002.

"The Myth of Robert Parker," Russ Bridenbaugh, *Wines & Vines*, June 2001.

"The Parker Principle," W. R. Tish, *LAT*, October 2004.

"The Powerful Palate," Nathan Cobb, *The Boston Globe Magazine*, Oct. 8, 1989.

"The Rich, Ripe Full-Bodied Life of Robert Parker," Peter Hann, *Business Week*, Aug. 18, 1986.

"The Saga of New World Wines," original musical production put on in 2003 in Napa Valley.

"The Standout," Jay McInerney, *House & Garden*, November 1999.

"The Trend-Watcher," Elin McCoy and John Frederick Walker, *Food & Wine*, April 1982.

"The Unbearable Lightness of Being," Ch'ng Poh Tiong, *Decanter*, July 2003.

"The Wine Advocacy of Robert Parker," Randal Barnett, *Warfield's*, December 1986.

"The Year California Won the Pennant," Moira Johnston, *New West*, August 2, 1976.

"Tim Mondavi on the Art and Business of Growing Wine," Cyril Penn, *WBM*, July 2001.

"Time Is Ripe for Change," Jordan Mackay, *LAT*, Sept. 15, 2004.

"Unfiltered Wines Gain Favor in California," Robyn Bullard, *WS*, June 15, 1994.

"Vintage and Market Report: The 1982 Vintage," Peter Allan Sichel.

"Vintage Report 2000," Peter Allan Sichel.

"Wet Dogs and Gushing Oranges," Sean Shesgreen, *The Chronicle of Higher Education*, Mar. 7, 2003.

"When He Spits, Vintners Tremble," Frank J. Prial, *NYT*, Dec. 6, 1989.

"Who is Robert Parker? And Why Should You Care?" Amy Virshup, *Smart Money*, September 1996.

"Why Wine Writers' Opinions Vary," James Suckling, *WS*, Oct. 10–15, 1983.

"Wine and Conversation: A New Look," Adrienne Lehrer, AAAS Annual Meeting, 2001.

"Wining and Dining, but with a hitch," Sara Engram, *Baltimore Sun*, Oct. 24, 2001.

"Wine: Corker of a Fracas," Joanna Simon, *The Sunday Times* (London), Jan. 17, 1988.

"Wine Critic Gets French Legion of Honor," Associated Press, *NYT*, June 23, 1999.

"Wine expert travels between two worlds," Jerry Shriver, *USA Today*, Sept. 3, 1999.

"Wine Etcetera," Elin McCoy and John Frederick Walker, *Food & Wine*, December 1981.

"Wine Gurus Get Brain Boost," *Reuters*, May 29, 2003.

"Wine Magnum Force," John Winthrop Haeger, *Los Angeles Magazine*, November 1998.

"Wine Quality Scores Part 1. Biology and the Wine Critics," *GVQ*, Nov. 1, 2002.

"Wine Talk," Terry Robards, *NYT*, Dec. 15, 1982; Jan. 19, 1983.

"Wine Talk," Frank J. Prial, *NYT*, Mar. 2, 1983; Mar. 9, 1983.

"Wine Threats End Book-Signing Tour," Dan Berger, *LAT*, Nov. 29, 1990.

"Wine Trends and Buzzwords," Nick Alabaster, wineanorak.com.

"Wine Writers Hedged for decades and wouldn't take a position. I do," Jane Macquitty, *The Times* (London), Dec. 5, 1987.

"Wine Writers: Squeezing the Grape for News," David Shaw, *LAT*, Part 1, Aug. 23, 1987, and Part 2, "Wine Critics Influence of Writers Can Be Heady," Aug. 24, 1987.

"The Zachy's Gazette: Parker's Hall of Fame," ad in *WS*, Nov. 30, 1990.

ACKNOWLEDGMENTS

I owe debts of gratitude to a long list of people who helped to make this book what it is.

I deeply appreciate the time Robert M. Parker, Jr., was willing to spend with me in person and on the telephone, in his home and on the road, and the hospitality he extended, despite knowing that he would have no say in what I wrote. Thanks are due him, too, for providing a number of requested photographs and allowing several to be reproduced in this book. His wife, Patricia Etzel Parker, was equally hospitable and forthcoming.

I also owe thanks to several people who work directly with Parker: his secretary, Joan Passman; his associate at *The Wine Advocate*, Pierre Rovani; his associates at eRobertParker.com; Mark Squires, who runs the site's bulletin board; his partner in Beaux Frères, Mike Etzel; and his literary agent, Robert Lescher.

During the thirty years I have covered the wine world I have had the opportunity to meet, talk, and taste with wine producers all over the world. They have welcomed me into their vineyards, their cellars, and often their homes, and I thank all for sharing their world, their wines, their knowledge, and their passion and points of view with me. The knowledge that I have gained from all of them informs every page of this book.

During the writing of this book my fellow wine writers and editors in the United States offered comments and stories, heated opinions, valuable information, suggestions of whom to talk to, and words of enthusiasm, and forwarded to me many appropriate articles I otherwise might not have seen. I'm particularly grateful to Gerald Asher, Dan Berger, Gerry Dawes, Alice Feiring, Robert Finigan, Eunice Fried, Mark Golodetz, Jim Gordon, Dana Jacobi, Harriet Lembeck, Jay McInerney, Peter Meltzer, Frank Prial, William Rice, Terry Robards, and W. R. Tish. Thanks are also due John Anderson, Michael Apstein, Mary Mulligan, Ed McCarthy, Robert Whitley, and Richard Elia.

I'm grateful to Marvin Shanken, the publisher and editor of *The Wine Spectator*, for a lengthy interview, and for access to early issues of the magazine.

My thanks, too, to Adam Strum, publisher of *The Wine Enthusiast*, for sharing information and early issues of that publication. Fabio Parasecoli of *Gambero Rosso* allowed me to copy many useful articles.

Robert Millman and Howard Kaplan offered historical perspective. Stuart Yaniger gave insight into the world of wine online. Dr. Linda Bartoshuk at the Yale Medical School shared her research on taste. Dr. Adrienne Lehrer discussed her work on wine language. Jerry Goodman and Charlie Rose provided tapes of their TV interviews with Parker.

Several dear friends also offered places to stay and meals during my travels: in Baltimore, Marguerite Thomas and Paul Lukacs; in Washington, D.C., Bob and Virginia Hurt; in London, Rosemary George and Christopher Galleymore; in New York, Mary Kay and Woody Flowers.

Many U.S.-based wine retailers and importers spoke to me at length about the wine business and Parker: Ruth Bassin, Elliot Burgess, Dick Caretta, Robert Chadderdon, Bill Deutsch, Joe Dressner, Robert Gourdin, Robert Haas, Daniel Johnnes, Robert Kacher, John Kapon, Mike Kapon, John Laird, Leonardo LoCascio, Kermit Lynch, Dr. Jay Miller, Mitchell Nathanson, Jeff Pogash, Ed Sands, Robert Schindler, Abdullah Simon, Ira Smith, Elliott Staren, Marvin Stirman, Guillaume Touton, and Jeff Zacharia. Emmanuel Berk of The Rare Wine Company loaned me an almost complete archive of Robert Finigan's newsletters and copied many early issues of *The Wine Advocate* for me.

Many thanks also to Dan Green, who first published Parker's books; to Carole Lalli, Parker's original editor at Simon & Schuster, who also worked with him—and me—during her tenure as editor in chief at *Food & Wine*; and to Phyllis Richman, former food editor of the *Washington Post*.

For logistical help, thanks are due to Beth Cotenoff and Pia Loavenbruck of Sopexa New York; Robin Kelley O'Connor of the Bordeaux Wine Bureau; Claire Contamine, formerly of the BIVB in Burgundy; and Florence Raffard of the CIVB in Bordeaux. Sheila Nicholas, Pam Hunter, and Tor Kenward were helpful in California. Anne Riives has been a constant source of information and suggestions, and Marsha Palanci, Delphine Boutier, Niki Singer, Margaret Stern, and the late Melissa Seré all provided vital information and paved the way for several important interviews. David Strada e-mailed articles and sent rare jazz CDs, Terry Ellis gave me information about Neil Young, and James Douglas Barron offered insight into Clement Greenberg.

Collector Park Smith shared memories of Parker. Peter M. F. Sichel gave perspective on the wine scene in Bordeaux. Henry Leung told about the tastings at Henry's Evergreen; Bob Lyster and Ed Kasselman recalled early Parker tastings.

In the United Kingdom, a number of wine professionals alerted me to useful articles and sent their own, spoke to me at length about the issues that Parker's power raises, made suggestions of whom to interview, and sometimes even interceded with those reluctant to talk. My thanks to Tim Atkin, Bill Baker, Michael Broadbent, Stephen Brook, Steve Browett, Clive Coates, Rosemary George, Malcolm Gluck, Anthony Hanson, Andrew Jefford, Jasper Morris, the late Dylan Paris, Roy Richards, Jancis Robinson and Nick Lander, Anthony Rose, Hugo Rose, Steven Spurrier, John Stimpfig, and Serena Sutcliffe. Thanks also to Amy Wislocki, the editor of *Decanter*, for allowing me to spend half a day in their office copying many pages from early issues of the magazine.

I have many people to thank in France for making time to be interviewed, beginning with wine writers Michel Bettane and Jean-Marc Quarin, and editor Christophe Tupinier of *Bourgogne Aujourd'hui*.

In Burgundy: Bertrand Ambroise, Bertrand Devillard, François Faiveley, Pierre Henri Gagey, Dominique Lafon, Jacques Lardière, Louis-Fabrice Latour, Floris Lemstra, François Lequin, Thierry Matrot, Allen Meadows, Pierre Meurgey, Etienne de Montille, Anne Parent, Nicolas Potel, Virginie Taupenot-Daniel, Aubert de Villaine (and his wife, Pamela), and especially Becky Wasserman, Russell Hone, and Peter Wasserman.

In Bordeaux: Hanna Agostini, Patrick Basedon, Olivier Bernard, Bill Blatch, Emmanuel Cruse, Pascal Delbeck, Marcel Ducasse, Laurent Ehrmann, May-Eliane de Lenquesaing, Gonzague Lurton, Fiona Morrison, Jean-Guillaume Prats, Michel Rolland, and Erik Samazeuilh. Also to: Bernard Audoy, Anthony Barton, Bruno Borie, Christopher Cannan, Florence and Daniel Cathiard, Jean-Michel Cazes, Jean-Marie Chadronnier, Laurent Cogombles, Jeffrey Davies, Baron Frédéric de Luze, Thierry Gardinier, Jean-Marc Guiraud, John Kolasa, Eugene Marly, Patrick Maroteaux, Edouard Moueix, Comte Stephan von Neipperg, Chantal Perse, Alain Raynaud, Dominique Renard, Jean-Luc Thunevin and Murielle Andraud, and Alexandre Thienpont.

Elsewhere in France: Michel Chapoutier, Jean-Luc Colombo, Etienne Hugel, Nicolas Joly, and Claude Taittinger.

I owe thanks to Angelo Gaja, Jacopo Biondi-Santi, Daniele Cernilli, and Riccardo and Renzo Cotarella in Italy, Pablo Alvarez Mezquiriz in Spain, and Adrian Bridge in Portugal. In Australia I'm grateful to Huon Hooke, Tim White, Grant Burge, and John Larchet; in New Zealand, Neil McCallum; and in Japan, Ernie Singer.

On the West Coast, a special thanks to Julie Garvey, James Hall, Anne

Moses, and Donald and Heather Patz, John and Doug Shafer, and Elias Fernandez, Ray Signorello, and Shari and Garen Staglin. Also, Stillman Brown and Dan Lewis, Jim Clendenen, Randy Dunn, Dennis Groth, Agustin Huneeus, David Lett, Leo McCloskey, Gwen McGill, Robert and Tim Mondavi, Manfred Krankl, Harvey Posert, Patrick Rooney, Ed Schwartz, Rob Sinskey, Clark Smith, and John Wetlaufer.

Thanks are due, too, to several libraries for their help. First and foremost, the library at the Culinary Institute of America in Hyde Park, and especially to librarian Christine Crawford-Oppenheimer, The New Milford, Connecticut, Public Library obtained books, and the staff at The New York Public Library was helpful in tracking down obscure tomes in both English and French at both their central research building and at the business library.

Dan Halpern, a true wine lover, has been the ideal editor—from his first enthusiasm for the proposal, to his patience in granting several deadline extensions, to his perceptive suggestions on the manuscript. At Ecco Press, I also thank his assistant, Robert Grover; E. J. Van Lanen; High Design for the wonderful cover; the book's designer, Jessica Shatan Heslin; and Lisa Wolff for careful copyediting.

My lawyer and agent, Alan Kaufman, another passionate wine lover, has offered wise advice and been a mainstay of support from the beginning.

I am blessed with dear close friends who listened with interest and enthusiasm throughout the writing of this book: Elene Kolb, Carroll Macdonald, Justine McCabe, Bridget Potter, and the late Elizabeth Crow.

I thank my son, Gavin McCoy Walker, for translating key articles in the Japanese press, for lively discussions about Bourdieu's theories and their relation to the Parker phenomenon, for his meticulous reading of the text, catching every misplaced and missing accent mark, and for his continued support.

My greatest thanks go to my husband, John Frederick Walker, with whom I first learned to love and write about wine. He has always been my surest sounding board and the ideas in this book have benefited greatly from his willingness to discuss them at the dinner table again, and again, and again. Most of all, I appreciate more than I could ever say his steadfast support.

Any shortcomings and errors that remain are, of course, my responsibility.

—ELIN McCOY

Photo Acknowledgments

Five photographs of Robert M. Parker, Jr., were provided by and are reproduced with the kind permission of Robert Parker. They are: Parker and Pat at Maxim's and Parker in 1978 at the launch of *The Wine Advocate* (taken by Patricia Parker); Parker at Château Rayas (taken by E. Junquenet) and Parker in Hermitage with Michel Chapoutier; and Parker and his family with Jacques Chirac.

The pictures of Parker spitting and Parker with his dogs originally appeared in *The Atlantic Monthly*, December 2000; they were taken by Christopher Barker, and I thank him for permission to reproduce them here.

The photograph of Parker and Hanna Agostini, taken by Jean-Jacques Saubi, is reproduced with the permission of photo agency MAXPPP in France.

Thanks are due to the officials of VinExpo for permission to use the photograph of Parker and Michel Rolland speaking at VinExpo New York in 2002, and also to Rolland and Parker for allowing it to be used.

The cartoon by Tony Husband first appeared in *Harpers Wine and Spirit Weekly*, April 2, 2002; I thank Tony Husband for allowing me to reproduce it.

Thanks are due François des Ligneris, owner of Château Soutard, who produced the "Vis Parker" poster, for providing it and for permission to reproduce it.

I am grateful to Laurent Cogombles for allowing me to reproduce the Château Bouscaut ad.

Dan Lewis kindly gave permission to reproduce the Jory Winery label.

The picture of Parker at Mark's Duck House is reproduced with the permission of Marty Katz.

INDEX

About the author

About the book

Insights,
Interviews
& More . . .

Read on

A Conversation with Elin McCoy

Richard Felber

Where were you born, Elin?

I was born in a suburb of Chicago, Illinois. As a child I lived in San Francisco; Washington, DC; Ann Arbor, Michigan; and Rochester, New York, before moving back to Chicago's North Shore, where I went to high school.

Tell me about your first name. How did you come by it? How is it pronounced?

The name "Elin" is Swedish; I was named for my grandmother on my mother's side, who was born and grew up in Sweden. It is pronounced as though the *E* were a long *A*: *Ā-lin.* One radio interviewer, amusingly enough, kept trying to remember the correct pronunciation before we went on the air and finally told me, "I'll just think of A-list."

What events from your childhood stand out?

There were so many; it's hard to single one out—lots involved reading, especially packing lunch in a backpack, loading it with books,

> ❝ The name 'Elin' is Swedish; I was named for my grandmother on my mother's side, who was born and grew up in Sweden. ❞

and climbing my favorite tree to sit and read all afternoon. I had quite a happy childhood and a tight-knit family. My parents and sister all loved to read and hike and travel. The highlight of my year was the summer, as I always spent a couple of hot-weather months in Michigan at a large lakeside summer cottage that also housed my grandparents, aunts, uncles, and six cousins. We sailed, swam, hiked the dunes, and water-skied; my older cousin and I bossed all the rest and commandeered the speedboat. Leaving the lake on Labor Day weekend was always the saddest day of the year.

What is your earliest memory of reading and being influenced by a book?

I was very small—maybe three and a half or four—and I suddenly realized I could read the words on the pages of a new book. I can still remember the feeling of power.

I was a voracious reader. My favorite book as a child was *Indian Captive* by Lois Lenski, in which Mary Jemison is captured in an Indian raid and lives the rest of her life with the tribe. It made me think that girls could do things, be brave, and have adventures. Nancy Drew was important too; also Jo in *Little Women* . . . I wanted to be a writer just as she was.

Did your parents drink wine? If so, did they drink "good" wine?

Yes, my parents drank wine, mostly when they gave dinner parties or when they ate out at the golf club or in restaurants. My father didn't collect wine or stock a cellar; the wines they drank were above jug, but hardly grand cru Bordeaux. A good martini (with an olive, not a twist) was the aperitif of choice. ▶

> 66 My father didn't collect wine or stock a cellar; the wines they drank were above jug, but hardly grand cru Bordeaux. 99

*When did you take your first sip of wine?
What do you recall about the experience?*

I can recall nothing about my first sip of wine, but I vividly recall the first time wine made an impression on me. I was fourteen and staying at home with my dad while my mother was away in Michigan with my younger sister. My father took me out to dinner at Jacques, a very elegant French restaurant in downtown Chicago. He ordered a bottle of Pouilly-Fuissé and asked the waiter to bring me a wine glass. The waiter reminded my father that he couldn't pour a glass for me; my father said he would serve me himself. I felt tremendously grown up and glamorous sipping and savoring my Pouilly-Fuissé.

Do you have any unusual or otherwise compelling anecdotes surrounding your collegiate experience?

Most important to me was going to France for my junior year. It was as though my eyes were suddenly opened wide to the world . . . and food and wine.

Name some jobs you've had, Elin. Anything unusual?

My first serious job was at an employment agency in Chicago the summer I was sixteen. I tried to match the people who came in with the jobs we had listed. I still remember interviewing women for a job as a Playboy Bunny, asking them if they would be willing to wear a costume with ears and a tail. The

66 My first serious job was at an employment agency in Chicago the summer I was sixteen. I still remember interviewing women for a job as a Playboy Bunny, asking them if they would be willing to wear a costume with ears and a tail. 99

interview process, however, did not include asking them to take off their clothes.

My first writing job (again a summer position) was at textbook publisher Scott Foresman, where I wrote reading exercises for *Think and Do* workbooks. These were companion volumes to the *Dick and Jane* readers; everyone learned to read with Dick and Jane back then.

How did you get into writing about wine?

I started coauthoring wine pieces with my husband for *New York* magazine after living on the West Coast and becoming enthused about California wine. Since then I've been a columnist and contributor on that subject (and many others) for a number of national magazines. In 2000 I became wine and spirits columnist for *Bloomberg Markets*.

Do you have any writerly quirks? When and where do you write? PC or pen?

I'm obsessed with notebooks and pens. I like fountain pens; unfortunately, they usually leak when you travel by plane. In researching this book I took all my interview notes in special small Spanish spiral-bound notebooks with squared paper that I buy in New York. I do tape every interview, but find taking notes more useful.

I generally compose directly on my computer, but occasionally write a rough draft in bigger spiral-bound notebooks. ▶

ёTheha

A Conversation with Elin McCoy *(continued)*

What do you rely upon for stimulation? Do you observe any particular beverage ritual?

The hardest part of writing is getting started. I have to have a big mug of cappuccino, but don't dare open a newspaper or check e-mail. I always stop writing each day when I still have a bit of a story or section to finish so I have something set to begin with in the morning.

What interests or enthusiasms do you have? Any hobbies or outdoor pursuits?

I am fascinated by archaeology. Working on digs is my idea of great fun. I love to hike, travel, garden (I'm passionate about lavender), listen to opera, and cook.

Town of residence?

Kent, Connecticut.

Husband, partner, children, pet(s)?

My husband, John Frederick Walker, writes primarily on natural history subjects and is an artist. Gavin Walker, my son, recently received an MA in Japanese studies from the University of Pennsylvania. I am temporarily without a dog, but won't be for long.

Finally, what's the greatest wine you've ever drunk?

A 1900 Château Margaux. But I'd have to say that Pinot Noir, especially red Burgundy, is my all-time favorite wine. ❧

> 66 I am fascinated by archaeology. Working on digs is my idea of great fun. 99

6

Aftertaste
Public Reaction to
The Emperor of Wine

LIKE EVERY WRITER who spends years laboring on a book, I hoped for lots of advance buzz to build interest and a slew of positive reviews after publication. But I didn't expect the fevered reaction from Parker's fan base that began while the book was still in the editing stage.

Months before publication, a copy of the manuscript somehow fell into the hands of one of Parker's acolytes, who leaked a few tantalizing tidbits from the book in a bulletin board post on Parker's Web site, eRobertParker.com. These postings included a story I had already decided to delete, largely because I felt it would lead to irrelevant French-bashing—which is, of course, exactly what happened in a long thread of heated discussion by people who hadn't even read the book.

What captivated Parker's fans far more, however, was a reference to a story that *is* in the book: Parker was at a black-tie dinner in Japan when a minor actress leaned provocatively toward him, exposing significant cleavage for the camera. This led to endless, juvenile postings like "wish I had a job causing big-breasted Japanese women to jump into my lap" and jokes about "vitisexual" events.

But the thread that most surprised me resulted from an early review that appeared on the Web site of British wine magazine *Decanter* a couple of weeks before the book went on sale in the United States. It began ▶

66 I didn't expect the fevered reaction from Parker's fan base that began while the book was still in the editing stage. 99

Aftertaste *(continued)*

"I've never shared a cabin with Robert Parker on a ten-month sea voyage and I won't need to after reading *The Emperor of Wine*" and wound up with "When you share a confined space with someone for a long time you get to know them very well indeed. . . ."

The result was a flurry of hilarious posts questioning whether Parker and I had had an affair. One read "The review seems to convey the very bizarre and scurrilous suggestion that Robert Parker was shacked up with McCoy." Parker himself eventually posted a response: "I spent all my time with McCoy fully dressed."

The chat room reactions were only the beginning. Once the book came out there were the usual and expected differences of opinion scattered throughout the one hundred plus mostly positive reviews that appeared in the United States and as far afield as England, Belgium, Chile, even India and Japan. One reviewer found my prose uncluttered, another thought it overly ornate; one thought there was too much about British wine writers, another not enough; and so on.

But there were far stronger pro and con reactions as well. The book in fact proved to be a Rorschach test for wine lovers who had already made up their minds about the most powerful man in wine. People who couldn't stand Parker and his influence thought I wasn't critical enough of him. A few Parker-hating reviewers even misread passages written with what I thought was evident sarcasm and irony to make it seem as though I thought Parker was a saint. People who worshipped Parker thought I was too critical;

> ❝ A review in the British wine magazine *Decanter* provoked a flurry of hilarious posts questioning whether Parker and I had had an affair. ❞

some of them were certain I'd set out to dig up dirt on their hero and do a hatchet job. A few accused me of purposely including several false stories about Parker in the book, but it turned out that they were alluding to two incidents where I gave both one person's version of an event and Parker's very different take.

On my book tour many wine retailers seemed eager to host signings, perhaps hoping customers who read the book would wean themselves from the dominance of a single palate's preferences when buying wine. But a couple of wineries that had expressed an interest in hosting an event backed out for fear of incurring Parker's wrath and having their scores lowered. This was yet more evidence, if I needed it, of just how powerful the Sage of Monkton had become.

One finger-wagging man at a book signing in Washington, DC, asked me aggressively, "Are you mean to Bobby in the book? I'm not going to buy it if you are." (He ended up buying two, perhaps because I patiently listened to his own stories about meeting "Bobby.") On the other hand, one wine writer said he refused to review the book in his publication because he hated Parker and his 100-point rating system too much to even acknowledge his existence.

But I also had a number of pleasant surprises on my book tour and met a remarkably diverse cross section of readers eager to delve into a story of ambition, wine, and global power. When the Nobel Prize–winning physicist Murray Gell-Mann was brought over to my table at the restaurant in The Little Nell in Aspen, Colorado, I was ▶

> " A couple of wineries that had expressed an interest in hosting a book event backed out for fear of incurring Parker's wrath and having their scores lowered. "

delighted to see he had my book under his arm.

All these varied and frequently passionate reactions to my book in the months following publication made me realize that it had touched a raw nerve in the wine world; attacks from both sides meant that I probably came close to achieving my goal of writing a fair and balanced biography.

In truth, I always saw the book in two parts: the first half, concerning Parker's rise to power, is sympathetic; the second half, about the wine world's reaction to his power, is critical of Parker. I don't see how anybody could have actually read my book and concluded that I belong in the Parker camp. I find ample evidence in the book of my conviction that his infallible palate is a myth. But I don't think Parker is the evil figure his detractors portray. I find his very American character appealing.

This ambivalence on my part has prompted a universal reaction from readers: they want to know what I *really* think of Parker. Do I like him or not? But Parker is now two people: an often likable man, and the personification of a dominant brand that stands for certain ideas and a particular taste profile. In the end, it's the brand and its increasingly negative effects on the wine world that trouble me. ❧

❝ There is ample evidence in the book of my conviction that Parker's infallible palate is a myth. But I don't think he is the evil figure his detractors portray. I find his very American character appealing. ❞

Elin McCoy on Writing About Robert Parker

What drew you to Parker as a subject for biography?

Parker's amazing rise to power, frankly. Everywhere I went on assignment, his name was on the lips of winemakers, retailers, consumers, and other journalists. I saw that Parker's life and career provided the keys to understanding the contemporary global wine industry.

And the tale of how Parker went from drinking soda to being the world's supreme wine judge was also a fantastic American story, the chronicle of a man who came from nowhere and succeeded beyond his wildest dreams. In the end, the book became a study in power.

What made you think you were the right person to write this biography?

I've covered the wine world for thirty years. I'm an insider, and I knew people would talk to me on the record. Many people who interviewed Parker didn't know much about wine and how that world works. They were seduced by the myth of the infallible superpalate. I wasn't.

Why is Parker so controversial?

Everyone questions the implications of one person having so much power and influence over an entire worldwide industry. Some people see him as the ultimate consumer advocate and the wine lover's best friend, ▶

> 66 I've covered the wine world for thirty years. I'm an insider, and I knew people would talk to me on the record. 99

fighting for improved wine quality everywhere. Others blame him for reshaping the taste of wine and are convinced winemakers are making wine just to please him. Even those who've benefited from his attention have a love-hate reaction to Parker's ascendency.

Are wineries really catering to his individual taste?

Are they ever! The truth is, taste follows power. Winemakers who hope to snare a top score actively craft wines to please his taste buds. Many new wineries, such as those in Spain's emerging wine regions, have built their success in America on following his lead. In France there's a word for it: *Parkerisé,* or "Parkerized." This refers to powerful wines made in a high alcohol, plush fruit style. Some vineyards give bonuses to workers when their wines get top Parker scores.

Does all this matter to the average wine lover?

Absolutely, and especially in America. Parker is responsible for the changing tastes in wine and his tastes are assumed to be what Americans like. That means that what's available in many wine stores is what Parker likes. It's often not what all consumers like.

What about the man?

Parker is definitely a classic type A entrepreneur in the best American tradition.

66 The truth is, taste follows power. Winemakers who hope to snare a top score actively craft wines to please Parker's taste buds. 99

He likes to say that he never went looking for power, but he's an ambitious guy who knew how to capitalize on circumstances and from early on worked at developing his Lone Ranger image. In reality, he's both generous and vindictive, down-home and arrogant. Parker is a champion of fledgling winemakers, yet is quick to tell people how they should be making wine and to denounce those who don't agree with him. He's bought into his own image.

How much time did you spend with Parker?

I have about thirty-five hours of taped conversations from conducting interviews at Parker's home, traveling with him to California wineries, and spending time on the phone. To get a full picture of his place in the wine world I spoke with winemakers, wine consultants, retailers, importers, and other journalists in France, England, the United States, Australia, New Zealand, Italy, Spain, and South America. It all adds up to well over two hundred hours of taped conversations.

Did Parker have any input?

Parker understood from the beginning that he would have no control over the book and that I wouldn't show him the manuscript before publication. I gave him a chance to respond to unflattering stories. It's a warts-and-all portrait of Parker and his influence, but I think it's fair and balanced. ▶

66 I spoke with winemakers, wine consultants, retailers, importers, and other journalists—well over two hundred hours of taped conversations. 99

Elin McCoy on Writing About Robert Parker
(continued)

What surprised you?

What I found surprising was the extent to which views of Parker were polarized. People were fiercely for or against his influence. There didn't seem to be much middle ground. I was also surprised by how much Parker's rise owed to America; in Europe he is seen as the embodiment of American taste. I realized that my book was also a social critique of America's fascination with experts and gurus, its embrace of ratings and scores, and of American power in the world marketplace.

What will happen after Parker?

Just about everyone who has interviewed me has asked this question. I deliberately ended the book without trying to make a prediction. When Parker finally fades from the scene, I think his one-man influence will fragment among critics specializing in one area (like Burgundy) and knowledgeable retailers, blogs, etc. Wines that depend heavily on Parker's scores will lose market share. I doubt very much there will ever again be an individual in the wine world with the power and global reach of Robert Parker. ᔫ

Author's Picks
A Mixed Case from
The Emperor of Wine

*Numerous wines are mentioned in the
book. I've selected a dozen here that figure
in Parker's story.*

Beaux Frères Pinot Noir (pp. 176–77)
Though not a fan of many Oregon Pinot
Noirs, Parker owns this winery in the
Willamette Valley with his brother-in-law
Mike Etzel. Recent vintages have become more
elegant, with very pure flavors of raspberries,
sassafras, and plums and a subtle hint of
earthiness. (The second label, Belles Soeurs,
is more affordable and more immediately
charming.)

Château Haut-Brion (p. 173)
Parker has the highest regard for this classic
first growth in the Pessac-Léognan appellation
of Bordeaux, and often points to that regard as
proof that he doesn't only like monster wines.
Distinctive and incredibly complex, the wine
has striking cigar box, mineral, and currant
flavors and a long, lingering finish.

Château Le Bon Pasteur (pp.137–38)
This estate in Bordeaux's Pomerol appellation
is owned by one of Parker's heroes, famed
consulting enologist and Merlot master
Michel Rolland, who believes in picking as late
as possible to obtain superripe fruit. The
resulting wines are very round, ripe, and richly
textured with almost a black cherry jam and
plum liqueur taste. ▶

> **66** Parker has the
> highest regard for
> Château Haut-
> Brion, and often
> points to that
> regard as proof
> that he doesn't
> only like monster
> wines. **99**

Author's Picks *(continued)*

Clarendon Hills Astralis (p. 236)

A handful of tiny producers in South Australia have benefited mightily from Parker's attention, mostly those who make supercharged wines from the Syrah (aka Shiraz) grape. Very highly rated by Parker, this extremely pricey wine is a rich and concentrated blockbuster that tastes of black raspberries, chocolate, and roasted coffee.

Costers del Siurana Clos de l'Obac (p. 234)

Spain is one of Parker's enthusiasms. Many of the wines from emerging regions, such as this one from Priorato, reflect his taste preferences. It is an unfiltered blend of Cabernet Sauvignon, Grenache, Syrah, Merlot, and Carignan. Parker championed early vintages, but the most recent ones have been lighter and more elegant and have earned lower scores.

Domaine du Pegau Châteauneuf-du-Pape (p. 224)

Parker has a special fondness for the wines of Châteauneuf-du-Pape, an appellation in the southern part of the Rhône Valley, and has done much to popularize them in the United States. Pegau's wines are in the superconcentrated style with fairly high alcohol levels. Their special bottling Cuvée da Capo is a Parker favorite.

Domaine Tempier Bandol Rosé (p. 121)

One of the most delicious and gulpable rosés from Provence, this has strawberry and spice flavors and plenty of character. Parker calls it "the world's most hedonistic rosé."

Guigal Côtes du Rhône (p. 225)

Guigal, based in Côte Rôtie in the northern Rhône, has received more 100-point scores from Parker than any other producer. These scores were awarded for Guigal's three extraordinary single-vineyard Côtes Rôties, La Mouline, La Landonne, and La Turque, all of which are dense, tannic, oaky, voluptuously textured, and wildly expensive. Guigal's simple, fruity, everyday Côtes du Rhône benefits handily by association with this Parker-produced reputation and sells for a tiny fraction of the price.

Morgante Don Antonio (p. 234)

The winemaking consultant at this tiny estate in Sicily is Riccardo Cotarella, whom Parker regards as Italy's Michel Rolland. Made from the local Nero d'Avola grape, the wine is very dark and powerful, with black fruit and chocolate flavors and a sensual texture.

Patz & Hall Alder Springs Chardonnay (p. 219)

This producer owns no vineyards. Patz & Hall instead purchases Pinot Noir and Chardonnay grapes from some of the best vineyards in California and bottles each separately. They produce several vineyard-designated Chardonnays; this one is all tropical fruit and silky texture, with hints of smoky oak.

Shafer Hillside Select Cabernet Sauvignon (p. 221)

This Napa Valley cult wine from the Stags Leap appellation is frequently on Parker's list of top California Cabernets. Smooth, rich, ▶

> 66 Guigal has received more 100-point scores from Parker than any other producer. 99

17

Author's Picks *(continued)*

and plush, it is a seamless, swashbuckling red that ages well.

Trimbach Clos Ste. Hune Riesling (p. 261)
Known for its whites, the Alsace region of northeastern France and its wines have been special for Parker since he first visited in 1967 with his then girlfriend—now wife—Pat. This sophisticated, rich, and intense white has marvelous underlying minerality.

Don't miss the next book by your favorite author. Sign up now for AuthorTracker by visiting www.AuthorTracker.com.